MYOCARDIAL DAMAGE

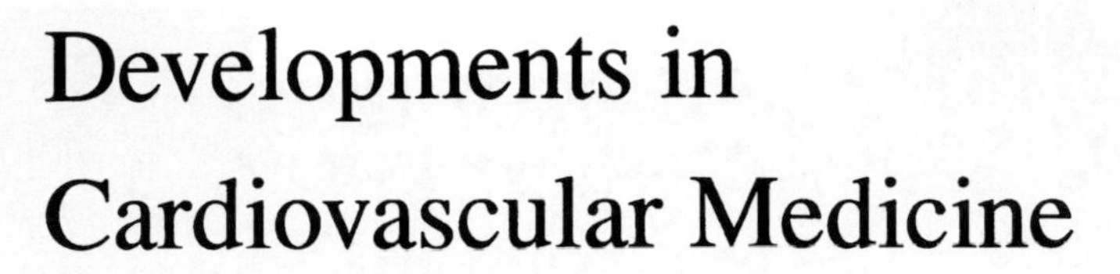

Developments in Cardiovascular Medicine

VOLUME 205

Myocardial Damage

Early Detection by Novel Biochemical Markers

edited by

JUAN CARLOS KASKI, MD, MRCP, FACC, FESC
Reader in Clinical Cardiology and Consultant Cardiologist,
Head of Coronary Artery Disease Research Group,
St George's Hospital Medical School, London, U.K.

and

DAVID W. HOLT, BSc, PhD, FRCPath
Director of Analytical Unit,
Department of Cardiological Sciences,
St George's Hospital Medical School, London, U.K.

KLUWER ACADEMIC PUBLISHERS
DORDRECHT / BOSTON / LONDON

A C.I.P. Catalogue record for this book is available from the Library of Congress.

ISBN 978-90-481-5056-4

Published by Kluwer Academic Publishers,
P.O. Box 17, 3300 AA Dordrecht, The Netherlands.

Sold and distributed in North, Central and South America
by Kluwer Academic Publishers,
101 Philip Drive, Norwell, MA 02061, U.S.A.

In all other countries, sold and distributed
by Kluwer Academic Publishers,
P.O. Box 322, 3300 AH Dordrecht, The Netherlands.

Printed on acid-free paper

Printed in the Netherlands

Table of contents

List of Contributors

Naab Al-Saady, MD, PhD
Lecturer/Hon. Senior Registrar, Department of Cardiological Sciences, St George's Hospital Medical School, London, UK

Jon R. Anderson, FRCS (CTh)
Senior Registrar, Department of Cardiothoracic Surgery, Hammersmith Hospital, London, UK

Fred S. Apple, PhD, DABCC
Director Clinical Laboratories, Hennepin County Medical Center, Minneapolis, MN, USA

Gary F. Baxter, PhD
Hon. Lecturer in Physiology, The Hatter Institute for Cardiovascular Studies, UCL Hospital and Medical School, London, UK

A. John Camm, BSc, MD, FRCP, FESC, FACC, CSt.J
Head of Department of Cardiological Sciences, St George's Hospital Medical School, London, UK

Paul O. Collinson, FRCPath, MD
Consultant Chemical Pathologist, Department of Chemical Pathology, Mayday University Hospital, Surrey, UK

Robbert J. de Winter, MD, PhD
Department of Cardiology, Academic Medical Centre, Amsterdam, The Netherlands

Norbert Frey, MD
Medizinische Universität zu Lübeck, Medizinische Klinik II, Lübeck, Germany

Jan F.C. Glatz, PhD
Department of Physiology, CARIM, Maastricht University, Maastricht, The Netherlands

Wim Th. Hermens, PhD
Department of Physiology, CARIM, Maastricht University, Maastricht, The Netherlands

David W. Holt, BSc, PhD, FRCPath
Director of Analytical Unit, Department of Cardiological Sciences, St George's Hospital Medical School, London, UK

Juan Carlos Kaski, MD, MRCP, FESC, FACC
Reader in Clinical Cardiology and Consultant Cardiologist, Head of Coronary Artery Disease Research Group, St George's Hospital Medical School, London, UK

Hugo A. Katus, MD
Medizinische Universität zu Lübeck, Medizinische Klinik II, Lübeck, Germany

Ernst-Georg Krause
Max Delbrück Zentrum für Molekulare Medizin, Molekulare Kardiologie, Berlin-Buch, Germany

Johannes Mair, MD
Institut für Medizinische Chemie & Biochemie, Innsbruck, Austria

Margit Müller-Bardorff, MD
Medizinische Universität zu Lübeck, Medizinische Klinik II, Lübeck, Germany

Mauro Panteghini, MD
1st Laboratory of Clinical Chemistry, Spedali Civili, Brescia, Italy

Mario Plebani, MD
Department of Laboratory Medicine, Central Laboratory, Padova, Italy

Bernd Puschendorf
Institut für Medizinische Chemie & Biochemie, Innsbruck, Austria

Jan Ravkilde, MD
Department of Cardiology B, Aarhus University Hospital, Aarhus N, Denmark

Peter Stubbs, MD, MBBS, MRCP
Department of Cardiology, Royal Free Hospital NHS Trust, London, UK

Alan H.B. Wu, PhD
Director, Clinical Chemistry Laboratory, Hartford Hospital, Hartford, UK

Acknowledgements

We are very grateful to

Boehringer Mannheim GmbH, Mannheim, Germany

and

Behring Diagnostics GmbH, Llederback, Germany

for their unrestricted educational grants which made this book possible.

We are also grateful to

Nicola H Tansey, Medical Education Co-ordinator, Cardiological Sciences, St George's Hospital, London

For co-ordinating and formatting this book and still smiling despite everything.

Foreword

The spectrum of unstable coronary syndromes has been the object of steadily increasing research particularly in respect of novel diagnostic and treatment modalities. Although the WHO criteria for the diagnosis of acute myocardial infarction have been known for decades, there is still much debate as to the proper use of these, especially the handling of biochemical markers. Traditional enzymes have limitations in diagnostic power and new markers have unclarified applicability with regard to standardisation of assays and decision limits. Furthermore, the growing understanding of the conception of minimal myocardial damage in the borderzone between unstable angina pectoris and myocardial infarction necessitates refinement in the definitions of the various entities within the kaleidoscope of acute myocardial ischaemia.

There has been a strong impetus within the European Society of Cardiology to present this subject to a broad audience of cardiologists, and as a consequence the theme of biochemical markers has become an important constituent of the Education and Training Programmes of the Society. This book has fundamental implications for the overall comprehension of this complex area.

This book on myocardial damage, compiled by international experts, will truly appeal to the basic scientist as well as the clinician searching for updated knowledge of the pathophysiology and risk stratification of unstable coronary syndromes.

Kristian Thygesen, MD, FESC, FACC
Chairman of the Education and Training Programme Committee
Vice-President of the European Society of Cardiology
The European Heart House
France

November 1997

Preface

The early detection of myocardial damage is one of the major challenges to contemporary cardiology. New biochemical markers have now emerged which appear to be highly sensitive and specific for the assessment of not only patients with myocardial infarction but those with unstable angina and prolonged chest pains. Some of these markers such as the troponins have been shown to have prognostic value in the context of acute chest pain. The incorporation of novel markers of myocardial damage to the routine diagnostic armamentarium is not without difficulty. The reasons for this include cost benefit implications, and the lack of definitive comprehensive publications dealing specifically with these issues. Technical difficulties with some of the novel markers are also a problem in some cases and this issue also needs to be carefully reviewed.

A critical analysis of the biochemical characteristics, sensitivity and specificity, as well as the potential clinical applications of the new markers, is required.

This monographic work addresses these issues and also sets up the basis for a redefinition of myocardial damage and myocardial infarction. The book will be of particular interest to biochemists, pharmacologists, cardiologists, general physicians and clinical and basic researchers. The important issue of myocardial damage in relation to pharmacological agents has been specifically addressed in the book and this topic will be of particular interest to both clinical pharmacologists and those working in the pharmaceutical industry.

We invited international authorities, whose original work and expertise in the field are widely recognised, to contribute chapters to the book.

Juan Carlos Kaski
and
David W. Holt

February 1998

Chapter 1

CELLULAR MECHANISMS OF ISCHAEMIC MYOCARDIAL DAMAGE

Gary F. Baxter

Despite the ancient Greek origins of the word (iscein = to restrict; aima = blood), a definition of "ischaemia" which is satisfactory and universally acceptable to clinicians and scientists remains elusive [1]; ancient provenance is no guarantee against controversy. Myocardial ischaemia develops when coronary blood supply to myocardium is reduced. Anoxia will develop when blood flow is completely impeded, whereas severe hypoxia will occur when there is some residual flow either through the occluded segment of the coronary artery or through collateral vessels. A pivotal factor in ischaemia is that oxygen supply to the mitochondria is inadequate to support oxidative phosphorylation [2, 3]. Uncoupling of oxidative phosphorylation rapidly produces profound biochemical and morphological changes within myocardium, the severity of which are ultimately determined by (a) the degree of oxygen deprivation and (b) the duration of impaired oxygenation. Generally, the injurious consequences of ischaemia are related to the degree of reduction in blood flow or to the duration of reduced blood flow, or both. Often, in experimental models and in clinical situations, ischaemia is followed by reperfusion, that is, the re-admission of oxygen and metabolic substrates. The process of reperfusion is associated with further biochemical and functional changes in myocardium, which we must also consider. In this chapter I shall review the key cellular events that occur during ischaemia and reperfusion and relate these to patterns of injury that may be either reversible or irreversible.

Experimental models of ischaemia and reperfusion

Usually, spontaneous coronary artery occlusion in humans is the result of thrombosis. The area of myocardium subtended by the occluded vessel is known as the myocardial risk zone. Unless there is some residual flow through the occluded vessel or some collateral blood flow, this risk zone will become severely ischaemic. Experimental models of coronary thrombosis are technically very difficult to employ. More commonly, experimental myocardial ischaemia in the intact heart is achieved by occlusion of a major coronary artery branch by ligation [4]. Usually the occlusion is temporary so that a period of reperfusion can be instituted. Coronary occlusion in this fashion may be conducted *in vivo*, in anaesthetised animals. Coronary occlusion is studied by this method in many species including rat, rabbit, pig and dog. Coronary occlusion may also be performed in isolated Langendorff-perfused hearts. The Langendorff preparation uses an intact heart, usually from a small animal species such as rat or rabbit, perfused retrogradely through the ascending aorta with a modified oxygenated Krebs' bicarbonate buffer. Although not subject to central innervation and circulating hormonal influences, the Langendorff-perfused heart will beat spontaneously for several hours and lends itself to investigations that might not be easily undertaken *in vivo*. For example, by stopping retrograde aortic flow, and hence flow through the entire coronary circulation, myocardial responses to global ischaemia can be investigated easily in the Langendorff preparation. Hence a range of whole heart experimental models, both *in vivo* and *in vitro*, allow us to induce, quantify and modify the various consequences of ischaemia-reperfusion: biochemical and metabolic changes; functional changes (contractile dysfunction and arrhythmia development); and structural changes (necrosis).

More 'reductionist' approaches to the study of myocardial ischaemia-reperfusion are also used. Isolated strips of myocardium may remain viable for several hours when superfused with oxygenated buffer solution and electrically stimulated to contract. Contractile responses to simulated ischaemia and subsequent re-oxygenation can be examined [5]. Isolated cardiac myocytes may be maintained in short-term culture conditions and various responses to simulated ischaemia studied. Isolated cell preparations are especially valuable for investigating the electrophysiological changes associated with ischaemia-reperfusion [6].

Functional consequences of ischaemia-reperfusion

Contractile dysfunction

The most obvious functional consequence of ischaemia is a rapidly-manifested decrease in myocardial contractility. Loss of sarcomere contraction is observed within seconds of the onset of ischaemia in experimental models. With regional ischaemia, the risk zone will become hypokinetic within a minute or so [7] and with global myocardial ischaemia complete arrest of the heart will occur. It is important to realise that reduced contractility occurs before the onset of tissue necrosis. Several factors, mechanical and biochemical, determine the reduced contractility in ischaemic myocardium. Soon after cessation of coronary flow there is a hydrostatically mediated decrease in myocardial contractility. Normal coronary perfusion pressure stretches sarcomeres and by Starling's law enhances tension development (the "garden hose" effect). Hence, the sudden reduction in coronary perfusion leads to sarcomere shortening and very rapid decline in developed force that is noticeable within seconds of cessation of coronary flow. In the ensuing minutes, metabolic changes will occur, which re-inforce and sustain the reduction in contractility (these are discussed in more detail below). Reduced adenosine triphosphate (ATP) availability for sarcomere contraction is likely to play a primary role in depressing contractility but in addition, decreased washout and accumulation of metabolites in the ischaemic zone may also be responsible. In particular, $[H^+]$, lactate and phosphate accumulation may inhibit the interaction of Ca^{2+} with troponin [8].

Depressed contractility in myocardial ischaemia is of major concern clinically. With a large risk zone, severely depressed contractility may result in acute ventricular failure. However, it is recognised that impaired contractility in the absence of necrosis may occur over prolonged periods in many clinical syndromes associated with repeated or chronic coronary hypoperfusion. This has been termed myocardial "hibernation" and probably represents a form of biochemical adaptation [9]. In the absence of necrosis, and with full reperfusion, myocardial contractility will recover completely (Figure 1.). However, full recovery may take several hours or days *in vivo*. This condition of incomplete but ultimately fully reversible contractility following reperfusion of viable tissue is called "myocardial stunning" [10] (Figure 1.).

Electrophysiological disturbances : ventricular arrhythmias

During ischaemia, profound metabolic and ionic perturbations may affect the development and propagation of action potentials in myocardium to the point where arrhythmias occur. These may range from isolated ventricular premature beats, salvos and runs of ventricular tachycardia, to ventricular fibrillation [11]. Electrophysiological changes related to disturbances of conduction, refractoriness and automaticity occur within a few minutes of the onset of ischaemia and account for early arrhythmias following coronary artery occlusion [12]. Several factors may augment and re-inforce the electrophysiological changes that predispose to arrhythmias during ischaemia. These include local release of catecholamines, the generation of reactive oxygen species, perturbances of the ionic milieu, especially efflux of intracellular K^+, and the generation of inflammatory mediators such as thromboxanes and prostanoids [13].

Reperfusion of myocardium following relatively brief periods of ischaemia may also precipitate arrhythmias [14]. Clinically these may be observed during thrombolysis [15] and after percutaneous transluminal coronary angioplasty (PTCA) [16]. It is clear that these arrhythmias are associated largely with the abrupt readmission of oxygen to ischaemic tissue [17], although other factors such as rapid changes in extracellular pH [18] and K^+ [19] may be important simultaneous triggers of arrhythmogenesis. The incidence and severity of reperfusion-induced arrhythmias are related to the duration of the preceding ischaemic period and also to the rate or "suddenness" of reperfusion. In experimental studies of coronary occlusion in the rat, the most severe reperfusion arrhythmias occur after a 10-15 minute occlusion [20], while graded reperfusion attenuates the severity of these arrhythmias [21]. With longer periods of ischaemia, necrosis will occur within the risk zone. Although reperfusion arrhythmias are less severe following prolonged ischaemia, infarcted and scar tissue cause inhomogeneity of conduction in the heart which predisposes to further arrhythmias at a later stage [22].

Reversible and irreversible cellular injury in ischaemia-reperfusion

The factors that affect the development of irreversible injury form the focus of the remainder of this chapter. Ultra-structural changes occur in myocardium soon after the onset of ischaemia. These may be considered reversible alterations if reperfusion of the tissue can be effected promptly. However, ischaemia lasting more than 20 minutes (without collateral flow) will cause irreversible cell injury, resulting in tissue necrosis. Often this form of ischaemic injury is referred to as "lethal" ischaemic injury. A number of "cardioprotective" agents have been examined in the experimental laboratory for their ability to enhance myocardial tolerance to ischaemia and limit the extent of necrosis during coronary occlusion. With many drugs, the benefit has been modest and it is clear that reperfusion is the *sine qua non* for tissue salvage. Reperfusion therapies in acute myocardial infarction (thrombolysis, PTCA, emergency coronary artery bypass grafting) have the primary aim of salvaging viable tissue, which may be reversibly injured, within the ischaemic risk zone and thereby limiting the extent of the necrotic injury. In myocardial infarction, mortality is closely related to the duration of unrelieved coronary occlusion. Simoons has very succinctly described this philosophy of treatment: "time is muscle and muscle is life" [23].

The factors that determine whether myocardium becomes irreversibly injured by ischaemia-reperfusion are impossible to identify with precision and the final steps that lead to cell death are unknown. It is generally assumed that cell death is related to morphological changes and these have been studied extensively. With the electron microscope it is possible to distinguish between those features of cell injury that may be defined as reversible with reperfusion, and those features that are associated with irreversible injury.

Histological and ultra-structural features of injury

Ultra-structural changes occur within 10 minutes of the onset of ischaemia and, while it is difficult to give a precise sequence of events, the progression of morphological damage is critically time-dependent [24-27]. Early changes after the onset of ischaemia are the loss of cytoplasmic glycogen granules [28]. Mitochondrial dense matrix granules disappear and the cristae become fragmented [29, 30]. Generalised cell swelling becomes apparent with ultrastructural features of membrane destabilisation: mitochondrial swelling occurs; the sarcolemma becomes distorted with evidence of subsarcolemmal oedema ("blebbing"); the sarcoplasmic reticulum becomes swollen [31-33]. As ischaemia progresses, mitochondrial cristae disruption increases [30] and the nucleus shows signs of chromatin aggregation and clearing of the nucleoplasm [34]. Even at this stage, with quite severe ultra-structural changes and a fragile sarcolemma, the injury may be reversible with prompt reperfusion so long as the sarcolemma remains intact [35].

Extending the period of ischaemia further will lead to irreversible cell injury. Most authorities agree that this point is reached after about 20 minutes of severe ischaemia. Probably the most important event leading to irreversible injury, and its distinguishing feature, is gross disruption of the sarcolemma [32]. This occurs as a result of increasing osmotic swelling and phospholipase activation and results in the critical loss of barrier function. The ionic gradient across the membrane is lost as electrolytes flow easily in either direction through the sarcolemmal disruptions. Of particular relevance, soluble proteins and enzymes leave the cell and, as we shall see in later chapters, the appearance of these in plasma has important diagnostic and prognostic implications. At this stage cells are irreversibly injured; myofibrillar proteolysis is evident, and many of the mitochondria are vacuolated and contain flocculent densities [23, 28, 30]. Such cells may be regarded as those destined to die. Reperfusion at this late stage merely hastens the process of cell death in irreversibly injured cells. However, there is some controversy concerning the potential for reperfusion to cause lethal injury in cells that are not already irreversibly injured; the concept of "lethal" reperfusion injury is discussed below.

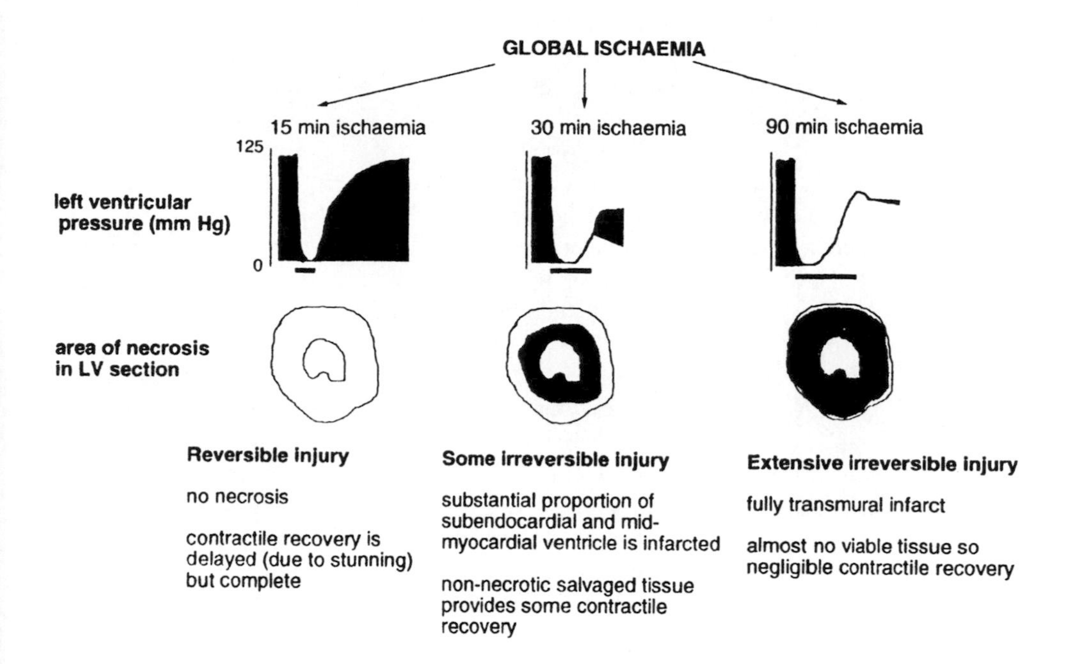

Figure 1. The time-dependency of ischaemic injury. Rabbit hearts were Langendorff perfused and subjected to global normothermic ischaemia for 15, 30 or 90 minutes followed by a period of reperfusion. For each time point, the top panel is a recording of the left ventricular pressure. The lower panel is a cross section of the left ventricle showing the extent of necrosis determined by triphenyltetrazolium staining (shaded zone). After a period of stabilisation, global ischaemia was induced and left ventricular contractions ceased rapidly. After 15 minutes ischaemia, reperfusion resulted in a gradual, but complete recovery of left ventricular contractility and there was no evidence of necrosis. During 30 minutes ischaemia, the end diastolic pressure increased markedly ("ischaemic contracture"). Reperfusion after 30 minutes ischaemia resulted in a partial recovery of left ventricular contractility. The subendocardium and midmyocardium was necrotic. During 90 minutes ischaemia, the development of ischaemic contracture was more pronounced. Reperfusion after 90 minutes resulted in negligible contractile recovery. Necrosis was fully transmural with all but a thin rim of epicardium failing to react with triphenyltetrazolium.

Gross pathology of irreversible injury and the 'wavefront of necrosis'

Permanent or prolonged coronary artery occlusion will result in extensive myocardial necrosis. However, the extent of necrosis and the rate at which it evolves are determined by the duration of ischaemia, the metabolic status of the heart during the ischaemic episode, and the residual myocardial blood flow. After epicardial coronary artery occlusion, myocardial ischaemia does not develop equally throughout the risk region. Under normal conditions there is preferential flow to the subendocardium to maintain the high metabolic rate of this zone due to high wall stresses. Therefore, the subendocardium is the area most likely to develop necrosis first, since the relative depth of ischaemia in this area is greater than elsewhere. Effectively, a transmural gradient of ischaemia is established and to this is related a transmural 'wavefront of necrosis' [36]. The process of necrosis begins in the subendocardial region and as the duration of ischaemia is extended, the area of necrosis extends into the midmyocardium and subepicardium.

Species that have a well-developed coronary collateral circulation display a slowed rate of necrosis relative to species with no preformed collateral vessels. For example, in the dog, which has a variably collateralised heart, coronary occlusion leads to the development of subendocardial necrosis within 40 minutes. After 3 hours of coronary occlusion, the zone of necrosis will include the midmyocardium of the risk zone. By 6 hours after coronary occlusion, the necrotic area will be fully transmural and maximal. The rate of progression of the wavefront is dependent on the degree of collateral flow. In species that have few coronary anastomoses and negligible collateral flow, such as the rabbit, rat and pig, essentially the same sequence of necrosis occurs but the infarction will be fully transmural within 40-60 minutes (see Figure 1). In humans, collaterals may be well developed in older patients with long-standing coronary disease [37] but it is important to bear in mind that while many patients presenting with acute myocardial infarction may have some collateral flow, in a substantial number of people sustaining their first coronary occlusion, death may occur before a collateral circulation can develop. In this respect, then, some humans will have a coronary anatomy resembling the dog's while others will be similar to the pig.

Biochemical determinants of ischaemia-reperfusion injury

It is possible here only to provide a brief overview of the major changes that occur in ischaemic and reperfused tissue and how these relate to the development of injury. The principal metabolic changes associated with ischaemia are the failure of adequate ATP generation by oxidative phosphorylation, the accumulation of by-products of anaerobic metabolism, the loss of ionic homeostasis, and the generation of reactive oxygen species.

High energy phosphate and nucleotide depletion

During ischaemia, the myocardium is mechanically quiescent (because of reduced contractility) but energy is still required to maintain homeostasis through the activity of essential enzyme reactions and the operation of ion channels. To maintain ATP levels, tissues require in addition to oxygen, a source of acyl coA and purine precursors. The generation of ATP under anaerobic conditions is possible through a switch from b-oxidation of fatty acids to glycolysis using glucose and glycogen stores [38, 39]. Anaerobic metabolism is an inefficient source of ATP and leads to the intracellular accumulation of H^+ and lactate as by-products. An alternative source of energy, which is available for a few minutes after the onset of ischaemia, is creatine phosphate. Creatine phosphate can donate high energy phosphate to adenosine diphosphate (ADP), a process that is catalysed by creatine kinase [8]. However the pool of creatine phosphate is rapidly depleted and, in addition, generation of ATP from creatine phosphate leads to the accumulation of inorganic phosphate. After anaerobic glycolysis and creatine phosphate, the final source of ATP is ADP. Two molecules of ADP can form one molecule each of ATP and adenosine monophosphate (AMP). AMP is rapidly dephosphorylated to adenosine, which leaves the cell and is broken down to xanthine. Thus during ischaemia, there is depletion of both high energy phosphates and the purine precursor of ATP. Although reperfusion provides a supply of nucleotides, depletion of the intracellular nucleotide pool may limit the rate at which ATP is generated after ischaemia and hence the rate of recovery of the tissue during reperfusion.

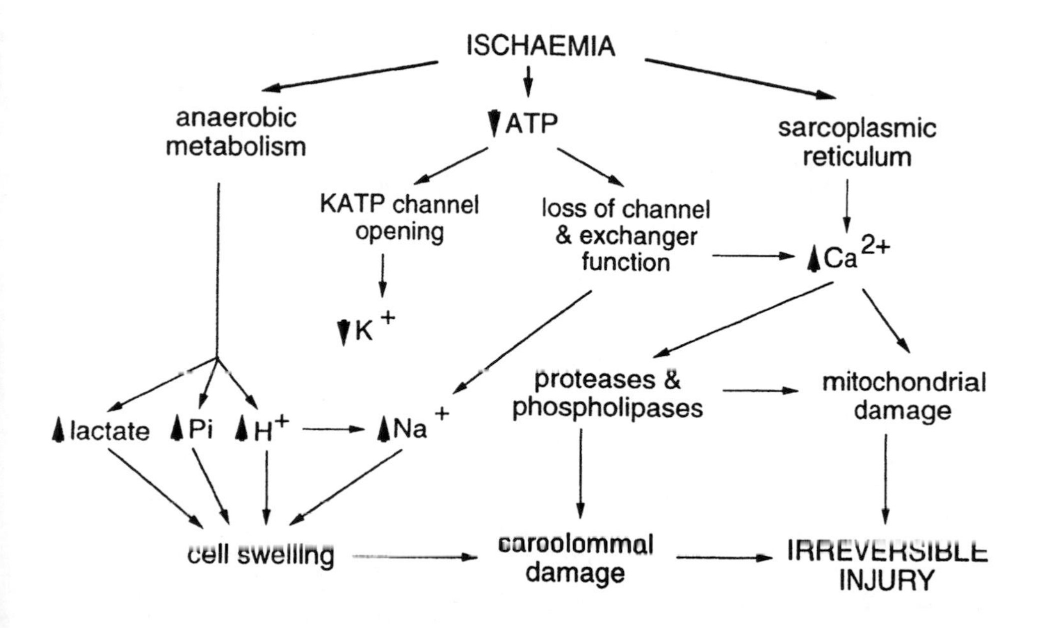

Figure 2. Biochemical and ionic changes during myocardial ischaemia. A series of intracellular changes occur after the onset of ischaemia. If ischaemia is of sufficient duration, alterations in metabolism and ion fluxes culminate in the development irreversible cell injury. For details, see text. Pi = inorganic phosphate.

Loss of ionic homoeostasis

Ionic homeostasis is lost within a few minutes of the onset of ischaemia. The regulation of various cation channels, especially Ca^{2+}, K^+ and Na^+ is ATP-dependent. Moreover, since the homeostasis of several ions is achieved by exchanger mechanisms (antiports and symports), perturbance in one ion during ischaemia may have effects on the regulation of others. An early ionic change after the onset of ischaemia is the efflux of K^+ indicated by a biphasic rise in extracellular K^+ [40]. The early phase of K^+ efflux is most likely to be mediated through opening of the ATP-sensitive K^+ channel (K_{ATP}). This outward rectifying channel opens in response to a fall in ATP and/ or a rise in ADP [41]. Co-transport of K^+ out of the cell with lactate and inorganic phosphate may also play a role [42]. Intracellular Na^+ rises progressively during ischaemia [42-44]. The mechanisms accounting for the early influx of Na^+ are not clear but may involve the activation of the Na/H antiporter which extrudes H^+ in exchange for Na^+. When ATP levels have declined drastically, Na-K-ATPase activity, which normally extrudes Na^+ in exchange for K^+, will decline thereby accentuating Na^+ influx. The combined accumulations of Na^+, H^+, lactate and phosphate during ischaemia lead to a marked rise in cytosolic osmolarity, which in turn accounts for cell swelling.

The accumulation of Ca^{2+} during ischaemia-reperfusion deserves special attention. Although during ischaemia of less than 60 minutes duration total tissue Ca^{2+} does not change substantially, cytoplasmic Ca^{2+} concentration begins to rise very soon after the onset of ischaemia [45, 46]. A 20-50% increase in Ca^{2+} concentration is detected after 10 minutes ischaemia in the Langendorff-perfused rat heart subjected to global ischaemia [46]. The major source of this early calcium overload is the sarcoplasmic reticulum. ATP decline during ischaemia reduces the uptake of Ca^{2+} into the sarcoplasmic reticulum by the Ca/Mg ATPase pump. In addition, stores of Ca^{2+} in sarcoplasmic reticulum may be released through ryanodine-sensitive Ca^{2+} release channels [47]. Although reduced availability of ATP would be expected to result in a reduced opening probability of these channels during ischaemia, activation of the protease calpain II causes disruption of junctional bridges ("feet') which contain these channels, resulting in impairment of their normal function [48]. As ischaemia is prolonged, the influx of Na will trigger the Na/Ca exchanger which extrudes three Na^+ in exchange for one Ca^{2+}. Intracellular Ca^{2+} rises further with the activation of several deleterious Ca^{2+} -dependent enzymes. These include Ca-ATPases, which hydrolyse ATP, causing further depletion of energy. Proteolysis and membrane disruption begins as Ca-dependent proteases and phospholipases are activated. Ca^{2+} overloading of mitochondria is apparent as dense areas on electron microscopy and effectively prevents mitochondrial respiration. The early moments of reperfusion may cause catastrophic Ca^{2+} overload in cells and accentuate the effects of ischaemia-induced Ca overload [49, 50]. Where membrane damage has occurred during ischaemia, the most important route for Ca^{2+} entry during reperfusion is through damaged areas of the sarcolemma. Unlimited amounts of Ca^{2+} may pour into the cell from the extracellular space through the disrupted sarcolemma. The L-type Ca^{2+} channel may be an important route of reperfusion Ca^{2+} overload during ischaemia and during reperfusion of less injured cells. Blockers of this channel may substantially attenuate rises in intracellular Ca^{2+} and limit the functional consequences of ischaemia-reperfusion but generally only if they are administered before the onset of ischaemia [51].

Generation of reactive oxygen species

Reactive oxygen species (oxygen free radicals; oxyradicals) are oxygen-containing chemical species with an unpaired electron, such as superoxide anion and hydroxyl radical. Hydrogen peroxide H_2O_2 is another important reactive oxygen species although it is not, in the strict sense, a free radical. Reactive oxygen species combine avidly with many biological molecules, especially with hydrogen in unsaturated fatty acid side chains and in protein thiol groups, leading to lipid peroxidation and protein denaturation [52]. There has been some interest recently in the possibility that reactive oxygen species may induce apoptosis [53], a potential mechanism of cell death discussed in more detail below. Reactive oxygen species are generated continuously under normal conditions and detoxified by intracellular and extracellular free radical scavengers and anti-oxidant enzymes. The most important of the intracellular anti-oxidants are catalase, which catalyses the breakdown of H_2O_2, superoxide dismutase (SOD), which reduces superoxide anions to H_2O_2, and reduced glutathione which converts H_2O_2 to water. During ischaemia, and to an even greater extent during reperfusion, the production of reactive oxygen species increases through several biochemical pathways [54, 55]. Normal anti-oxidant and scavenging activity may be overwhelmed or depressed under these conditions.

The unchecked activity of reactive oxygen species is likely to play an important role in the genesis of ischaemia-reperfusion injury. Evidence for the involvement of reactive oxygen species in ischaemia-reperfusion injury comes from several sources. First of all, detection methods such as electron spin resonance have shown in experimental studies that radicals are generated during ischaemia [56]. There is generation of enormous quantities during the first minutes of reperfusion and this burst of oxygen radicals gradually subsides over several hours [57]. There is evidence that these events in experimental models are also observed under clinical conditions [58]. Secondly, the exogenous administration of free radical generating mixtures such as xanthine and xanthine oxidase (which generates superoxide) to isolated hearts and cardiac myocytes results in functional and electrophysiological changes similar to those seen in post-ischaemic reperfusion and thought to be due to Ca^{2+} overload [59]. Thirdly, the exogenous administration of free radical scavengers and anti-oxidant enzymes at reperfusion can attenuate the functional consequences of post-ischaemic reperfusion, namely arrhythmias [60] and myocardial stunning [61]. However, the ability of these scavengers and enzymes to modify the morphological changes associated with ischaemia-reperfusion and thereby limit the extent of necrosis is not proven. There have been numerous experimental studies to assess the possibility that these agents, given just before and during reperfusion, can prevent irreversible cell injury occurring as a result of the free radical burst. Some of these studies have reported limitation of infarct size, suggesting that reperfusion causes lethal cell injury. Other studies have reported no limitation of infarct size with these agents suggesting that lethal cell injury occurs only during ischaemia. These conflicting results [see reference 62] highlight an area of controversy, which is discussed in detail below.

Lethal ischaemic injury versus lethal reperfusion injury

It is clear that reperfusion is associated with prolonged depression of contractility (myocardial stunning) and the development of arrhythmias (see above). Both these forms of reperfusion dysfunction may be regarded as reversible in as much as cell death is not a feature of the conditions. The existence of lethal (i.e. cytotoxic) reperfusion injury has been debated extensively and remains controversial [63-65]. Reperfusion of ischaemic myocardium is necessary for halting the process of necrosis. Moreover, it is required for the recovery of reversibly injured myocytes. However, as we have seen, the process of reperfusion imposes a sudden oxidative stress on already damaged tissue and reperfusion may be a cause of injury (see Figure 3). Although massive efflux of enzymes, which is related to membrane damage, occurs predominantly during reperfusion it does not automatically follow that irreversible injury is occurring as a result of reperfusion. During ischaemia there is little, if any, flow through the injured tissue and washout of enzymes does not occur until flow is restored. Therefore, reperfusion is required not only to halt ischaemic injury but may also be necessary to detect irreversible ischaemic injury.

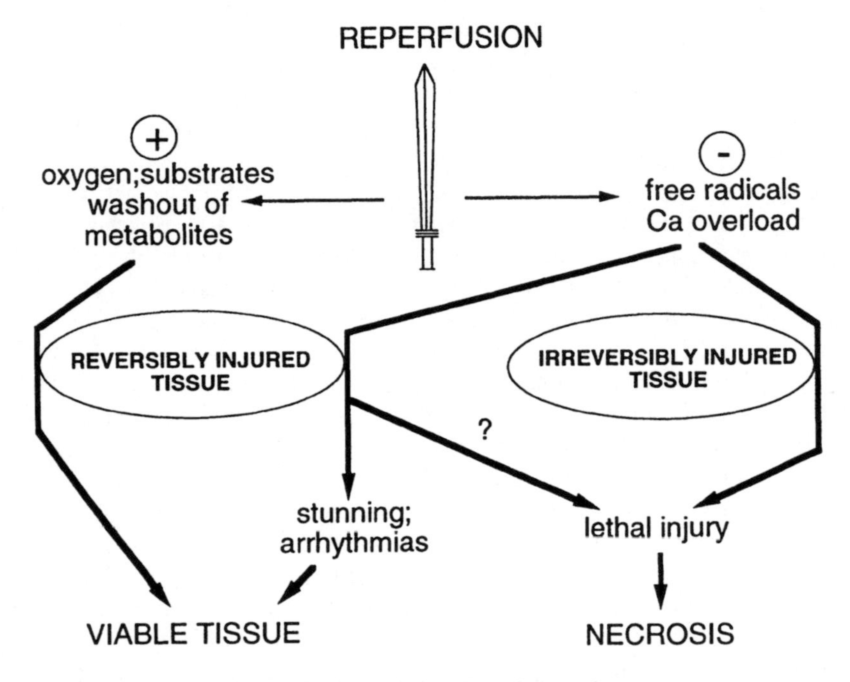

Figure 3. The "double-edged sword" of reperfusion. Reperfusion of ischaemic myocardium has both beneficial components (restoration of oxygen and metabolic substrate supply and washout of accumulated metabolites) and deleterious components (generation of oxygen free radicals and calcium overload). The beneficial components of reperfusion are necessary for the salvage of tissue that has been reversibly injured during ischaemia. However, tissue that has already been irreversibly damaged by ischaemia can not be salvaged by reperfusion. The deleterious aspects of reperfusion may hasten the death of tissue that was irreversibly injured during ischaemia. Reperfusion may also have adverse effects on reversibly injured tissue. Myocardial stunning and arrhythmias may occur during reperfusion and are manifestations of 'reperfusion injury'. The notion that reperfusion may cause lethal reperfusion injury in tissue that was not irreversibly damaged by ischaemia is controversial. For further details, see text.

Direct evidence in support of lethal reperfusion injury is sparse but theoretically this form of injury could result at the myocyte level from biochemical perturbations that occur with the abrupt re-introduction of molecular oxygen, particularly the generation of reactive oxygen species. Additionally there could be other non-myocyte contributors to reperfusion injury including endothelial damage, platelet aggregation, and neutrophil adhesion and activation, but limitations of space preclude discussion of these here. The existence of lethal reperfusion injury is at present unresolved largely because of methodological difficulties. First, it is not possible to quantitate tissue necrosis following relatively short periods of ischaemia with electron or light microscopy. The classical tetrazolium staining techniques, which are used to quantitate experimental necrosis, require a period of reflow for enzyme and co-factor washout from irreversibly injured cells. Secondly, it is impossible to assess the phenomenon of reperfusion properly without a preceding period of ischaemia, the duration of which determines the severity of cellular perturbations that occur during reperfusion. Thirdly, reperfusion of myocardium and the evolution of an infarct *in vivo* are dynamic and complex events involving a host of cell types and many cytokine and hormonal mediators.

Thus, some authors question if reperfusion can cause cell death directly, arguing that restoration of flow accelerates or manifests any injury that occurred during the preceding ischaemic period - the tissue that appears to undergo necrosis during reperfusion is in fact "condemned to die" during ischaemia [65]. On the other hand, some experimental studies, in which the effects of 'simulated reperfusion' (without ischaemia) have been examined in isolated myocytes, suggest that the free radical components of reperfusion can directly cause irreversible cell injury [66]. The design of experiments *in vivo* to examine the existence of lethal reperfusion injury is more difficult. The effects of substances that could modify one or more components of the reperfusion process have been studied. Unfortunately many of these studies have been flawed in design for several reasons: (i) the agent was given before the onset of ischaemia or given during ischaemia in a collateralised species, thereby modifying cell injury during ischaemia; (ii) hearts were subjected to extremely prolonged periods of ischaemia so that extensive lethal injury was sustained before reperfusion; (iii) hearts were reperfused only for a short period of time which could either invalidate the estimation of necrosis by tetrazolium staining or, if an index of contractility was being used, could prevent discrimination between myocardial stunning and irreversible injury. Nevertheless, studies in which modifying agents were given just before reperfusion, or during the early moments of reperfusion, provide indirect evidence that additional injury may occur during reperfusion. Several recent studies evaluating magnesium as an adjunct to reperfusion are good examples of this approach and provide evidence, persuasive for this author at least, that reperfusion per se causes irreversible cell injury that leads to necrosis [67].

Apoptosis in the mechanism of ischaemia-reperfusion injury

How does apoptosis differ from necrosis?

Apoptosis is a special form of cell death. It is often called "programmed cell death" because the activation of apoptotic processes is a necessary component of tissue regulation in almost every area of biology [68]. However, apoptosis is known to occur in many pathological conditions and there has been some interest recently that it may occur in addition to ischaemic necrosis in the ischaemic and reperfused myocardium. There are several critical distinctions between cell death due to necrosis and cell death due to apoptosis. These distinctions begin with the early ultra-structural changes. We have already seen that during the evolution of ischaemic injury that results in necrosis, cell swelling occurs with organelle disruption, amorphous chromatin clumping, and eventually sarcolemmal breakdown. During apoptosis a very different pattern of ultrastructural changes occurs. Cell shrinkage is observed with maintenance of sarcolemmal integrity. Nuclear chromatin becomes fragmented and compacted, and is observed as dense packages closely associated with the nuclear membrane. Cytoplasmic blebs are formed which round off into discrete membrane-bound apoptotic bodies, containing densely packed but intact organelles. These apoptotic bodies are phagocytosed by neighbouring cells [69]. There are many stimuli for apoptosis and a number of these are physiological stimuli. There are also a number of known inhibitors of apoptosis and under normal physiological conditions there is a finely balanced control mechanism for regulating programmed cell death.

Evidence for apoptosis in ischaemia-reperfusion injury

The apoptotic process occurs very rapidly and, because apoptosed cells are destroyed by phagocytosis, no inflammatory reaction occurs. Hence, detection of apoptosis by conventional histological approaches may be very difficult. However, apoptosis can be detected, using specialised techniques that detect DNA fragmentation, in acute myocardial infarction showing that apoptosis occurs during ischaemia-reperfusion in addition to necrotic cell death [70-72]. However, it is not clear at what stage apoptosis occurs in relation to necrosis. It is also unclear whether it occurs predominantly during ischaemia or during reperfusion. Studies of permanent coronary artery occlusion in the rat (i.e. ischaemia without reperfusion) suggest that apoptosis represents the major form of myocyte death [73]. In contrast, rabbit hearts subjected to ischaemia-reperfusion display signs of apoptosis only during reperfusion and not in ischaemic myocytes [74]. Another study, which complicates this dichotomy still further, examined rat hearts subjected to ischaemia and reperfusion [75]. Two and a quarter hours of coronary occlusion was necessary to produce evidence of apoptosis in the ischaemic myocytes. However, myocardium made ischaemic for only 45 minutes and then reperfused showed evidence of apoptosis after only 1 hour of reperfusion. However, the amount of apoptosis in reperfused myocardium was less than in permanently ischaemic myocardium, suggesting that reperfusion reduces the total number of cells undergoing apoptosis but increases the rate at which apoptosis occurs in irreversibly injured cells. Investigation of apoptosis as a mechanism of cell death in myocardial ischaemia-reperfusion is at an early stage but it is likely that we shall hear far more about this process in the next few years as more sophisticated and reliable markers for its detection become available.

References

1. Hearse DJ. Myocardial ischaemia: can we agree on a definition for the 21st century? Cardiovasc Res 1996;28:1737-1744.
2. Opie LH. Myocardial metabolism in ischemia. In: Heusch G (ed) Pathophysiology and Rational Pharmacotherapy of Myocardial Ischemia. New York: Springer-Verlag, 1990, pp 37-57.
3. Ganz P, Braunwald E. Coronary blood flow and myocardial ischemia. In: Braunwald E (ed) Heart Disease: A Textbook of Cardiovascular Medicine. Philadelphia: WB Saunders, 1997, pp 1161-1183.
4. Ytrehus K, Downey JM. Experimental models assessing the physiology of myocardial ischemia. Curr Opin Cardiol 1993;8:581-588.
5. Walker DM, Walker JM, Pugsley WB, Pattison CW, Yellon DM. Preconditioning in isolated superfused human muscle. J Mol Cell Cardiol 1995;27:1349-1357.

6. Pinson A (ed). "The Heart Cell in Culture", CRC Press, Boca Raton, 1987.
7. Tennant R, Wiggers CJ. The effect of coronary occlusion on myocardial contraction. Am J Physiol 1935;112:351-361.
8. Allen DG, Orchard CH. Myocardial contractile function during ischemia and hypoxia. Circ Res 1987;60:153-168.
9. Ross J Jr. Myocardial perfusion-contractrion matching. Implications for coronary heart disease and hibernation. Circualtion 1991;83:1076-1083.
10. Braunwald E, Kloner RA. The stunned myocardium: prolonged, post-iscehmic ventricular dysfunction. Circulation 1982;66:1146-1149.
11. Curtis MJ, Macleod BA, Walker MJ. Models for the study of arrhythmias in myocardial ischaemia and infarction: the use of the rat. J Mol Cell Cardiol 1987;19:399-419.
12. Janse MJ, Wit AL. Electrophysiological mechanisms of ventricular arrhythmias resulting from myocardial ischemia and infarction. Physiol Rev 1989;69:1049-1169.
13. Parratt JR (ed). Early Arrhythmias Resulting from Myocardial Ischaemia: Mechanisms and Prevention by Drugs. London: MacMillan, 1982.
14. Manning AS, Hearse DJ. Reperfusion-induced arrhythmias: mechanisms and prevention. J Mol Cell Cardiol 1984;16:497-518.
15. Goldberg S, Greenspon AJ, Urban PL, Muza, Berger B, Walinsky P, Maroko PR. Reperfusion arrhythmia: a marker of restoration of antegrade flow during intracoronary thrombolysis for acute myocardial infarction. Am Heart J 1983;105:26-32.
16. Taggart P, Sutton P, Holdright D, Swanton H. Reperfusion arrhythmia in man: a rare but real event. J Mol Cell Cardiol 1996;28:A24 (abstract).
17. Manning AS. Reperfusion-induced arrhythmias: do free radicals play a critical role? Free Radical Biol Med 1988;4:305-316.
18. Avkiran M, Ibuke C, Shimada Y, Haddock PS. Effects of acidic reperfusion on arrhythmias and Na^+ - K^+ -ATPase activity in regionally ischemic rat hearts. Am J Physiol 1996;270:H957-H964.
19. Curtis MJ, Hearse DJ. Ischaemia-induced and reperfusion-induced arrhythmias differ in their sensitivity to potassium: implications for the mechanisms of initiation and maintenance of ventricular fibrillation. J Mol Cell Cardiol 1989;21:21-40.
20. Crome R, Hearse D, Manning A. Relationship between cellular cyclic AMP content and the incidence of ventricular fibrillation upon reperfusion after varying periods of ischaemia. J Mol Cell Cardiol 1983;15 (suppl 1):180 (abstract).
21. Corr PB, Witowski FX. Potential electrophysiological mechanisms reponsible for dysrhythmias associated with reperfusion of ischaemic myocardium. Circulation 1983;68 (suppl I):I-16 - I-24.
22. Misier AR, Opthof T, van Hemel NM, Vermeulen JT, de Bakker JM, Defauw JJ, van Capelle FJ, Janse MJ. Dispersion of 'refractoriness' in noninfarcted myocardium of patients with ventricular tachycardia or ventricular fibrillation after myocardial infarction. Circulation 1995;91:2566-2572.
23. Simoons ML, Boersma E, Maas ACP, Deckers JW. Management of myocardial infarction: the proper priorities. Eur Heart J 1997;18:896-899.
24. Schaper J, Hein S, heinrichs CM, Weihrauch D. Myocardial injury and repair. In: Parrat JR (ed) Myocardial response to Acute Injury. London: MacMillan, 1992, pp 1-16.
25. Jennings R, Ganote C. Structural changes in myocardium during acute ischemia. Circ Res 1974; 34/35 (suppl III): III-156-III-172.
26. SchaperJ. Unltrastructure of the myocardium in acute ischemia. In: Schaper W (ed) The Pathophysiology of Myocardial Perfusion. Amsterdam: Elsevier, 1979, pp 581-674.
27. Schaper J. Ultrastructural changes of the myocardium in regional ischemia and infarction. Eur Heart J 1986;7: 3-9.
28. Jennings R, Baum J, Herdson P. Fine structural changes in myocardial ischemic injury. Arch Pathol 1965;79:135.
29. Jennings RB, Shen AC, Hill ML, Ganote CE, Herdson PB. Mitochondrial matrix densities in myocardial ischemia and autolysis. Exp Mol Pathol 1978;29:55-65.
30. Jennings RB, Ganote CE. Mitochondrial structure and function in acute myocardial ischemic injury. Circ Res 1976;38 (suppl I): I-80 - I-91.
31. Whalen DA Jr, Hamilton DG, Ganote CE, Jennings RB. Effect of a transient period of ischemia on myocardial cells. I. Effects on cell volume regulation. Am J Pathol 1974;74:81-398.
32. Jennings RB, Reimer KA, Steenbergen C Jr. Myocardial ischemia revisited. The osmolar load, membrane damage, and reperfusion. J Mol Cell Cardiol 1986;18:769-780.
33. Steenbergen C Jr, Hill ML, Jennings RB. Cytoskeletal damage during myocardial ischemia. Circ Res

1987;60:478-486.
34. Schaper J, Mulch J, Winkler B, Schaper W. Ultrastructural, functional and biochemical criteria for estimation of reversibility of ischemic injury: a study on the effects of global ischemia on the isolated dog heart. J Mol Cell Cardiol 1979;11:521-541.
35. Herdson PB, Sommers HM, Jennings RB. A comparative study of the fine structure of normal and ischemic dog myocardium with special reference to early changes following temporary occlusion of a coronary artery. Am J Pathol 1965;46:367-386.
36. Reimer KA, Jennings RB. The 'wavefront phenomenon' of myocardial ischemic cell death. II. Transmural progression of necrosis within the framework of ischemic bed size (myocardium at risk) and collateral flow. Lab Invest 1979;40:633-644.
37. Schaper W, Gorge G, Winkler B, Schaper J. The collateral circulation of the heart. Prog Cardiovasc Dis 1988;31:57-77.
38. Morgan H, Neely J. Metabolic regulation and myocardial function. In: Hurst J (ed) The Heart. New York: McGraw-Hill, 1990, pp91-105.
39. Opie LH. Metabolism of free fatty acids, glucose and catecholaminmes in acute myocardial infarction. Am J cardiol 1975;36:938-953.
40. Dresdner K. Ionic fluxes in ischemia and infarction. In: Rosen MR, Janse MJ, Wit AL (eds) Cardiac Electrophysiology: A Textbook. New York: Futura, 1990, pp 695-718.
41. Edwards G. Weston AH. The pharmacology of ATP-sensitive potassium channels. Annu Rev Pharmacol Toxicol 1993;33:597-637.
42. Kleber A. Resting membrane potential, extracellular potassium activity, and intracellular sodium activity during acute global ischemia in isolated perfused guinea-pig hearts. Circ Res 1983;52:442-450.
43. Malloy C, Buster D, Castro M, Geraldes C, Jeffrey F, Sherry A. Influence of global ischemia on intracellular sodium in the perfused rat heart. Mag Res Med 1990;15:33-44.
44. Pike M, Kitakaze M, Marban E. ^{23}Na-NMR measurements of intracellular sodium in intact perfused ferret hearts during ischemia and reperfusion. Am J Physiol 1990;259:1767-H1773.
45. Marban E, Kitakaze M, Kusuoka H, Porterfield J, Yue D, Chacko V. Intracellular free calcium concentrations measured with 19F NMR spectroscopy in intact ferret hearts. Proc Natl Acad Sci USA 1987;84:6005-6009.
46. Steenbergen C, Murphy E, Levy L, London RE. Elevation in cytosolic free calcium concentration early in myocardial ischemia in perfused rat heart. Circ Res 1987;60:700-707.
47. Feher JJ, LeBolt WR, Manson NH. Differential effect of global ischemia on the ryanodine-sensitive and ryanodine-insensitive calcium uptake of cardiac sarcoplasmic reticulum. Circ Res 1989;65:1400-1408.
48. Rardon DP, Cefali DC, Mitchell RD, Seiler SM, Hathaway DR, Jones LR. Digestion of cardiac and skeletal muscle junctional sarcoplasmic reticulum vesicles with calpain II. Effects on Ca^{2+} release channel. Circ Res 1990;67:84-96.
49. Hearse DJ, Tosaki A. Free radicals and calcium: simultaneous interacting triggers as determinants of vulnerability to reperfusion-induced arrhythmias in the rat heart. J Mol Cell Cardiol 1988;20:213-223.
50. Altschuld RA, Ganote CE, Nayler WG, Piper HM. What consitutes the calcium paradox? J Mol Cell Cardiol 1991;23:65-767.
51. Nayler WG. Calcium Antagonists. London: Academic Press, 1988.
52. Halliwell B, Gutteridge JM, Cross CE. Free radicals, antioxidants, and human disease: where are we now? J Lab Clin Med 1992;119:598-620.
53. McConkey DJ, Orrenius S. Signal transduction pathways in apoptosis. Stem Cells 1996;14:619-631.
54. Loesser K, Kukreja R, Kassiha S, Jesse R, Hess M. Oxidative damage to the myocardium: fundamental mechanisms of myocardial injury. Cardioscience 1991;2:199-216.
55. Ferrari R, Ceconi C, Curello S, Cargnoni A, Alfieri O; Pardini A; Marzollo P; Visioli O Oxygen free radicals and myocardial damage: protective role of thiol-containing agents. Am J Med 1991;91(3C):95S-105S.
56. Rao PS, Cohen MV, Mueller HS. Production of free radicals and lipid peroxides in early experimental myocardial ischaemia. J Mol Cell Cardiol 1983;15:713-716.
57. Bolli R, Jeroudi MO, Patel BS, DuBose CM, Lai EK, Roberts R, McCay PB. Direct evidence that oxygen derived free radicals contribute to post-ischemic myocardial dysfunction in the intact dog. Proc Natl Acad Sci USA 1989; 86:4695-4699.
58. Grech ED, Dodd NJ, Bellamy CM, Perry RA, Morrison WL, Ramsdale DR. Free radical generation during angioplasty reperfusion for acute myocardial infarction. Lancet 1993;341:990-991.

59. Coetzee WA, Opie LH. Effects of oxygen free radicals on isolated cardiac myocytes from guinea-pig ventricle: electrophysiological studies. J Mol Cell Cardiol 1992;24:651-663.
60. Woodward B, Zakaria MN. Effect of some free radical scavengers on reperfusion induced arrhythmias in the isolated rat heart. J Mol Cell Cardiol 1985;17:485-493.
61. Bolli R, Jeroudi MD, Patel BS, Aruoma OI, Halliwell B, Lai EK, McCay PB. Marked reduction of free radical generation and contractile dysfunction by antioxidant therapy begun at the time of reperfusion. Evidence that myocardial "stunning" is a manifestation of reperfusion injury. Circ Res 1989;65:607-622.
62. Yellon DM, Downey JM. Current research views on myocardial reperfusion and reperfusion injury. Cardioscience 1990;1:89-98.
63. Braunwald E, Kloner RA. Myocardial reperfusion: a double-edged sword? J Clin Invest 1985;76:1713-1719.
64. Opie LH. Reperfusion injury and its pharmacologic modification. Circulation 1989; 80:1049-1062.
65. Jennings RB, Reimer KA. Lethal reperfusion injury: fact or fancy? In: Parratt JR (ed), Myocardial Response to Acute Injury, London: MacMillan, 1992, pp 17-34.
66. Vanden Hoek TL, Shao Z, Li C, Zak R, Schumacker PT, Becker LB. Reperfusion injury on cardiac myocytes after simulated ischemia. Am J Physiol 1996;270:H1334-H1341.
67. Baxter GF, Sumeray MS, Walker JM. Infarct size and magnesium: insights from experimental studies into LIMIT-2 and ISIS-4. Lancet 1996;348:1424-1426.
68. Nagata S. Apoptosis by death factor. Cell 1997;88:355-365.
69. Wyllie AH, Kerr JF, Currie AR. Cell death: the significance of apoptosis. Int Rev Cytol 1980;68:251-306.
70. Misao J, Hayakawa Y, Ohno M, Kato S, Fujiwara T, Fujiwara H. Expression of bcl-2 protein, an inhibitor of apoptosis, and Bax, an accelerator of apoptosis, in ventriuclar myocytes of human hearts with myocardial infarction. Circulation 1996;94:1506-1512.
71. Olivetti G, Quaini F, Sala R, Lagrasta C, Corradi D, Bonacini E, Gambert SR, Cigola E, Anversa P. Acute myocardial infarction in humans is associated with activation of programmed myocyte death in the surviving portion of the heart. J Mol Cell Cardiol 1996;28: 2005-2016.
72. Saraste A, Pulkki K, Kallajoki M, Henriksen K, Parvinen M, Voipio-Pulkki LM. Apoptosis in human acute myocardial infarction. Circulation 1997;95:320-323.
73. Kajstura J, Cheng W, Reiss K, Clark WA, Sonnenblick EH, Krajewski S, Reed JC, Olivetti G, Anversa P. Apoptotic and necrotic myocyte cell deaths are independent contributing variables of infarct size in rats. Lab Invest 1996;74:86-107.
74. Gottlieb RA, Burleson KO, Kloner RA, Babior BM, Engler RL. Reperfusion injury induces apoptosis in rabbit cardiomyocytes. J Clin Invest 1994;94:1621-1628.
75. Fliss H, Gattinger D. Apoptosis in ischemic and reperfused rat myocardium. Circ Res 1996;79:949-956.

Chapter 2

CREATINE KINASE ISOFORMS

David W. Holt

In the late 1980's and early 1990's the most sensitive commercially available marker of myocardial damage was serum measurement of the activity of the MB isoenzyme of creatine kinase (CK). Although relatively specific for myocardial muscle the measurement of this analyte was not without its drawbacks, including poor sensitivity for acute myocardial infarction (AMI) in the time period 8-12 hours after the onset of symptoms and poor specificity in the context of skeletal muscle damage [1,2]. Sensitivity was improved by the introduction of assays for CK-MB mass [3,4], and these assays were also less prone to instability of the analyte and to interference caused by haemolysis [5].

Against this background an assay systems were introduced for the separation of CK isoforms. These techniques held out the promise of an early and sensitive diagnostic aid for AMI, allowing accurate triage of patients entering the Emergency Room. Subsequent work in this laboratory and by others identified clinical settings in which the separation of these isoforms could provide interesting information on myocardial damage. Subsequently, events rather overtook the implementation of these assays as more cardio-specific analytes became available, which could be measured by analytical techniques which were less labour intensive.

This chapter will record the work on the separation of CK isoforms, and put it in context with the measurements which have tended to replace these techniques.

Analytical Techniques

The cytosolic enzyme creatine kinase (EC 2.7.3.2) catalyses the reversible transfer of a phosphate group from ATP to creatine. It exists in the cytoplasm of the cells of human tissue as a dimer consisting of two monomers, designated M and B [6]. These subunits combine to produce three dimeric isoenzymes termed CK-BB, CK-MB, and CK-MM, readily distinguishable by electrophoretic mobility [7]. The CK-MB isoenzyme is found, predominantly, in cardiac tissue. The CK-MM isoezyme is widely distributed throughout skeletal and cardiac muscle.

The CK-MM and CK-MB isoenzymes of CK exist in blood as a collection of isoforms; CK-MM has three isoforms numbered 1, 2 and 3, whilst CK-MB has two isoforms, 1 and 2. Two isoforms are found only in tissue, CK-MM3 and CK-MB2. When the tissue isoforms are released into the blood stream, from damaged skeletal muscle or myocardial tissue, they are converted to their respective isoezyme isoforms by the enzyme carboxypeptidase-N. The isoforms differ only by a single carboxyterminal lysine residue on the M sub-unit. Thus, the CK-MM3 isoform is converted, successively to CK-MM2 and CK-MM1 by the cleavage of the two lysine residues. The breakdown of the CK-MB2 isoform to CK-MB1 is shown diagrammatically in Figure 1.

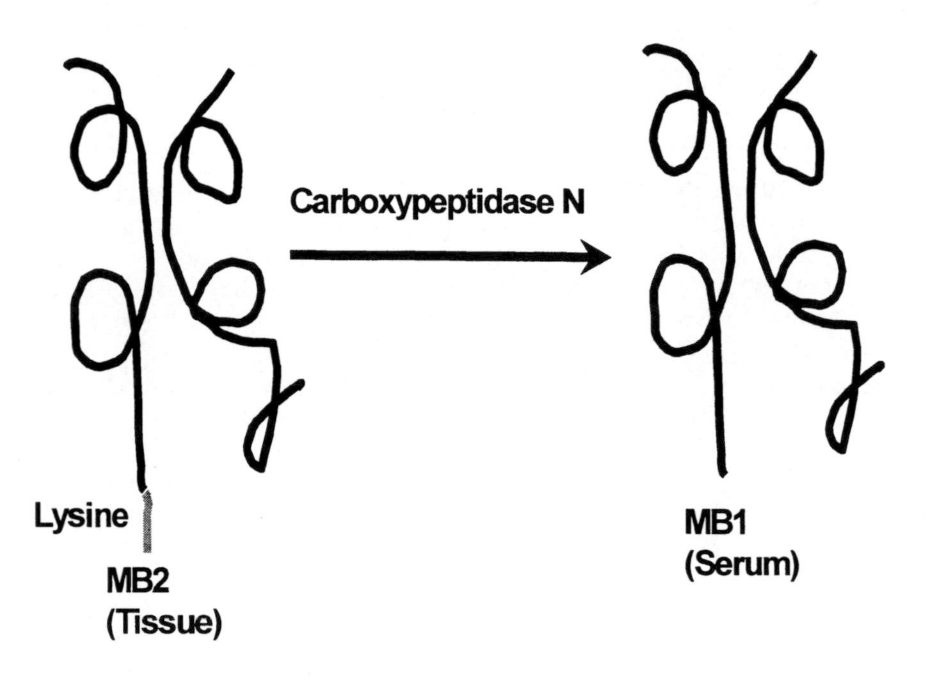

Figure 1. Diagrammatic representation of CK-MB isoforms.

Early published studies related to the electrophoretic or anion-exchange liquid chromatographic separation of the MM isoforms [8,9]. The sensitivity of these techniques was not sufficient to allow the separation and quantitation of the MB isoforms at relatively low CK-MB activities, during the early period following myocardial infarction.

Much of the later published data refer to the electrophoretic separation of the various isoforms using a rapid electrophoresis assay (REP) which was developed for the CK isoforms by Helena Laboratories (Beaumont, Texas, USA). This technique is based on high voltage, temperature controlled, electrophoresis of the sample on a specially designed agarose gel. Separation time for the CK isoforms is less than 20 minutes. Application of the plasma sample and substrate are automated, together with the incubation and development of the fluorescencent product. Quantitation is by densitometric scanning. This method is both accurate and reproducible at CK-MB activity values which fall well within the normal reference interval and can be used at plasma CK-MB activities of less than 5U/L [10]. Typical scans for the MM and MB isoforms are shown in Figure 2.

The original apparatus developed by Helena Laboratories was improved on by them and a system designed specifically for CK-MB isoform separation was produced

(Cardio REP). This automated system has a total run time, from sample application to automatic scanning and editing, of under 20 minutes [11].

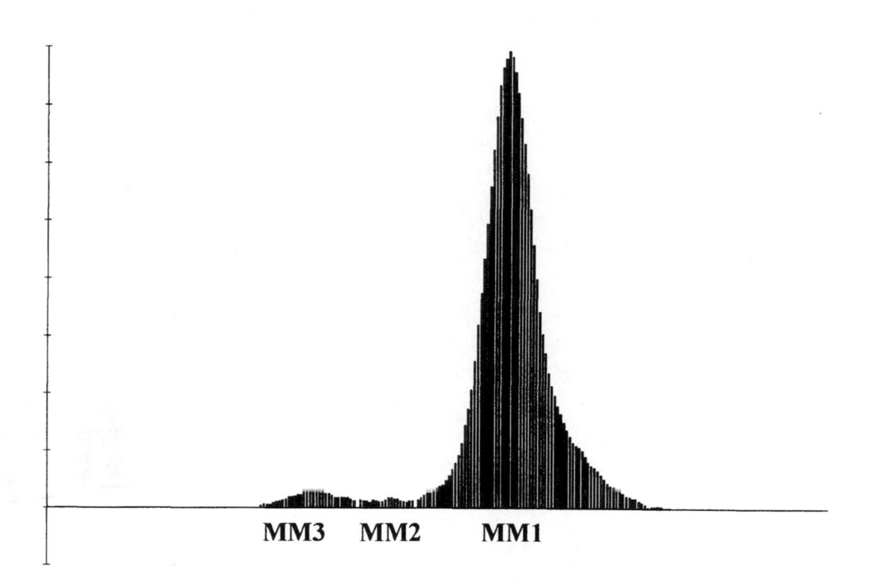

Figure 2a. Densitometric scan showing separation of CK-MM isoforms in a plasma sample from a healthy human subject. Isoforms were separated using Helena REP apparatus. The ratio CK-MM3/CKMM1 was 0.03. The total CK activity was 90U/L. Data: unpublished, Analytical Unit.

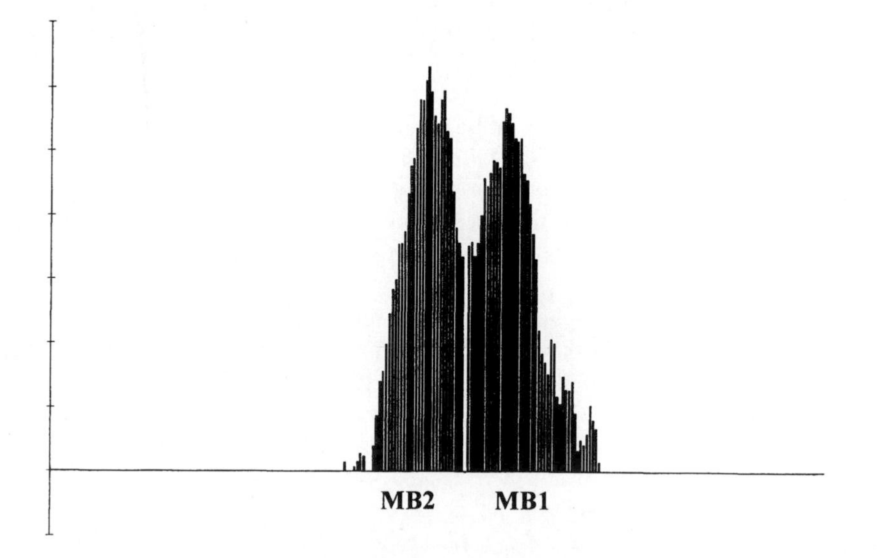

Figure 2b. Densitometric scan showing separation of CK-MB isoforms in a plasma sample from a healthy human subject. Isoforms were separated using Helena REP apparatus. The ratio CK-MB2/CKMB1 was 0.96. The CK-MB activity was 5U/L. Data: unpublished, Analytical Unit.

One group has reported on the use of another electrophoretic technique for the separation of CK-MB isoforms - isoelectric focusing [12]. Using this technique a tissue form and three other isoforms without a carboxyterminal lysine were separated. The clinical significance of these additional isoforms of MB is not clear and does not seem to have been of any particular value [13].

In addition, immunoassays for the measurement of CK-MB2 have been described. One possesses a high sensitivity for the MB2 isoform (0.2μg/L) but the assay is a little cumbersome to use [14,15]. A specific antibody to the B-subunit is used in association with a capture antibody to the CK-M + lysine subunit. Quantitation is by difference, using a second assay to measure total CK-MB mass. Another immunoassay technique is based on an immunochemical extraction procedure specific from CK-MB and CK-MB2. Automated assays of both analytes are performed following the extraction procedure [16].

One problem associated with the separation and measurement of the CK isoforms is that of analyte stability. The enzyme carboxypeptidase-N continues the *in vitro* breakdown of the tissue isoforms in blood samples after collection. To avoid this it is necessary to collect the samples into specific collection tubes prior to separation. A variety of inhibitors of the enzyme have been used, including 15mmol EDTA [17].

Applications

The ratio between the tissue isoforms and the final plasma modified isoforms can be used as a measure of recent tissue damage. Thus, the ratios CK-MM3/MM1 and CK-MB2/MB1 have been used as markers of myocardial damage. Increases in these ratios indicate a rise in the tissue form of the respective isoform, caused by recent entry of the tissue isoform into the circulation.

Wu *et al* [8] demonstrated an increase in the relative proportion of CK-MM3, in comparison with CK-MM1, at time points before total CK and CK-MB activities exceeded the normal reference interval in patients with AMI. In apparently healthy subjects the CK-MM3/MM1 ratio was in the range 0.03-0.29. Typical results following AMI were for this ratio to exceed 0.37 within 5 hours after the onset of chest pain and to peak by about 9 hours. In contrast, CK-MB and total CK peaked at about 18 and 25 hours, respectively. Sensitivity for the diagnosis of AMI was 90% in the period 7-9 hours after the onset of pain, whereas this sensitivity was not achieved by CK-MB activity until 13-15 hours.

The lack of specificity for myocardial muscle injury associated with the CK-MM isoenzyme prompted interest in the development of systems with sufficient sensitivity to detect changes in CK-MB isoform ratios at CK-MB activities which were well within the normal reference range. Puleo *et al* [10] reported on such a system, showing that either isoform could be detected at activities as low as 1.25U/L.

These authors had already established data for healthy subjects by measuring the CK-MB2/MB1 ratio in 56 healthy volunteers and 50 hospitalised patients without signs of myocardial ischaemia [1]. The mean (±SD) ratio in the two groups was 0.94±0.39 and 1.09±0.39. These are very similar values to those generated in this laboratory following the analysis of 248 samples collected from 50 apparently healthy subjects. Table 1 summarises these data and shows that the ratio was calculated from samples with low mean values for both total CK and CK-MB mass. Allowing 3 SD values, the upper limit of the reference range for the ratio is 1.6.

Table 1. Mean value for CK-MB2/MB1 isoform ratio in 248 samples from 50 healthy human volunteers, in comparison with mean values for total CK and CK-MB mass. Data: unpublished, Analytical Unit.

Analyte	Mean	SD
CK (U/L)	70	20
CK-MB (μg/L)	1.3	0.84
CK-MB2/MB1	0.84	0.25

Using the Helena REP system Puleo *et al* [2] showed that the CK-MB2/MB1 ratio could be used to diagnose or rule out AMI within the first 6 hours after the onset of symptoms. The isoform assay showed a sensitivity for the diagnosis of AMI of 95.7% within 6 hours of the onset of symptoms, compared with only 48% for CK-MB activity, as illustrated in Figure 3.

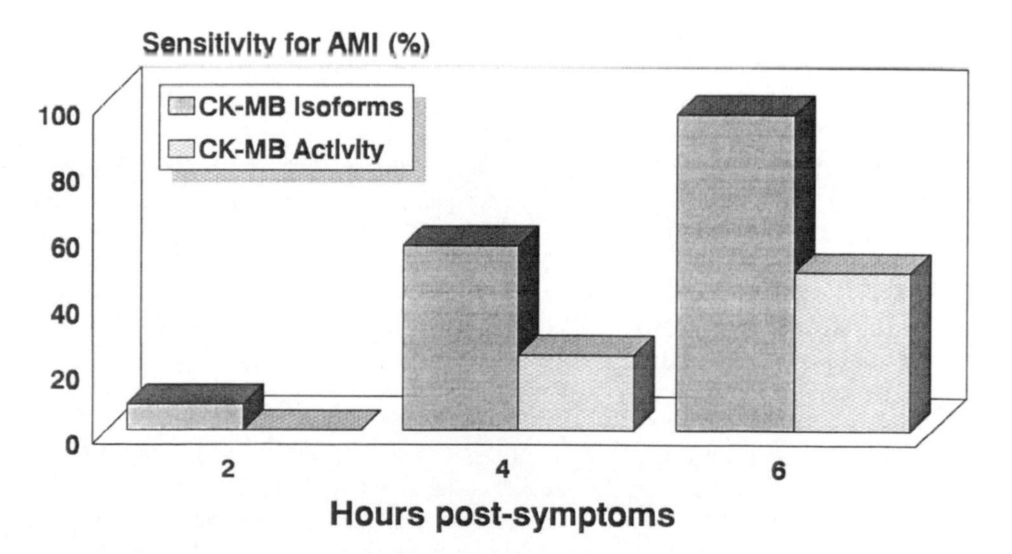

Figure 3. Sensitivity for AMI by means of CK-MB isoform separation and CK-MB activity measurement during the first 6 hours after the onset of symptoms. Based on Puleo *et al* [2].

However, the results of such studies can be markedly influenced by the discriminator values set for the markers, and the choice of CK-MB activity as the comparative marker, whilst appropriate at the time, was already being superseded by the more sensitive measurement of CK-MB mass [3,4].

Clearance of CK-MB is relatively rapid after infarction and the separation of CK-MB isoforms is of little clinical value for samples from patients who have presented with AMI late after the onset of symptoms. Table 2 demonstrates this point in samples from 4 patients who had presented to hospital at least 18 hours after the onset of chest pain. Whilst one sample contained only the CK-MB2 isoform, the other samples showed ratios for CK-MB2/MB1 within the reference range. Troponin I, which has slower release kinetics was, however, still of clinical value in all of these cases, being well above the discriminatory concentration for myocardial damage (0.1µg/L Sanofi-Pasteur assay).

Table 2. CK-MB2/MB1 isoform ratios in samples from 4 patients with a diagnosis of AMI admitted to hospital late after the onset of symptoms. The isoform ratios are compared with cardiac troponin I measurements made using the Sanofi-Pasteur assay. Data: unpublished, Analytical Unit.

Time Post-pain (h)	CK-MB2/MB1	Troponin I (µg/L)
18	100% CK-MB2	5.1
24	0.9	4.5
35	1.4	3.1
36	0.5	1.4

Both CK-MM and CK-MB isoform analysis have been used for the non-invasive detection of vessel patency following thrombolytic therapy [18-21]. A comprehensive study of 207 patients with AMI treated with plasminogen activator examined serial measurements of CK-MB mass, CK-MB2/MB1 ratio and CK-MM3/MM1 ratio [21]. In comparison with angiographically confirmed flow status, at 90 minutes following thrombolytic therapy, there was considerable overlap for all of these markers between those with patent infarct-related vessels (TIMI 2/3 flow) compared with those having closed vessels (TIMI 0/1 flow). It was concluded that, although significant differences could be shown for these markers with regard to outcome, they were of limited clinical value in comparison with an angiographic assessment of vessel patency.

At this institution interest in the use of CK-MB isoform measurements has focused on more subtle myocardial damage than that encountered in AMI [22].

The measurement has proved useful in the detection of peri-operative myocardial damage during cardiac surgery, particularly when comparing myocardial preservation techniques, such as those provided by various cardioplegic solutions and ventricular fibrillation with intermittent cross-clamping [23]. Figure 4 shows the time course of the changes in CK-MB2/MB1 ratio from induction of anaesthesia to 5 days after aortic cross-clamp release. The kinetics of CK-MB2 release is such that the ratio normally returns to below the upper reference value within 24 hours of surgery. This allows for the possibility of detecting acute myocardial events subsequent to surgery, at time points when the troponins have still not returned to normal values. It was concluded that CK-MB isoform separation could be useful for studies which require sensitive and discriminating end points for myocardial damage.

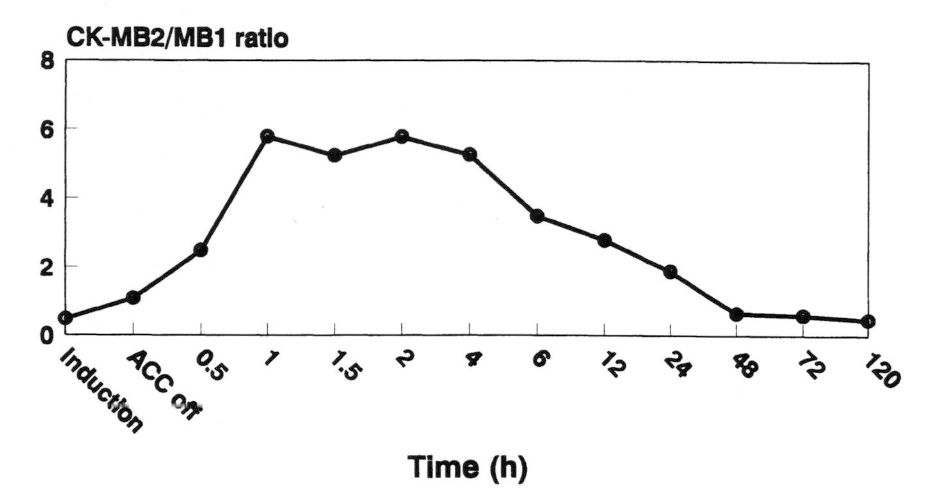

Figure 4. Changes in CK-MB isoform ratio during coronary artery bypass operation and the early post-operative period. The time axid is not linear, to emphasise the changes during the first 24 hours. Induction = induction of anaesthesia, ACC off = aortic cross-clamp removal. Data: unpublished, Analytical Unit.

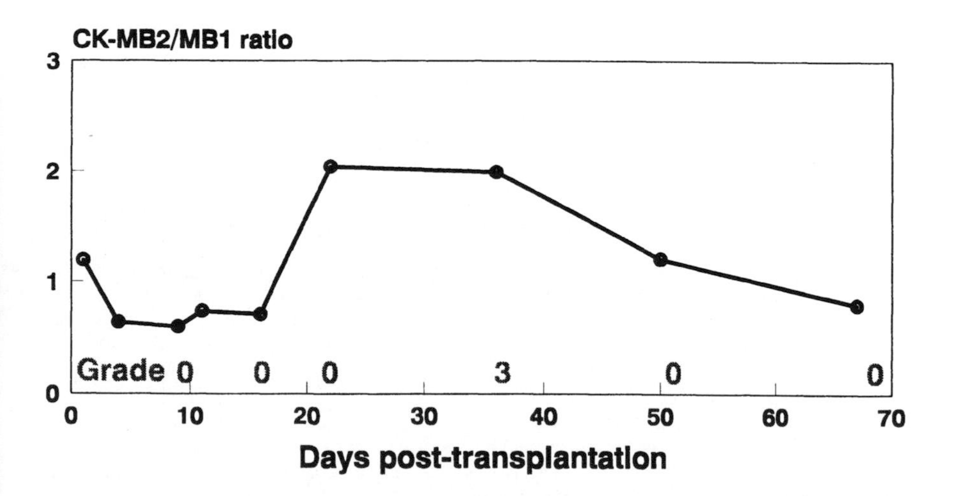

Figure 5. Changes in CK-MB isoform ratio following orthotopic heart transplantation. Grade indicates the grade of rejection diagnosed by endomyocardial biopsy. Data: unpublished, Analytical Unit.

Another interesting application was in the detection of heart graft rejection, following orthotopic transplantation [24,25]. There was a trend for the CK-MB2/MB1 ratio to rise before biopsy signs of rejection were apparent. Typical data are shown in Figure 5, from which it can be seen that the isoform ratio increases at least 15 days before the diagnosis of a serious, grade 3, rejection. The ratio was increasing at time points when CK-MB activity was well within normal limits and when routine biopsies were still detecting no signs of graft rejection (grade 0). Although the specificity and sensitivity for the diagnosis of rejection was of the order of 60%, better than other non-invasive tests, the measurement was not reliable enough to replace the gold standard of endomyocardial biopsy.

Other applications of CK-MB isoform separation have included the detection of minor myocardial damage following invasive cardiovascular procedures, such as coronary angioplasty [22], as a marker of deteriorating heart function in dilated cardiomyopathy[26] and radio-frequency ablation of accessory conduction pathways [27], as well as the pre-clinical applications outlined in Chapter 16 of this book.

Conclusions

Separation and quantitation of CK isoforms is a sensitive and rapid method for detecting myocardial damage soon after the release of the tissue isoforms from damaged muscle. Within the context of the time the methodology first appeared in a commercially viable form, it was a significant advance. However, the development of methods for the measurement of the troponins, and data on their prognostic value, has largely confined the technique to research settings. Because the kinetics of CK-MB release are faster in comparison with the troponins, isoform separation may still play a useful role, in association with measurements of the more cardio-specific markers.

References

1. Puleo PR, Guadagno PA, Roberts R, *et al.* Early diagnosis of acute myocardial infarction based on assay for subforms of creatine kinase-MB [see comments]. Circulation 1990; 82: 759-764.
2. Puleo PR, Meyer D, Wathen C, *et al.* Use of a rapid assay of subforms of creatine kinase-MB to diagnose or rule out acute myocardial infarction [see comments]. New England Journal of Medicine 1994; 331: 561-566.
3. Mair J, Artner-Dworzak E, Dienstl A, *et al.* Early detection of acute myocardial infarction by measurement of mass concentration of creatine kinase-MB. American Journal of Cardiology 1991; 68: 1545-1550.
4. Collinson PO, Rosalki SB, Kuwana T, *et al.* Early diagnosis of acute myocardial infarction by CK-MB mass measurements. Annals of Clinical Biochemistry 1992; 29: 43-47.
5. Hossein-Nia M, Johnston A, Holt DW. Performance of the Tandem-E CKMB II assay for the measurement of creatine kinase MB isoenzyme concentration. Commun Lab Med 1993; 7-11.
6. Perryman MB, Strauss AW, Buettner TL, Roberts R. Molecular heterogeneity of creatine kinase isoenzymes. Biochimica et Biophysica Acta 1983; 747: 284-290.
7. Lang H, Wurzburg U. Creatine kinase, an enzyme of many forms. Clinical Chemistry 1982; 28: 1439-1447.
8. Wu AH, Gornet TG, Wu VH, Brockie RE, Nishikawa A. Early diagnosis of acute myocardial infarction by rapid analysis of creatine kinase isoenzyme-3 (CK-MM) sub-types. Clinical Chemistry 1987; 33: 358-362.
9. Apple FS, Sharkey SW, Werdick M, Elsperger KJ, Tilbury RT. Analyses of creatine kinase isoenzymes and isoforms in serum to detect reperfusion after acute myocardial infarction. Clinical Chemistry 1987;

33: 507-511.
10. Puleo PR, Guadagno PA, Roberts R, Perryman MB. Sensitive, rapid assay of subforms of creatine kinase MB in plasma. Clinical Chemistry 1989; 35: 1452-1455.
11. Secchiero S, Altinier S, Zaninotto M, Lachin M, Plebani M. Evaluation of a new automated system for the determination of CK-MB isoforms. Journal of Clinical Laboratory Analysis 1995; 9: 359-365.
12. Kanemitsu F, Okigaki T. Creatine kinase MB isoforms for early diagnosis and monitoring of acute myocardial infarction. Clinica Chimica Acta 1992; 206: 191-199.
13. Kanemitsu F, Okigaki T. Characterization of human creatine kinase BB and MB isoforms by means of isoelectric focusing. Clinica Chimica Acta 1994; 231: 1-9.
14. Laurino JP, Fischberg-Bender E, Galligan S, Chang J. An immunochemical mass assay for the direct measurement of creatine kinase MB2. Annals of Clinical & Laboratory Science 1995; 25: 252-263.
15. Laurino JP, Bender EW, Kessimian N, Chang J, Pelletier T, Usategui M. Comparative sensitivities and specificities of the mass measurements of CK-MB2, CK-MB, and myoglobin for diagnosing acute myocardial infarction. Clinical Chemistry 1996; 42: 1454-1459.
16. McBride JH, Schotters SB. Immunochemical extraction and automated measurement of plasma creatine kinase MB isoenzyme and creatine kinase MB2 isoform. Journal of Clinical Laboratory Analysis 1997; 11: 163-168.
17. Davies J, Reynolds T, Penney MD. Creatine kinase isoforms: investigation of inhibitors of in vitro degradation and establishment of a reference range. Annals of Clinical Biochemistry 1992; 29: 202-205.
18. Puleo PR, Perryman MB, Bresser MA, Rokey R, Pratt CM, Roberts R. Creatine kinase isoform analysis in the detection and assessment of thrombolysis in man. Circulation 1987; 75: 1162-1169.
19. Christenson RH, Ohman EM, Vollmer RT, *et al.* Serum release of the creatine kinase tissue-specific isoforms MM3 and MB2 is simultaneous during myocardial reperfusion. Clinica Chimica Acta 1991; 200: 23-33.
20. Jaffe AS, Eisenberg PR, Abendschein DR. Conjoint use of MM and MB creatine kinase isoforms in detection of coronary recanalization. American Heart Journal 1994; 127: 1461-1466.
21. Christenson RH, Ohman EM, Topol EJ, *et al.* Creatine kinase MM and MB isoforms in patients receiving thrombolytic therapy and acute angiography. TAMI Study Group. Clinical Chemistry 1995; 41: 844-852.
22. Hossein-Nia M, Kallis P, Brown PA, *et al.* Creatine kinase MB isoforms: sensitive markers of ischemic myocardial damage. Clinical Chemistry 1994; 40: 1265-1271.
23. Anderson JR, Hossein-Nia M, Kallis P, *et al.* Comparison of two strategies for myocardial management during coronary artery operations [see comments]. Annals of Thoracic Surgery 1994; 58: 768-72; discussion 772-3.
24. Hossein-Nia M, Kallis P, Brown P, Murday AJ, Treasure T, Holt DW. Creatine kinase MB2 isoform release as a marker of peri-operative myocardial damage during cardiac transplantation. Clin Biochem 1994; 494-497.
25. Anderson JR, Hossein-Nia M, Brown P, Corbishley C, Murday AJ, Holt DW. Creatine Kinase MB isoforms: a potential predictor of acute cardiac allograft rejection. J Heart Transplant 1995; 14: 666-670.
26. Hossein-Nia M, Baig K, Goldman JH, *et al.* Creatine kinase isoforms as circulating markers of deterioration in idiopathic dilated cardiomyopathy. Clin Cardiol 1997; 20: 55-60.
27. Katrisis D, Hossein-Nia M, Anastasakis A, *et al.* Use of troponin-T concentration and creatinine kinase isoforms for quantitation of myocardial injury induced by radiofrequency catheter ablation. European Heart Journal 1997; 18: 1007-1013.

Chapter 3

MYOCARDIAL DAMAGE: THE ROLE OF TROPONIN T

Norbert Frey, Margit Müller-Bardorff, Hugo A. Katus

The troponin complex is located on the thin filament of the contractile apparatus (Figure 1). It consists of the subunits-troponin T (39 kD), troponin I (26 kD), and troponin C (18 kD) [1]. These subunits are the products of different genes and are not related to each other in protein structure [2].

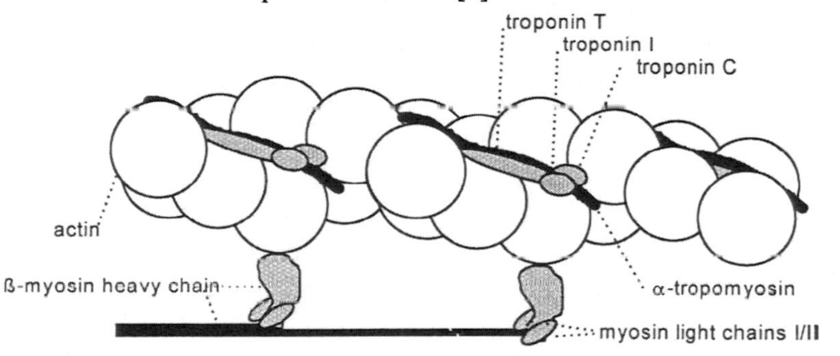

Figure 1. The sarcomere containing the troponin complex

The troponin complex plays a fundamental role in the transmission of calcium signal for acto-myosin-interaction. While troponin C acts as a molecular calcium sensor, troponin T and I regulate the strength and velocity of contraction by their isoform composition and phosphorylation status [3,4].

Distribution of troponin isoforms in different muscle types

The regulatory troponin complex is found exclusively in striated muscle. Within striated cardiac and skeletal muscle different isoforms of troponin T, I, and C are expressed, resulting in a muscle type specific pattern. This pattern may be altered when the muscle is exposed response to different workloads, stimulation protocols, and exogenous stimuli [5, 6]. The isoforms of the troponins result from transcription of genes which are specific for the respective muscle type. For troponin C, one specific gene encodes for the fast skeletal muscle isoform and another one for the slow cardiac troponin C isoform. By contrast troponin T and I are encoded by different genes in cardiac muscle, slow and fast twitch skeletal muscle. Only during fetal development in cardiac troponin

T expressed in skeletal muscle and *vice versa* skeletal muscle troponin I is expressed in cardiac muscle [7]. The skeletal and cardiac isoforms of troponin T and I differ markedly in their protein structure. However, for troponin proteins a high interspecies homology has been found [8].

Plasticity of isoform expression of troponin T and I subunits in muscle

Differentiated muscle retains a certain degree of plasticity by transcription of different genes, translational modifications by differential splicing of exonic sequencies, and posttranslational modification by different degrees of phosphorylation. In skeletal muscle, multiple isoforms of skeletal troponin T are observed, which differ by up to 11 aminoacids. The presence of up to four isoforms of cardiac troponin T has also been reported to be present in cardiac tissue as a consequence of myocardial hypertrophy [9]. This isoform composition may affect the contractile behaviour of the respective muscle types. However, the significance of the presence of multiple cardiac troponin T isoforms in human myocardium is still unknown. Mesnard *et al.* [10] found a reduced cardiac Troponin T isoform in only 50% of samples of failing human ventricles regardless of the cause of failure whereas another troponin T isoform was unexpectedly decreased compared to controls.

Both troponin T and I possess multiple sites which are accessible for phosphorylation. However, in viable myocytes the phosphorylation rate of troponin T is low. It is mediated via protein kinase C-ζ and mainly restricted to two, as yet, undetermined sites. In contrast, at least 7 sites of troponin I are modified by variable degrees of phosphorylation transducted by both protein kinase A and different isoforms of protein kinase C. For troponin I the phosphorylation sites are distributed throughout the molecule at serine 78, 43, 23/24 and threonin 144 and two more as yet unidentified sites [11]. The phosphorylation of troponin T and I results in a decrease in actomyosin ATPase activity and/or calcium sensitivity and eventually in a lower rate of cross bridge cycling.

Isoform expression of troponins

So far a reexpression of cardiac troponin T has not been observed in skeletal muscle. However, due to its fetal expression in skeletal muscle it may be that troponin T could be reexpressed in diseased and/or regenerating skeletal muscle. In fact, there is experimental evidence for reexpression of cardiac troponin T in regenerating skeletal muscle of rat [12]. In respect of expression of myofibrillar proteins in humans, the rat is not a good animal model since marked differences in tissue composition of isoforms of myofibrillar proteins have been observed between rat and human. It has been also argued that the elevations of troponin T observed in patients with skeletal muscle injury may indicate reexpression of cardiac troponin T. However, it has been shown that, with a more specific assay, cardiac troponin T is not elevated after skeletal muscle injury or in patients with skeletal muscle diseases. Furthermore, it has been reported that cardiac troponin T is elevated in skeletal muscle of Duchenne patients and patients with endstage renal disease [13,14,15]. The antibodies employed in the immunological analysis by these studies reacted with multiple bands of myofibrillar proteins other than troponin T on Westernblot analysis. Using more specific antibodies, we were not able to reproduce these findings, even when we used highly sensitive PCR-methods to detect a re-expression of cardiac troponin T in skeletal muscle [16]. Thus, at present, there is no clear evidence for a significant reexpression of cardiac troponin T in differentiated skeletal muscle.

Compartmentation and degradation of troponin subunits in myocardium

The troponins are myofibrillar proteins which are complexed in the three dimensional network of the contractile machinery. A minor fraction of troponin T and troponin I of approximately 6 to 8 % is found in the cytosolic pool [17]. This cytosolic pool may serve as a precursor pool for myofibrillar assembly.

However, the true concentration of the myofibrillar proteins, compartmented in the myocyte, is difficult to determine, as all solubilisation methods may lead to extraction of denatured or complexed proteins. This may result in altered immunological detection [18, 19, 20]. As observed for many different cardiac proteins, the relative concentration of the contractile proteins may change with workload and wall-stress. This could further complicate the assessment of infarct size when serum concentrations of the troponins are measured

Following the loss of membrane integrity the cytosolic pool of troponins may be released from damaged myocytes. This has been shown experimentally for troponin T using isolated Langendorff preparations damaged by the "Calcium Paradox" [21]. In this model, calcium overload leads to a massive release of cTnT without disturbance of the ordinated structure of the myofibrils. It may be speculated that troponin I could behave similarly in this experimental model.

The modes of liberation of troponins

The mode of liberation of troponins from the myofibrils are unclear at present unclear. It is believed, that the myofibrils are degraded after activation of proteolytic enzymes such as calpain, following a rise in intracellular calcium concentration. Others have shown, that a shift in pH, by itself, may induce a dissociation of troponins from the myofibrils. There are only few studies available following the disappearance of troponins from the myofibrils in experimental infarctions. In one such study troponin I was rapidly degraded, while unaltered troponin T was shown to persist for up to 6 days after the onset of myocardial infarction [22].

The troponin fragments which originate during the degradation process and, finally, gain access to the circulation are undetermined. Gorza and co-workers [23], reported that ischemia may disturb the ordered structure of the myofibrils and expose antigenic epitopes not normally accessible in nonischemic tissue. They also observed that following myocardial ischemia of troponin T and troponin I containing complexes of 250 kD and 65 kD, are formed. Wu *et al* identified circulating complexes of troponin I and troponin C, homodimers of troponin I, unaltered troponin I, and fragments of troponin I of many different sizes. By contrast, for troponin T, only one species could be identified in circulation. Wu and *et al* also observed, that phosphorylation of troponin I may interfere with recognition of the epitope by different monospecific antibodies [personal communication]. Since troponin T and I do not occur in peripheral blood as single unaltered molecules, it is likely that different assays using different antibodies may not identify some of the fragments, complexes or phosphorylated isoforms.

Analytical methods for the detections of troponins

For cardiac troponin T determinations only one assay [24] is currently available (Boehringer Mannheim, FRG). This troponin T assay has undergone several steps of improvement [25, 26, 27, 28] (Figure 2). At present, the laboratory based methods use an immunoelectric technique employing two cardiac specific antibodies. The measurement can be completed within 10 minutes with an analytical sensitivity of 0,04 µg/l using the ELECSYS-analyser. In addition to the laboratory based methods, a test strip assay was developed for bedside analysis of whole blood samples. This assay has a detection limit of 0.08 µg/l and uses an automated test strip reader. This reader converts the intensity of a positive line via a CCD camera and a computer program into quantitative results within a concentration range of 0.1 to 3.0 µg/l.

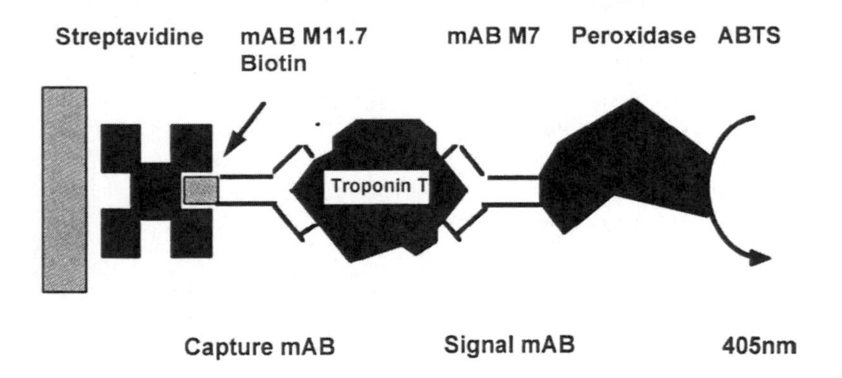

Figure 2.A. Principle of troponin T assays: troponin T ELISA

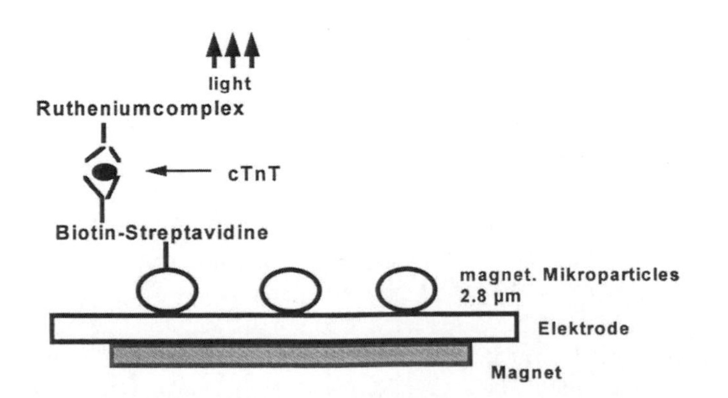

Figure 2.B. Principle of troponin T assays: electrochemiluminiscence assay

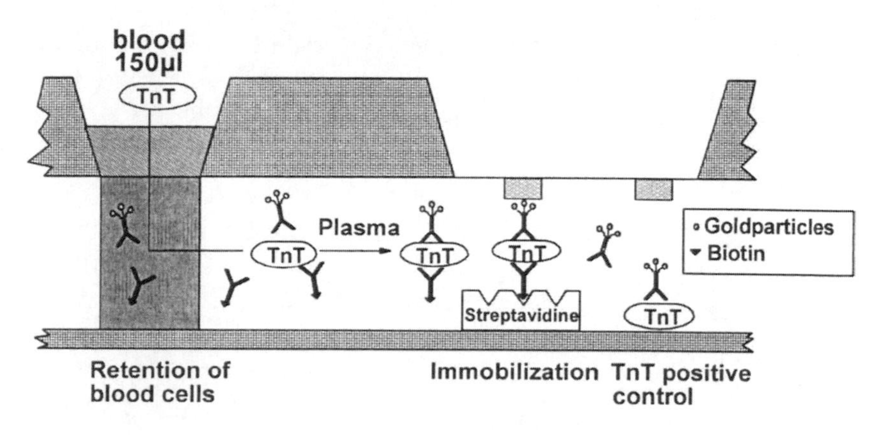

Figure 2.C. Principle of troponin T assays: troponin T bedside assay

The antibodies utilised in the Boehringer Mannheim test are well characterized. The two epitopes of the selected monoclonal antibodies are only 6 aminoacid residues apart [25,29]. This part of the troponin T molecule differs markedly in protein structure in cardiac and skeletal muscle.

By contrast, several troponin I assays [30] are available by different manufacturers. These assays use either mono- or polyclonal antibodies, which are directed against different antigenic determinants. As a consequence, these assays have different analytical sensitivities and discriminator values. They are therefore difficult to standardise. However, standardisation of the different assays is urgently needed to allow a comparative analysis of the different troponin I concentrations observed in patients with myocardial damage.

A bedside assay is available for troponin I which uses a goat polyclonal antibody and a mouse monoclonal antibody (Spectral Diagnostic). This qualitative assay has a detection limit of 0.1 µg/l and may be comparable in its diagnostic performance to the second generation of the troponin T bedside assay [31].

Analytical sensitivity of troponin T and I and discriminator values in clinical practice

In table 1 information regarding currently available troponin assays is given. As this is a rapidly evolving field, it is impossible to cover all available assays and analytical evaluations of the many different methods. However, it should be stressed that the clinical value of new diagnostic methods strongly depends on both the properties and performance of the assay systems used and the analyses conducted.

In healthy blood donors, troponin T values are below the detection limit of the assay. In 4595 patients without cardiac diseases seen by general practioners, 99.6 % of all samples were below 0.06 µg/l [25]. It is unclear whether the minor increase in troponin T levels above 0.06 µg/l found in some patients was due to the presence of minor myocardial damage or whether it is explicable by the analytical imprecision of the assay at the detection limit . Cardiac troponin T is also undetectable in healthy blood donors, but it was found to be slightly elevated in patients with heart failure. It is conceivable that, in the latter group of patients, there is myocardial cell damage severe enough to cause release of traces of troponin T. Thus at present, it is believed, that troponin T is either not normally present or escapes detection by available assays. The absence of the marker in blood is one of the requirements for a high sensitivity of a diagnostic assay.

Molecule	Company	Instrument	Analyt. Sens. [µg/l]	Cut-off [µg/l]
bed side assays				
cTnT	Boehringer M	-	0.08	n.a.
cTnT	Boehringer M	Cardiac Reader	<0.08	0.1
cTnI	Spectral	-	0.1	0.1
Laboraty based methods				
cTnT	Boehringer M	ELECSYS 2010	0.01	0.1
cTnI	Dade	STRATUS II	0.35	1.5
cTnI	Behring M	OPUS	0.5	2.0
cTnI	Sanofi	ACCESS	0.03	0.1

Table 1. Features of some of the available troponin assays

Furthermore, the absence of detectable levels raises the possibility that the release of what are strictily intracellular molecules into the blood stream may cause induction of auto-antibody formation. In fact Bohner *et al* [32] showed the presence of auto-antibodies directed against troponin I in patients with proven reinfarctions. In these patients they observed extremely low levels of troponin I but high levels of CKMB. However, auto-antibodies were not reported for troponin T in the same patient.

Diagnosis of acute myocardial infarction

The diagnostic performance of troponins has been tested in patients with myocardial infarction against the "gold standard" of CKMB and ECG changes. Due to this comparative approach CKMB must perform at least as well as the troponins in this setting. It is currently accepted, that the diagnosis of myocardial infarction cannot be maintained if appropriately collected blood samples are negative for troponins on repeated measurements.

Both the kinetics of troponin appearance and its persistence in circulation must be considered when the diagnostic efficiency of troponins is evaluated. In patients with definite myocardial infarction, troponin T appears in the circulation 3-4 hours after the onset of pain in circulation. However, in some patients with myocardial infarction troponin T or I may not become elevated before 8 hours after the onset of symptoms. It is, therefore, recommended that a second test for troponins is performed at 8 to 24 hours after the onset of symptoms in order to definitely exclude the presence of myocardial infarction.

In the German Multicenter Trial [22] it was observed that 48 % of all patients with myocardial infarction were positive for troponin T on admission. In the GUSTO III-Trial only 18 % of the chest pain patients were positive, when tested with the rapid bed side assay for troponin T, which has a discriminator value of 0.22 μg/l [33]. Thus troponin T is clearly not suitable for the early diagnostic work up of patients with myocardial infarction.

Troponins remain elevated for a prolonged time following the onset of acute myocardial infarction. The prolonged elevation of troponin T is due to the continued release of this molecule from necrotising cardiomyocytes, since the serum half life of troponin T is only two hours [34]. The serum half life of troponin I is unknown at present.

Due to its prolonged release, troponin T remains elevated for more than 2 weeks in some patients. In a direct comparison cardiac troponin T was elevated longer than troponin I and was significantly more sensitive on the 7th day of acute myocardial infarction [35].

For data exist on the value of troponins to assess noninvasively the size of acute myocardial infarction. In experimental studies troponin T levels on day 3 correlated with histological estimates of infarct size [36]. For troponin I a comparable relationship was not observed, in fact Cummins and co-workers reported that troponin I levels do not correlate with infarct size in an experimental canine model [37]. In clinical studies it was reported, that both troponin T and I are correlated to the size of thallium scan perfusion defects.

When myocardial cell damage is induced during radiofrequency ablation the troponins appear in circulation and rise to much higher levels than those found for CKMB [38, 39]. In an experimental setting of isolated Langendorff heart preparations with hypoxemia periods of 20, 40, and 60 minutes, troponin T and CK appeared in the coronary sinus effluent at similar time intervals [40].

The added diagnostic value of troponins in patients with definite myocardial infarction is limited, since the diagnosis of acute myocardial infarction is defined by WHO criteria. Thus, troponins do not aid in the diagnostic classification of these patients. Furthermore troponins are not helpful in the selection of treatment strategies in patients who present very early after the onset of symptoms. In these patients a negative troponin T value only indicates that the patient is presenting early after the onset of pain and that they belong to the group of patients which will benefit most from reperfusion therapy. Thus, due to the timing of its appearance in the circulation, cardiac tropins are not suitable for the selection of treatment strategies. However, available data indicate, that patients with troponin T elevations on admission have a much higher mortality than patients who present with a negative troponin T test and myocardial infarction. Thus troponin T may be used in the early phase of a definite myocardial infarction to assess the risks of the individual patient [50,53].

Assessment of reperfusion of the myocardial infarction zone

The appearance of cardiac markers in blood depends on reperfusion of the myocardial infarction zone. Consequently the serum concentration changes of cardiac markers may indicate the efficiency and time delay from reperfusion of the infarct zone. For troponin

T, a biphasic release was observed following recanalisation of infarct related arteries, while a monophasic rise in serum concentration can be observed in patients with a permanent occlusion. The observed differences can be explained by the variable washout of the unbound cytostolic pool of cTnT. Different indices can be delineated from the serum concentration curves in order to predict non-invasively reperfusion of the infarct zone. We have proposed a ratio of the troponin T levels at 14 to 38 hours [41]. Others have used the difference between troponin T levels at baseline and 90 minutes after the onset of treatment [42].

Troponin I release kinetics are also altered by reperfusion. However, the differences are not as striking as those observed for troponin T. Troponin I release resembles that of CK in patients with reperfused acute myocardial infarction [43].

Minor myocardial damage

In 1990 we [44] first reported on the presence of troponin T in patients with unstable angina but no other evidence for acute myocardial infarction, based on ECG recordings and cardiac enzymes. These patients were a high risk subgroup for further cardiac events as several of the troponin T positive patients progressed to acute myocardial infarction or death. This early study triggered great interest and a large number of trials subsequently confirmed that approximately 30 % of all patients with angina at rest have positive results for troponin T [22,45-55]. It was also shown, that troponin T positive patients have a survival rate in hospital similar to that of patients with definite acute myocardial infarction treated with thrombolytic agents. In the GUSTO IIa trial, troponin T elevations and reversible ST segment depression were independend predictors of cardiac events, while CKMB mass elevations were not [53]. These findings essentially confirmed the results shown in the European Multicenter Trial [45]. Several investigators asessed the prognostic value of troponin T measurements at admission in patients with unstable angina. In the Scandinavian Multicenter trial [46] no difference was observed in the 6 months' complication rate when Troponin T positive unstable angina patients and patients with definite myocardial infarction were compared. By contrast, unstable agina patients without troponin T elevations had an outcome similar to that in patients with atypical chest pain. In the FRISC trial [54] cardiac events were also similar at 5 months follow up in patients with acute myocardial infarction and patients with troponin T positive unstable angina. The FRISC study group for the first time indicated that troponin T may also aid in the selection of treatment modalities [56]. This trial showed that dalteparin treatment was only effective in patients with elevated troponin T (Figure 3).

By contrast troponin T negative patients had a similar cardiac event rate regardless of dalteparin treatment. When dalteparin treatment was stopped in the troponin T positive group at 4 weeks, the event rate rose to levels observed in the troponin T positive placebo group at the end of 5 months. This indirectly shows that thrombus formation in unstable plaques, and embolization, may cause micronecroses which are detectable as troponin T elevations in blood.

Stubbs *et al* [49] reported, that patients with cTnT positive unstable angina at the index event remain a high risk subgroup well beyond 1 year follow up. At present it is unclear what causes the persistence of an increased risk for cardiac events in patients with troponin T positive angina, even after appropriate treatment was begun. Troponin I elevations were also reported to indicate a high risk complication rates with increasing troponin I levels in blood [57]. These findings were supported subclass of unstable angina. Antmann and co-workers found an increase in by the reports of different groups using different troponin I assays.

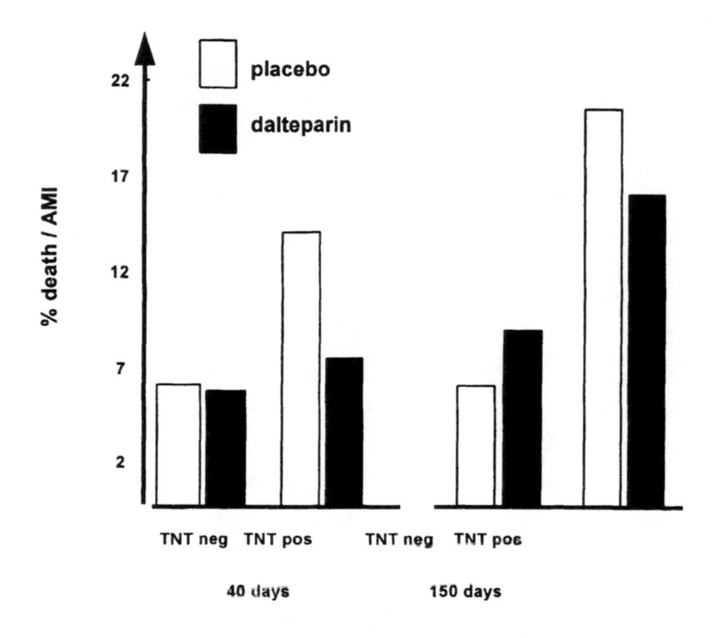

Figure 3. FRISC-trial: dalteparin treatment in troponin T positive and negative patients with acute coronary syndromes. Dalteparin was stopped at 30 days. Note the "catch-up"-phenomenon in complications in the troponin T positive group treated with dalteparin for 30 days.

Four trials compared troponin T and I in unstable angina patients. In the GUSTO IIa studies, troponin I was not independently predictive when troponin T was forced into a logistic regression model first. However, troponin T remained predictive even when troponin I was considered first [58]. In the TRIM study there was no statistical difference in the power of risk stratifications between troponin T and I. However, troponin T tended to be more predictive than Troponin I [59]. The comparison of the performance of the rapid bedside assay of tropoinin T and I, which was published recently [31] is not designed to allow a comparative assessment of the performance of both markers. Hamm and coworkers tested an outdated low sensitivity troponin T assay, which is no longer available. Furthermore they used troponin I measurements that were not performed under field, but under laboratory conditions. In contrast to troponin T troponin I was measured from fresh plasma instead of whole blood samples by a single experienced laboratory technician during regular working hours. Furthermore the physicians were blinded only for troponin I but not for troponin T results. Thus, this comparison between both assays was highly biased in favour of troponin I. Under these unbalanced conditions there was a tendency towards a better predictive power of the cTnI assay which, however, was not statistically significant.

The comparative analysis of troponin T and I assays is difficult since different cut off levels for troponin I varying from 1.5 to 0.03 μg/l were used by different groups. The liberal use of different troponin I discriminator levels hinders a meaningful comparison of troponin T and I performances. Furthermore, it is necessary to record changes in specificity, which may result in modifications in cut off levels. For example even a conservative cut-off value of 1.0μg/l could lead to false positive cTnI test results [60]. Thus, in future studies, floating discriminator values for both troponin I and troponin T could be used to test the predictive power of each marker molecule.

The high sensitivity of troponin measurements has also lead to their use to assess troponin T in patients with myocarditis and for monitoring graft rejection in patients following cardiac transplantation. Lauer *et al* [61] found troponin T to be the best predictor for the presence of myocarditis and could demonstrate a relationship between troponin T levels and histological grades of interstitial infiltrates. Dengler *et al* [62] showed, that troponin T levels greater than0.04 μg/l have a high negative predictive value for a rejection episode of 96 %. However, the positive predictive value was only about 85 %. Thus troponin T negative patients can be followed conservatively, while troponin T positive patients need to be followed by repetitive endomyocardial biopsies, which is the routine procedure for monitoring graft rejection in most centres.

Summary

The troponins have become the markers of choice in patients with suspected acute coronary syndromes. However, in patients with definite myocardial infarction cardiac enzymes and ECG recordings are sufficient for the diagnosis. In the group of patients with chest pain the troponins not only serve to rule out acute myocardial infarction but also to perfom a risk stratification in these patients. In the future the troponins may also be used for the selection of treatment regimens in patients with acute coronary syndromes.

The relative value of troponin I as compared to troponin T will depend on the analytical performance of the assay systems since both markers are similar in many biological respects. The scientific evaluations of this comparative analysis requires that both markers are analysed under identical conditions using samples from well defined subgroups of patients.

References

1. Potter JB. Preparation of troponin and ist subunits. Methods Enzymol. 1982;85: 241-263.
2. Wade R, Kedes L. Developmental regulation of contractile protein genes. Ann Rev Physiol 1989; 51:179-88.
3. Ruegg JC. Troponin, the on-off switch of muscle contraction in striated muscle. In: Ruegg ed. Calcium in Muscle Activation. Springer Verlag; 1986; 83-94.
4. Farah CS, Reinach FC. The troponin complex and regulation of muscle contraction. FASEB J 1995;9:755-67.
5. Gao L, Kennedy JM, Solaro RJ. Differential Expression of TnI and TnT Isoforms in Rabbit Heart during the Perinatal Period and during Cardiovascular Stress. J Mol Cell Cardiol 1995;27:541-50.
6. Müller-Bardorff M, Katus HA, Lange R, Jahn L. Cardiomyoplasty: Effects of permanent electrical stimulation of skeletal muscle on the isoform pattern of myofibrillar proteins and the expression of muscle specific transcription factors. Circulation 1995; 95: I-48.

7. Mesnard L, Logeart D, Taviaux S, Diriong S, Mercadier JJ, Samson F. Human cardiac troponin T: cloning and expression of new isoforms in the normal and failing heart. Circ Res 1995;76:687-692.
8. Mesnard L, Samson F, Espinasse I, Durand J, Neveux JY, Mercadier JJ. Molecular cloning and developmental expression of human cardiac troponin T. FEBS Lett 1993;328:139-44.
9. Gulati J, Akella AB, Nikolic SD, Starc V, Siri F. Shifts in contractile regulatory protein subunitstroponin T and troponin I in cardiac hypertrophy. Biochem Biophys Res Commun 1994;202:384-90.
10. Mesnard-Rouiller L, Mercadier JJ, Butler-Browne G, Heimburger M, Logeart D, Allan PD, Samson F. Troponin T mRNA and protein isoforms in the left ventricle: pattern of expression in failing and control hearts. J Mol Cell Cardiol 1997;29:3043-55.
11. Jideama NM, Noland A, Raynor RL, Blobe GC, Fabbro D, Kazanietz MG, Blumberg PM, Hannun YA, Kuo JF. Phosphorylation Specificities of Protein Kinase C Isozymes for Bovine Cardiac Troponin I and Troponin T and Sites within these Proteins and Regulation of Myofilament Properties. J Biol Chem 1996;271:23277-23283.
12. Sabry MA, Dhoot GK. Identification and pattern of transitions of cardiac, adult slow and slow skeletal muscle like embryonic isoforms of troponin T in developing rat and human skeletal muscles. J Muscle Res Cell Motil 1991;12:262-70.
13. Bodor GS, Survant L, Voss EM, Smith S, Porterfield D, Apple FS. Cardiac troponin T composition in normal and regenerating human skeletal muscle. Clin Chem 1997;43:476-84.
14. Braun SL, Pongratz DE, Bialk P, Liem S, Schlotter B, Vogt W. Discrepant results for cardiac troponin T and troponin I in chronic myopathy depending on instrument and assay generation. Clin Chem 1996; 42: 2039-41.
15. Akagi M, Nagake Y, Makino H, Shikata K, Ota Z. A comperative study of myocardial troponin T levels in patients undergoing hemodialysis. Jpn J Nephrol 1995; 37: 639-43.
16. Haller C, Zehelein J, Remppis A, Müller-Bardorff M, Katus HA. Cardiac troponin T in patients with end stage renal disease: absence of expression in truncal skeletal muscle. Clin Chem [in press].
17. Katus HA, Remppis A, Scheffold T, Diederich KW, Kübler W. Intracellular compartmentation of cardiac troponin T and ist release kinetics in patients with reperfused and nonreperfused myocardial infarction. Am J Cardiol 1991;67:1360-7.
18. Lavigne L, Waskiewics D, Pervaiz G, Fagan G, Whiteley G. Investigation of serum troponin I hetcrogeneity and complexation to troponin T. Clin Chem 1996; 42: S312.
19. Katruhka A, Petterson K, Lovgren T, Mitrunen K, Beresnikova A, Bulargina T, Esakova T, Severina M. Cardiac troponin I cardiac troponin C complex in serum of patients with acute myocardial infarction. Proc XVI Inter Cong Clin Chem 1996: 221.
20. Katruhka A, Petterson K, Lovgren T, Mitrunen K,Mykkanen P, Beresnikova A, Esakova T. Cardiac troponin C influences the binding of monoclonal antibodies to cardiac troponin I. Proc XVI Inter Cong Clin Chem 1996: 221.
21. Remppis A, Scheffold T, Greten J, Haass M, Greten T, Kubler W, Katus HA. Intracellular compartmentation of troponin T: release kinetics after global ischemia and calcium paradox in the isolated perfused rat heart. J Mol Cell Cardiol 1995;27:793-803.
22. Katus HA, Remppis A, Neumann FJ, Scheffold T, Diederich KW, Vinar G, Noe A, Matern G, Kübler W. The diagnostic efficiency of troponin T measurements in acute myocardial infarction . Circulation 1991;83:902-12.
23. Gorza L, Menabo R, Vitadello M, Bergamini CM, Di Lisa F. Cardiomyocyte Troponin T Immunreactivity is modified by cross-linking resulting from intracellular calcium overload. Circulation 1996;93:1896-1904.
24. Katus HA, Looser S, Hallermayer K, Remppis A, Scheffold T, Borgya A, Essig U, Geuss U. Development and in vitro characterization of a new immunoassay of cardiac troponin T. Clin Chem; 38: 386-393.
25. Müller-Bardorff M, Hallermayer K, Schröder A, Ebert C, Borgya A, Gerhardt W, Remppis A, Zehelein J, Katus HA. Improved troponin T ELISA specific for cardiac troponin T isoform: assay development and analytical and clinical validation. Clin Chem 1997;43458-466.
26. Müller-Bardorff M, Freitag H, Scheffold T, Remppis A, Kübler W, Katus HA. Development and characterization of a rapid assay for bedside determinations of cardiac troponin T. Circulation 1995; 92: 2869-2875.
27. Müller-Bardorff M, Kampmann M, Rauscher T, Ebert J, Zehelein J, Hallermayer K, Klein G, Katus HA. Characterization and clinical evaluation of a rapid, sensitive and highly specific electrochemiluminiscence immunoassay for cardiac troponin T. Eur Heart J 1997; 18: P3676.

28. Müller-Bardorff M, Rauscher T, Kampmann M, Schoolmann S, Mangold D, Zerback R, Remppis A. Quantitative assessment of the bedside troponin T assay with an automated reader. Circulation 1997; 96: 1516A.
29. Macht M, Fiedler W, Kürzinger K, Przbylski M. Mass spectrometric mapping of protein epitope structures of myocardial infact markers myoglobin and troponin T. Biochemistry 1996;35:15633-15639.
30. Bodor GS, Porter S, Landt Y, Ladenson JH. Development of monoclonal antibodies of cardiac troponin I and preliminary results in suspected cases of myocardial infarction. Clin Chem 1992; 38: 2203-14.
31. Hamm CW, Goldmann BU, Heeschen C, Kreymann G, Berger J, Meinertz T,. Emergency room triage of patients with acute chest pain by means of rapid testing for cardiac troponin T or troponin I. N Engl J Med 1997;337:1648-1653.
32. Bohner J, von Pape KW, Hannes W, Stegmann T. False-nagative immunoassay results for cardiac troponin I probably due to circulating troponin I antibodies. Clin Chem 1996; 42: 2046.
33. Ohman EM, Armstrong PW, Weaver WD, Gibler WB, Bates ER, Stebbins AL, Hochmann JS. Prognostic value of whole-blood qualitative troponin T testing in patients with acute myocardial infarction in the GUSTO-III trial. Circulation 1997, 96: I-216.
34. Katus HA, Remppis A, Looser S, Hallermeier K, Scheffold T, Kübler W. Enzyme linked immunoassay of cardiac troponin T for the detection of acute myocardial infarction in patients. J Mol Cell Cardiol 1989; 21: 1349-53.
35. Mair J, Genser N, Morandell D, Maier J, Mair P, Lechleitner P, Calzolari C, Larue C, Ambach E, Dienstl F, Pau B, Puschendorf B. Cardiac troponin I in the diagnosis of myocardial injury and infarction. Clin Chim Acta 1996; 245:19-38.
36. Zimmermann R, Zehelein J, Licka MB, Tiefenbacher CP, Dengler TJ. Troponin T level 72 hours after acute myocardial infarction as serologic estimate of infarct size. Circulation 1997; 96: I-274.
37. Cummins B, Cummins P. Cardiac-specific troponin release in canine experimental myocardial infarction: Development of a sensitive enzyme-linked immunoassay. J Mol Cell Cardiol 1987; 19: 999-1010.
38. Del Rey JM, Madrid AH, Novo L, Sanchez A, Martin J, Ruby J, Pallares E, Kais J, Willer BG, Manzano JG, Silvestre I, Jimenez A, Ripoll E, Moro C. Evaluation of biochemical markers of myocardial lesion after radiofrequency ablation. Value of troponin I. Rev Esp Cardiol 1997; 50:552-60.
39. Katritsis D, Hossein-Nia M, Anastasakis A, Poloniecki I, Holt DW, Camm AJ, Ward DE, Rowland E. Use of troponin T concentration and kinase isoforms for quantitation of myocardial injury induced by radiofrequency catheter ablation. Eur Heart J 1997;18:1007-13.
40. Yamahara Y, Asayama J, Kobara M, Ohta B, Matsumoto T, Miyazaki H, Tatsumi T, Ishibashi K, Inoue D *et al.* Basic Res Cardiol 1994; 89:241-9.
41. Remppis A, Scheffold T, Karrer O, Zehelein J, Hamm C, Grünig E, Bode C, Kübler W, Katus HA. Assessment of reperfusion of the infarct zone after acute myocardial infarction by serial cardiac troponin T measurements in serum. Br Heart J 1994; 71: 242-8.
42. Laperche T, Steg G, Dehoux M, Bessiano J, Grollier G, Aliot E, Mossard JM, Aubry P, Coisne D, Hanssen M, Iliou MC, the PERM Study Group. A study of biochemical markers of reperfusion early after thrombolysis for acute myocardial infarction. Circulation 1995;92:2079-86.
43. Bertinchant JP, Larue C, Pernel I, Ledermann B, Fabbro-Peray P, Beck L, Calzolari C, Trinquier S, Nigond J, Pau B. Release kinetics of serum cardiac troponin I in ischemic myocardial injury. Clin Biochem 1996; 29: 587-594.
44. Katus HA, Kübler W. Detection of myocardial cell damage in patients with unstable angina by serodiagnostic tools. In: Unstable Angina; Bleifeld W, Braunwald E and Hamm C [Editors]. Springer Verlag Heidelberg 1990; 92-100.
45. Hamm CW, Ravkilde J, Gerhardt W, Jorgensen P, Peheim E, Ljungdahl L, Katus HA. The prognostic value of serum troponin T in unstable angina. N Engl J Med 1992;327:146-50.
46. Ravkilde J, Hørder M, Gerhardt W, Ljungdahl L, Pettersson T, Tryding N, Møller BH, Hamfeldt Å, Graven T, Asberg A, Helin M, Penttilä I, Thygesen K. Diagnostic performance and prognostic value of serum troponin T in suspected acute myocardial infarction. Scand J Clin Lab Invest 1993; 53: 677-685.
47. Seino Y, Tomita Y, Takano T, Hayakawa H. Early identification of cardiac events with serum troponin T in patients with unstable angina. Lancet 1993; 342:1236-1237.
48. Burlina A, Zaniotto M, Secchiero S, Ruben D, Accorsi F. Troponin T as a marker of ischaemic myocardial injury. Clin Biochem 1994;27:113-21.

49. Stubbs P, Collinson P, Moseley D, Greenwood T, Noble MIM. Prospective study of the role of cardiac Troponin T in patients admitted with unstable angina. Br Med J 1996a; 313: 262-264.
50. Stubbs P, Collinson P, Moseley D, Greenwood T, Noble M. Prognostic significance of admission troponin T concentrations in patients with myocardial infarction. Circulation 1996b; 94: 1291-1297.
51. De Winter RJ, Koster RW, Schotveld JH, Sturk A, van Straalen JP, Sanders GT. Prognostic value of troponin T, myoglobin, and CK-MB mass in patients presenting with chest pain without acute myocardial infarction. Heart 1996; 75: 235-239.
52. Wu AHB, Abbas SA, Green S, Pearsall L, Dhakam S, Azar R, Onoroski M, Senaie A, McKay RG, Waters D. Prognostic value of cardiac troponin T in unstable angina pectoris. Am J Cardiol 1995; 76: 970-2.
53. Ohman EM, Armstrong PW, Christenson RH, Granger CB, Katus HA, Hamm CW, O'Hanesian MA, Wagner GS, Kleiman NS, Harrell FE, Califf RM, Topol EJ for the GUSTO-IIA Investigators. Cardiac troponin T levels for risk stratification in acute myocardial ischemia. N Engl J Med 1996; 335: 1333-1341.
54. Lindahl B, Venge P, Wallentin L. Relation between troponin T and the risk of subsequent cardiac events in unstable coronary artery disease. Circulation 1996; 93: 1651-1657.
55. Pettijohn TL, Doyle T, Spiekerman AM, Watson LE, Riggs MW, Lawrence ME. Usefulness of positive troponin-T and negative creatine kinase levels in identifying high-risk patients with unstable angina pectoris. Am J Cardiol 1997; 80: 510-511.
56. Lindahl B, Venge P, Wallentin L, for the Fragmin in Unstable Coronary Artery Disease [FRISC] Study Group. Troponin T identifies patients with unstable coronary artery disease who benefit from long-term antithrombotic protection. J Am Coll Cardiol 1997b; 29: 43-48.
57. Antman EM, Milenko J, Tanasijevic MD, Thompson B, Schactman M, McCabe CH, Cannon CP, Fischer GA, Fung AY, Thompson C, Wybenga D, Braunwald E. Cardiac-specific troponin I levels to predict the risk of mortality in patients with acute coronary syndromes. N Engl J Med 1996; 335: 1342-9.
58. Ohman EM, Christenson RH, Peck S, Pieper K, Katus HA, Newby K, Armstrong P. Comparison of troponin I and T for risk stratification in patients with acute coronary syndromes. Circulation 1996; 94: I-323.
59. Lüscher MS, Thygesen K, Ravkilde J, Heickendorff L, for the TRIM study group. Applicability of cardiac troponin T and I for early risk stratification in unstable coronary artery disease. Circulation 1997; 96: 2578-2585.
60. Wright SA, Sawyer DB, Sacks DB, Chyu S, Goldhaber SZ. Elevation of troponin I levels in patients without evidence of myocardial injury. JAMA 1997; 278: 2144.
61. Lauer B, Niederau C, Kühl U, Schannwell M, Pauschinger M, Strauer BE, Schultheiss HP. Cardiac troponin T in patients with clinically suspected myocarditis. JACC 1997; 30: 1354-9.
62. Dengler TJ, Braun K, Zehelein J, Bärle S, Szabo G, Zimmermann R. Elevated troponin T serum levels in acute graft rejection after cardiac transplantation. Circulation 1996; 94: I-290.

Chapter 4

TROPONIN I: STRUCTURE, PHYSIOLOGY AND IT'S ROLE IN RISK STRATIFICATION OF ANGINA PATIENTS

Mario Plebani and Mauro Panteghini

Ischaemic heart disease is one of the most common and serious health problems in contemporary society and acute myocardial infarction (AMI) is the major cause of cardiovascular morbidity and death. Progress in the field of thrombolytic therapy has significantly affected the management of patients with AMI. Timely reperfusion for an evolving infarction often prevents immediate mortality, and reduces the risk of severe infarct, thus preserving left ventricular function. Although the familiar triad of clinical history, electrocardiogram (ECG), and serum enzyme analysis is of importance in the diagnosis of AMI, it does not always allow an early diagnosis, particularly in cases of so-called non-Q-wave infarction and minor myocardial necrosis. Moreover, although specialist techniques such as planar radioisotope imaging, single photon emission computed tomography (SPECT), position emission tomography (PET), echocardiography, ventriculography and nuclear magnetic resonance (NMR) imaging have been used to confirm the diagnosis in patients with equivocal symptoms and ECG findings, most of these procedures are not widely available to patients due to very high costs and long turnaround time. Among the various bichemical markers which have been suggested to be suitable for the detection of myocardial injury, only a very small number can be singled out as being of particular significance. A serious limitation of most of these markers is the lack of cardiac specificity, which results in a high percentage of false-positive results. For this reason, research in the past decade has focused on the structural and functional proteins of the cardiac muscle, which are highly specific for the myocardium. Initially the search started by choosing the cardiac myosin-antimyosin system; cardiac myosin being present in the myocardium at a relatively high intracellular concentration. However, the highly complex situation regarding the specificity of cardiac myosin for myocardial tissue, together with its isoform switching at various developmental stages and under a number of pathological conditions, has jeopardised its position as an ideal cardiospecific marker. More recently, the characterization of the troponin complex and the development of immuno assays for the measurement of these proteins in serum represent a major advance in the detection of myocardial damage.

The troponin complex: biochemistry and physiology

Troponin, together with tropomyosin, belongs to the group of structural proteins involved in the regulation of striated and cardiac muscle contraction [1,2]. The troponin complex, located on the thin filament of striated muscle, controls the interaction of thick and thin filaments in response to alterations in intracellular Ca^{2+} concentrations. It is composed of three protein components, troponin C (TnC), troponin I (TnI), and troponin T (TnT), each of which performs a specific function (Figure 1).

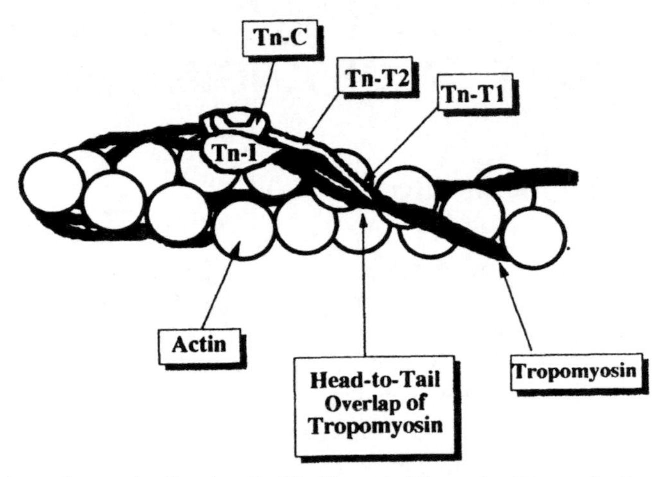

Figure 1. The troponin complex found on the thin filament of the contractile complex in muscle tissue.

TnC is a 18.5 kD, Ca^{2+}-binding protein containing four metal-binding sites. Two sites located in the C-terminal globular domain of TnC bind both Ca^{2+} (K_{ass} 10^7 L/mol) and Mg^{2+} (K_{ass} 10^3 L/mol), whereas the other two sites, located in the N-terminal domain, specifically bind Ca^{2+} with a K_{ass} of 10^5 L/mol. TnI, a 23.9 kD protein, inhibits actomyosin ATPase activity and this inhibition is reversed by the addition of Ca^{2+}-saturated TnC. TnI and TnC closely interact with each other, the strength of their interaction depending on the saturation of Ca^{2+}-binding sites of TnC [3]. In the presence of Ca^{2+}, the K_{ass} value for the TnI-TnC complex is 10^8-10^9 L/mol [4]. Multiple sites of TnI-TnC interaction have been localized and, supposedly, in the presence of Ca^{2+}, TnI wraps itself around the central helix of TnC, forming contacts with both N- and C-terminal globular domains containing Ca^{2+}-binding sites [5,6].

The troponin complex plays a fundamental role in the transmission of intracellular calcium signals into actin-myosin interaction. Following membrane depolarization, myoplasmic calcium occupies the calcium-binding sites of TnC, leading to a conformational change in this troponin subunit. This conformational change is transmitted mainly via TnT to tropomyosin and TnI, this process allowing the interaction between actin and myosin. Cross-bridges are then formed between the thick and thin filaments and this results in muscle contraction. The strength and velocity of contraction are modulated by the numbers and affinities of Ca^{2+}-binding sites of TnC and by the structure, phosphorylation, and steric arrangement of TnT, TnI, and tropomyosin [7]. A current unresolved issue is just what is regulated when Ca^{2+} binds to TnC. Ca^{2+} may act as a switch and promote the transition from the weak to the strong state in an "all or none" fashion, or Ca^{2+} may increase the rate of transition from the weak to the strong state in a "graded" fashion [8]. In terms of activation of the thin filament whether an "all or none" or a "graded" response, there is an emerging picture of how the reaction of Ca^{2+} and TnC is transduced to affect the reactivity of actin by triggering the actin-myosin crossbridge reaction and putting the thin filament in "on" and "off" states. The "off" state is associated with low cytosolic Ca^{2+}. There are weak interactions between TnC and TnI and possibly between TnC and TnT; TnI binds to actin strongly. The "off" state may also be characterized by a poorly understood interaction of TnI with TnT and a potential interaction between TnT and actin [9]. These interactions, which are transmitted to tropomyosin through TnT, hold tropomyosin in a conformational state or location that alters actin activity so that actin-crossbridge reaction is impeded. The strong interaction between TnI and actin in relaxing conditions is also likely to inhibit the reaction of crossbridges and to produce the blocked state. The inhibitory reactions are reversed when Ca^{2+} binds to TnC, resulting in an activation of myofilament activity (the "on" state) [10].

Distribution of troponin subunits in different muscles

The regulatory troponin complex is not present in smooth muscle. Contraction in this type of muscle is regulated by myosin phosphorylation/dephosphorylation and Ca^{2+}-dependent caldesmon binding [11,12]. The proteins of the regulatory troponin complex, which are reported to exist in multiple isoforms, are expressed in a tissue-specific and developmentally regulated manner [13]. All the troponin complex proteins are single polypeptides and show a relatively simple tissue distribution, compared to myosin or tropomyosin. Three isoforms, encoded by separate genes, have been reported for TnI, which is a single polypeptide chain of 211 amino acids without any evidence of heterogeneity in its aminoacid sequence [14,15]. These isoforms are associated with cardiac, slow skeletal, and fast skeletal muscle and may contribute to the functional differences between the different muscle types. The TnI isoforms have been found to be tissue-specific, but species non-specific. Studies in a variety of different mammals, including rabbit, cow, baboon, monkey, and man have indicated that the structural relationship of cardiac TnI (cTnI) is strongly preserved in all mammalian species. A comparative study of cTnI from different species by Berson *et al* [16] showed that all cTnI appeared to have similar molecular weights, but had different overall charges. These different molecules have the same inhibitory effect on skeletal actomyosin ATPase activity and the same ability to be phosphorylated.

In skeletal muscle, the TnI isoforms (sTnI) are located in different types of cells in the myofibrils. The fibres of the type-1 contain only the slow isoform, whereas the fast isoform is present in type-2 cells. Vice versa, the distribution pattern of the cardiac isoform (cTnI) is uniform throughout the atrial and ventricular chambers possibly due to the presence of only one cell type in the heart tissue; furthermore, cTnl is uniquely located in the myocardium [17].

Comparative studies on the sequences of cTnI and sTnI show that their inhibitory regions are almost identical in the two proteins, although the former contains 32 to 33 extra amino acids at the N-terminal region which are not present in sTnI [14,18]. This sequence includes two adjacent serine residues at positions 22 and 23 or 23 and 24, depending on the animal species, that can be phosphorylated by protein kinase A [19]. The extent of cTnI phosphorylation seems to correlate with the degree of positive inotropy of the heart as stimulated by catecholamines [20]. Several lines of experimental evidence suggest that phosphorylation of cTnI may have a unique and important regulatory role in controlling cardiac function. Physiologically, the phosphorylation of cTnI appears to influence the Ca^{2+} regulation of cardiac myofibrils through a change in the interaction between cTnI and TnC [21]. The absence of the N-terminal serine residues confers reduced responsiveness of the heart to b-adrenergic stimulation. The amino acid sequence from 104-115 is the peptide imparting on ATPase inhibitory activity of the TnI molecule. This region has also been found to be important for TnI interaction with TnC and tropomyosin and any alteration in this sequence results in a loss of, or reduction in, the inhibitory activity of TnI. Recently Larue *et al* [22] carried out epitope analysis of cTnI by 40 apparently cTnI-specific monoclonal antibodies showing at least six different epitopes on the cTnI molecule; some of these were highly specific for cTnI and were not found in sTnI. Wilkinson and Grand have previously shown that cTnI has 40% sequence dissimilarity from the skeletal troponins [23]. This unique amino acid sequence makes cTnI an ideal candidate for the laboratory detection of myocardial injury and has facilitated the development of monoclonal antibodies that do not react with sTnI.

Developmental characteristics of cTnI

Both cardiac and skeletal TnI are expressed in the human ventricle throughout fetal development, but sTnI is the predominant isoform at all fetal stages. After birth, there is a switch to the expression of cTnI alone, the phenotype of adult myocardium, this process being complete a few months later. In particular, sTnI protein and mRNA are absent by the ninth postnatal month, as demonstrated by Sasse *et al* [24]. This developmental transition in troponin gene expression has functional implications for the contractility of the developing myocardium and raises questions as to the mechanisms regulating the expression of these genes. In fact, as already mentioned, the extended N-terminal sequence of cTnI contains serine residues that are selectively phosphorylated when hearts are perfused with adrenergic agonists or when purified TnI is treated with cAMP-dependent protein kinase. Indeed, the transition in expression from sTnI to cTnI during neonatal life may account in part for the differing response of fetal and neonatal myocardium to β-adrenergic stimulation [25,26]. The molecular mechanisms of this transition are unknown, although it is believed that the developmental changes in hormones, particularly thyroid hormones, are directly involved in the process [13]. However, more recent data suggest that thyroid hormones do not significantly affect cTnI expression [27]. The present hypothesis is that, during skeletal muscle development, neural contact plays an important role in the development of the mature phenotype. Denervation or cross innervation between fast and slow skeletal muscle types have a profound effect on isoform transition [28]. cTnI is not expressed by developing or diseased human skeletal muscle. Immunohistochemical studies conducted on chicken embryos and human fetal tissues have found no evidence of cTnI expression by normal animal and human embryonic or adult skeletal muscle [29]. These observations, together with the previously reported data on TnI switching in developing heart, confirm that sTnI is the early embryonic TnI in human muscle. For these reasons cTnI cannot be detected in skeletal muscle biopsy samples of patients with polymyositis, Duchenne muscular dystrophy, or in other degenerative processes of the muscle, even when the synthesis of the monomer B of creatine kinase is increased [30]. The absence of cTnI in diseased as well in nondiseased human skeletal muscle further explains the excellent cardiac specificity of cTnI measurement for the diagnosis of myocardial damage.

Kinetics of cTnI release

The mechanisms of troponin degradation in the early stages after AMI have been studied. Cummins and Auckland [31], who developed the first radioimmunoassay for cTnI measurement in serum, showed that the release of the protein resembles that of creatine kinase MB for its early release, but a biphasic profile was evident, with serum elevations within 4-6 h, a peak concentration at 18 h, and cTnI concentrations above normal for up to 6-8 days post-infarction. Many factors that contribute to alterations in the intracardiac environment may directly affect the stability of troponin or act indirectly by activating proteolytic enzymes. The anaerobic activation of proteolytic enzymes, like cathepsins, reduction in tissue ATP, accumulation of lactate, elevation of pH, and changes in the intracellular distribution of electrolytes, were suggested to be responsible for this degradation [32]. The relatively early release of cTnI could be the result of rapid myofibrillar breakdown but may also reflect the presence of free cTnI within the sarcoplasm. cTnI tightly interacts with two other troponin components (TnC and TnT) and the separation of these components can be achieved only in the presence of a high concentration of urea or other denaturing agents [33]. Recent data, obtained using different pairs of cTnI-specific monoclonal antibodies and other sophisticated experimental conditions, indicate that in AMI the largest proportion of cTnI is liberated in the form of a complex, mostly together with TnC, whereas only a small proportion of cTnI circulates in the bloodstream as the free form [34]. The ratio of total to free cTnI in the serum varies with time after AMI and is different in serum samples from different patients. As a rule, the ratio $(cTnI)_{total}/(cTnI)_{free}$ is low at the beginning of the post-AMI period, it then increases to its maximum value and, at the end of the observation period, returns to its initial value. At its peak concentration, total cTnI is 5-12 times higher than the corresponding concentration of free cTnI. The release kinetics of total cTnI is similar to that previously described in the literature [35]: the peak of the total cTnI concentration is observed 15-25 h after the onset of chest pain and the concentration of the protein remains increased for at least 80-100 h.

In the light of these findings, the standardization of different methods for cTnI measurement in serum seems mandatory. In fact, differences in currently available assays could be explained by the different specificity of antibodies, pairs of antibodies, purification and type of antigens (purified free cTnI or cTnI-cTnC complex) and reaction conditions. However, from a clinical point of view, cTnI is a specific and sensitive marker for myocardial damage as shown by recent studies. There is increasing interest in introducing it into the clinical setting.

cTnI for risk stratification in patients with acute coronary syndromes

From the pathophysiological point of view, acute coronary syndromes (ACS) have a common mechanism: the atherosclerotic plaque ruptures and a mural or occlusive thrombus forms, impairing or interrupting the perfusion of myocardial tissue [36]. Large plaque ulceration can lead to total occlusion of a coronary artery resulting in AMI, but also smaller, nonocclusive, plaques may produce minor myocardial injury associated with irreversible damage in clinically defined unstable angina (UA) patients [37]. The boundaries between UA and AMI are then overlapping and UA is thought to be the prelude to AMI in many cases. The rationale for the use of highly sensitive and specific biochemical markers, such as cardiac troponins, in ACS patients is based on the possibility to detect accurately the presence of even minor degrees of myocardial damage and, consequently, ascertain whether a progression of UA to AMI has begun [38].

In recent years increasing evidence indicates that abnormal troponin measurements identify a sub-group of patients (approximately one fifth to one third of all patients presenting with a diagnosis of UA) who have an increased risk of major cardiac events. This links cardiac troponin elevation to adverse outcome in patients with ACS [39-42]. The first reports on the role of troponin in UA, as assessed by cTnT, the first commercially available test and showed that the short-term (<6 months) risk of cardiac death or AMI was increased in patients who had abnormal cTnT values [39,43,44]. cTnI is the newer of the two troponin markers, and the studies on risk stratification were initiated some years later. Unlike cTnT, for which there is only one manufacturer of a commercial assay, the results for cTnI have been obtained using different cTnI assays and therefore different cut-off concentration points. Despite these facts, the diagnostic and prognostic value previously showed for cTnT, has now been shown also for cTnI.

In a first study by Wu *et al* [45], cTnI measurements (Opus Behring; cutoff value, 1.6 μg/L) were used for risk stratification in 74 consecutive non-AMI patients with chest pain. With the use of odds ratios, these authors showed that poor outcome was significantly more frequent in the high-cTnI group than in the low-cTnI group; a substantial improvement compared with the prognastic value of creatine kinase MB. Using the same method (Opus Behring) but a different cut-off point (0.5 μg/L), Panteghini *et al* [46] confirmed that the short-term incidence of cardiac events in UA patients was significantly related to the increase in serum cTnI. Interestingly, the same group recently obtained identical clinical results by using a different assay (Access Sanofi Diagnostics Pasteur; cutoff value, 0.03 μg/L), showing that the prognostic value of cTnI determination in UA patients is not method-dependent [47]. In another study of 44 patients with UA, positive cTnI results (ELISA Sanofi Diagnostics Pasteur; cutoff value, 0.1 μg/L) were detected in 45% of patients [48]. These authors found a 1-year cardiac event rate of 33% for patients with and have for those without increased cTnI concentrations. Furthermore, increased cTnI was a significant univariate predictor of cardiac death or fatal AMI.

The first large study that assessed the prognostic value of a single measurement of cTnI in serum at the time of hospital admission was reported on patients in the Thrombolysis in Myocardial Ischemia Phase IIIB (TIMI IIIB) trial by Antman *et al*

[41]. In a retrospective evaluation, these authors measured cTnI (Stratus Dade; cutoff value, 0.4 μg/L) in 1404 patients with UA and non-Q-wave AMI. Elevated serum cTnI concentrations predicted a significantly increased mortality at 42 days (Figure 2).

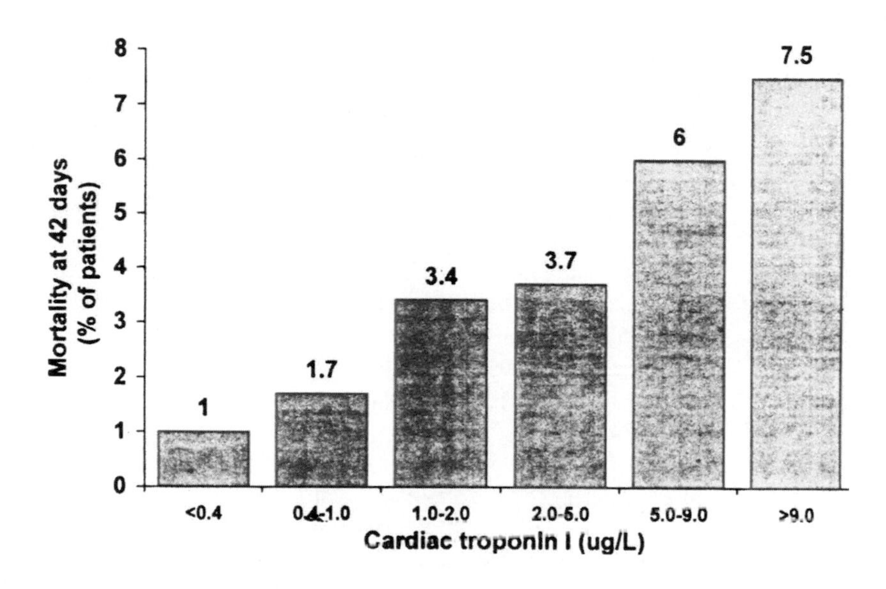

Figure 2. Mortality rates at 42 days according to cardiac tropinin I concentrations in patients of the TIMI IIIB trial (adapted from ref. 41)

In particular, the incidence of death was 1% in patients with undetectable cTnI concentrations and 3.7% in patients with increased cTnI. Each increment of 1 μg/L in cTnI concentrations was associated with a significant increase (+3%) in the risk ratio for death even after adjustement for other factors influencing mortality. Although interesting, this report has some limitations, such as its retrospective approach and a somehow confusing classification of the chest pain episodes. However, we do not believe that these limitations significantly affect the importance of its main message i.e., that cTnI provides prognostic information beyond that supplied by demographic characteristics or electrocardiographic findings in patients presenting with ACS. Confirming these findings, Galvani *et al* [42] prospectively studied 91 UA patients in whom non-Q-wave AMI was excluded. Their data confirmed that cTnI (Stratus Dade; cutoff value, 3.1 μg/L) is an important and independent prognostic variable in UA patients, predicting short- (30-day) and long-term (1-year) patient outcome. Finally, three additional preliminary studies [49-51], presented during 1997, further confirmed the prognostic potential of cTnI to predict cardiac events.

So far, only one prospective study has compared the prognostic value of cTnT with that of cTnI in the same cohort of ACS patients [52]. In 516 patients suspected of having ACS, both cTnT (ES 300 Boehringer; optimal cutoff, 0.1 μg/L) and cTnI (Opus Behring; optimal cutoff value, 2.0 μg/L) provided independent prognostic information with regard to major cardiac events at 30-days follow-up period. Also, in the subset of patients who were retrospectively categorized as UA (n=309), both cTnT and cTnI mantained comparable prognostic information regarding the occurence of cardiac death/AMI. In particular, among the patients with elevated cTnI concentration, 3.2% died during the follow-up period, which is comparable to the 3.7% mortality rate that Antman *et al* [41] found in their cTnI-positive group. The event rate in patients with low concentrations of cTnI was also concordant in the two studies: 0.7% in this study versus 1.0% in the study by Antman *et al* [41]. A close correlation of cTnT and cTnI values was found, which further supports the hypothesis that release of these proteins reflects the same pathological process in the myocardium. Interestingly, the predictabilities of the two troponins varied at different cutoff values and were already significant at 0.05 μg/L for cTnT and 1.5 μg/L for cTnI. Recently, three additional preliminary studies confirmed these findings, by using a different cTnI assay [53] or different cutoff values [54,55].

Optimal blood sampling for risk stratification using cTnI determination

In the study of Antman *et al* [41], the sample for cTnI taken on enrollment provided suitable prognostic information regarding cardiac events. Conversely, in their study, Luscher *et al* [52] used the highest value within the first 6 h after admission. When only the admission sample was used for risk stratification in this study, a considerable number of patients who were suffering from significant myocardial damage were overlooked. On the basis of these and other published data [47], it seems that an in-hospital observation period of at least 12 h gives an accurate identification of the majority of the UA patients at greater risk of cardiac events. A few samples, i.e., two or three, collected in the first 12 h after admission, are probably enough to obtain this information. Hamm *et al* [56] showed that the events rate in patients with negative cTnI test in this period was only 0.3%.

Clinical implications

As discussed previously, the measurement of troponins in serum allows the identification of a new group of patients with unstable ACS characterized by minor myocardial injury, who are at high risk of developing short- and long-term adverse cardiac events. Recent evidence suggests that patients with an ACS and documented abnormal release of cardiac troponins must be regarded as “high risk”, independently of their clinical characteristics.

Having demonstrated that troponins are useful for risk stratification, the challenge is now to reduce the high risk of subsequent ischemic events in ACS patients with an elevated troponin test result. The applicability of the cardiac troponins as prognostic indicators may also provide a useful tool for selecting optimal treatment. A study has recently shown that elevation of cTnT identifies a subgroup of patients in whom prolonged antithrombotic treatment with a low-molecular-weight heparin can improve prognosis [57]. However, additional prospective outcome studies are needed to document whether antithrombotic agents and/or aggressive revascularization treatments can improve prognosis in troponin-positive patients.

	Group cTnI- (n=15)		Group cTnI+ (n=7)
Mean luminal stenosis, %	82±13		85±12
Multivessel disease	7		4
Lesion morphology :			
• type A lesions, no.	9		1
• type B1 lesions, no.	8	**P=0.041**	6
• type B2 lesions, no.	4		7

Lesion morphology classified according to American Heart Association criteria.

Figure 3. Angiographic characteristics for 22 patients with unstable angina, according to whether they were negative or positive for cardiac troponin I (adapted from ref. 47).

The possibility that elevated cardiac troponin concentrations may be associated with dangerous vulnerable plaques, responsible for AMI and death, needs also to be evaluated prospectively in studies correlating the morphologic features of plaques with cardiac troponin concentrations in patients with ACS. In this respect [58], it is surprising that Antman *et al* [41] did not find significant differences in the incidence of coronary thrombi on angiography between patients who had increased cTnI and those who did not. More recent studies seem to show however a significant association of the cTnI increase with the severity of lesion morphology at angiography [46,47,59,60]. In one of these studies [46], UA patients with and without measurable amounts of cTnI (Opus Behring; cutoff value, 0.5 μg/L) did not differ with respect to the quantitative degree of coronary disease at angiography, but the type of lesion morphology, classified according to American Heart Association (AHA) criteria [61], was significantly different between the two groups (P=0.0019); ulcerated lesions or thrombus formation were present in all cTnI-positive patients. Similar results were also obtained by the same group of researchers, using a different cTnI assay (Access Sanofi Diagnostics Pasteur; cutoff value, 0.03 μg/L) (Figure 3) [47]. In addition, Benamer *et al* [60], using multivariate analysis, showed that cTnI increase (Stratus Dade; cutoff value, 0.3 μg/L) within 24 h from hospital admission can predict the complexity (presence of thrombus or AHA type C lesions) of the coronary lesions in UA patients (odds ratio: 3.6).

Conclusion

One of the most important challenges in the management of ACS patients is risk stratification. The possibility of identifying patients who are at high risk of subsequent ischemic events and who could benefit by more aggressive medical treatment or invasive revascularization procedures appears to be closer now with the troponin assays. The measurement of cTnI concentrations in serum may provide a rational and cost-effective basis to individualize these subjects, guiding therapy and controlling its effectiveness. In particular, the following points should be considered in relation to cTnI measurement in patients with suspected ACS:

1. High concentrations of cTnI predict cardiac events (death, AMI) better than low concentrations.
2. Early measurements (<12 hours after admission) are sufficiently predictive.
3. High cTnI concentrations correlate with the presence of disrupted coronary plaque morphology.

References

1. Leavis P, Gergely J. Thin filament proteins and thin filament-linked regulation of vertebrate muscle contraction. CRC Crit Rev Biochem 1984;16:235-305.
2. Zot AS, Potter JD. Structural aspects of troponin-tropomyosin regulation of skeletal muscle contraction. Annu Rev Biophys Chem 1987;16:535-59.
3. Heeley D, Golosinska K, Smillie LB. The effects of troponin fragments T1 and T2 on the binding of nonpolymerizable tropomyosin to F-actin in the presence or absence of troponin I and troponin C. J Biol Chem 1987;262:9971-8.
4. Ingraham RH, Swenson CA. Binary interactions of troponin subunits. J Biol Chem 1984;259: 9544-8.
5. Farah CS, Miyamoto CA, Ramos CHI, *et al.* Structural and regulatory functions of the NH_2-and COOH-terminal regions of skeletal muscle troponin I. J Biol Chem 1994;269:5230-40.
6. Olah GA, Trewhella J. A model structure of the muscle protein complex $4Ca^{2+}$-troponin C-troponin I derived from small-angle scattering data: implication for regulation. Biochemistry 1994;33:12800-6.
7. Solaro RJ, Pan BS. Control and modulation of contractile activity of cardiac myofilaments. In: Sperekalis N, editor. Physiology and pathophysiology of the heart. Boston: Kluwer Academic Publishers, 1988: 291-3.
8. Solaro RJ. Troponin I-Troponin C interactions and molecular signalling in cardiac myofilaments. In: Sideman S, Beyar R, editors. Molecular and subcellular cardiology: effects of structure and function. New York: Plenum Press, 1995: 109-15.
9. Heeley DH, Smillie LB. Interaction of rabbit skeletal muscle troponin T and F-actin at physiological ionic strenght. Biochemistry 1988;27:8227-31.
10. Lehrer S. The regulatory switch of the muscle thin filament: Ca^{2+} or myosin heads? J Mus Res Cell Motility 1994;15:232-6.
11. Marston SB, Smith CWJ. The thin filament of smooth muscles. J Muscle Res Cell Motil 1985;6:669-708.
12. Kamm KE, Stull JT. Regulation of smooth muscle contractile elements by second messengers. Ann Rev Physiol 1989;51:299-313.
13. Haider KH, Stimson WH. Cardiac troponin-I: a biochemical marker for cardiac cell necrosis. Disease Markers 1993;11:205-15.
14. Vallins WJ, Brand NJ, Dabhade N, Butler-Browne G, Yacoub MH, Barton PJR. Molecular cloning of human cardiac troponin I using polymerase chain reaction. FEBS Lett 1990;270: 57-61.

15. MacGeoch C, Barton PJR, Vallins WJ, Bhavsar P, Spurr NK. The human cardiac troponin locus: assignement to chromosome 19p13.2-19q13.2. Human Genet 1991;88:101-4.
16. Berson G, Samuel JL, Swynghedauw B. A comparative study of the cTnI factor from mammalians. Pfugers Arch 1978;374:277-83.
17. Humphrey JE, Cummins P. Regulatory protein of the myocardium. Atrial and ventricular tropomyosin and Tn-I in the developing and adult bovine and human hearts. J Mol Cell Cardiol 1984;16:643-57.
18. Leszyk J, Dumaswala K, Potter JD, Collins JH. Amino acid sequence of bovine cardiac troponin. Biochemistry 1988;27:2821-27.
19. Moir AJG, Solaro RJ, Pery SV. The site of phosphorylation of troponin I in the perfused rabbit heart. Biochem J 1980;185:505-13.
20. Zhang R, Zhao J, Mandveno A, Potter JD. Cardiac troponin I phosphorylation increases the rate of cardiac muscle relaxation. Circ Res 1995;76:1028-35.
21. Liao R, Wang CK, Cheung HC. Effect of phosphorylation of cardiac troponin I on the fluorescence properties of its single tryptophan as determined by picosecond spectroscopy. SPIE Proc. 1990;1204:600-705.
22. Larue C, Defacque-Lacquement H, Calzolari C, Le Nguyen D, Pau B. New monoclonal antibodies as probes for human cardiac troponin I: epitopic analysis with synthetic peptides. Mol Immunol 1992;29:271-8.
23. Wilkinson JM, Grand RJA. Comparison of amino acid sequence of troponin I from different striated muscles. Nature 1978;271:31-5.
24. Sasse S, Brand NJ, Kyprianou P *et al.* Troponin I gene expression during human cardiac development and in end-stage heart failure. Cir Res 1993;72:932-8.
25. Anderson PAW. Fetal and neonatal physiology and pharmacology. Curr Opin Cardiol 1990; 5:3-16.
26. Artman M, Kithas PA, Wike JS, Strada SJ. Inotropic responses change during postnatal maturation in rabbit. Am J Physiol 1988;255:H335-42.
27. Ausoni S, de Nardi C, Moretti P, Gorza L, Schiaffino S. Developmental expression of rat cardiac troponin I mRNA. Development 1991;112:1041-51.
28. Pette D, Vrbova G. Neural control of phenotypic expression in mammalian muscle fibres. Muscle Nerve 1985;8:676-89.
29. Toyota N, Shimada Y. Differentiation of troponin in cardiac and skeletal muscles in chicken embryos as studied by immunofluorescent analysis. J Cell Biol 1981;91:497-504.
30. Bodor GS, Porterfield D, Voss EM, Smith S, Apple FS. Cardiac troponin-I is not expressed in fetal and healthy or diseased adult human skeletal muscle tissue. Clin Chem 1995;41:1710-5.
31. Cummins P, Auckland ML. Cardiac specific radioimmunoassay of troponin-I in the diagnosis of myocardial infarction. Clin Sci 1983;64:42-6.
32. Katagiri T, Kobayashi Y, Sasai Y, Toba K, Niitani H. Alterations in cardiac troponin subunits in myocardial infarction. Jap Heart 1981;22:653-64.
33. Greaser ML, Gergerly J. Purification and properties of the components from troponin . J Biol Chem 1973;248:2125-33.
34. Katrukha AG, Bereznikova AV, Esakova TV, *et al.* Troponin I is released in bloodstream of patients with acute myocardial infarction not in free form but as complex. Clin Chem 1997; 43:1379-85.
35. Bhayana V, Henderson AR. Biochemical markers of myocardial damage. Clin Biochem 1995;28:1-29.
36. Fuster V, Fallon JT, Nemerson Y. Coronary thrombosis. The Lancet 1996;348:S7-10.
37. Falk E. Unstable angina with fatal outcome: dynamic coronary thrombosis leading to infarction and/or sudden death: autopsy evidence of recurrent mural thrombosis with peripheral embolization culminating in total vascular occlusion. Circulation 1985;71:699-708.
38. Wu AHB. Use of cardiac markers as assessed by outcomes analysis. Clin Biochem 1997;30:339-50.
39. Hamm CW, Ravkilde J, Gehrardt W, *et al.* The prognostic value of serum troponin T in unstable angina. N Engl J Med 1992;327:146-50.
40. Lindahl B, Venge P, Wallentin L, for the FRISC Study Group. Relation between troponin T and the risk of subsequent cardiac events in unstable coronary artery disease. Circulation 1996;93:1651-7.
41. Antman EM, Tanasijevic MJ, Thompson B, *et al.* Cardiac-specific troponin I levels to predict the risk of mortality in patients with acute coronary syndromes. N Engl J Med 1996;335:1342-9.
42. Galvani M, Ottani F, Ferrini D, *et al.* Prognostic influence of elevated values of cardiac troponin I in patients with unstable angina. Circulation 1997;95:2053-9.
43. Ravkilde J, Horder M, Gehrardt W, *et al.* Diagnostic performance and prognostic value of serum troponin T in suspected acute myocardial infarction. Scad J Clin Lab Invest 1993;53:677-85.

44. Burlina A, Zaninotto M, Secchiero S, Rubin D, Accorsi F. Troponin T as a marker of ischemic myocardial injury. Clin Biochem 1994;27:113-21.
45. Wu AHB, Feng YJ, Contois JH, Azar R, Waters D. Prognostic value of cardiac troponin I in patients with chest pain. Clin Chem 1996;42:651-2.
46. Panteghini M, Bonora R, Pagani F. Rapid and specific immunoassay for cardiac troponin I in the diagnosis of myocardial damage. Int J Clin Lab Res 1997;27:60-4.
47. Panteghini M, Bonora R, Pagani F, Buffoli F, Cuccia C. Rapid, highly sensitive immunoassay for determination of cardiac troponin I in patients with myocardial cell damage. Clin Chem 1997;43:1464-5.
48. Bertinchant JP, Larue C, Pernel I, *et al.* Release kinetics of serum cardiac troponin I in ischemic myocardial injury. Clin Biochem 1996;29:587-94.
49. Benamer H, Benessiano J, Gaultier CJ, *et al.* Prognostic markers in unstable angina: relationship between troponin I and C-reactive protein. J Am Coll Cardiol 1994;29(suppl A):233-4.
50. Kano S, Nishimura S, Tashito Y, *et al.* Cardiac troponin-I in diagnosis and prognosis of unstable coronary artery disease. Clin Chem 1997;43:S157.
51. Willmore TM, Gallahue F, Cramner H, *et al.* Influence of cardiac troponin I in emergency department patients. Circulation 1997;96(suppl I):364.
52. Luscher MS, Thygesen K, Ravkilde J, Heickendorff L, for the TRIM Study Group. Applicability of cardiac troponin T and I for early risk stratification in unstable coronary artery disease. Circulation 1997;96:2578-85.
53. Lindahl B, Venge P, Wallentin L. Comparison of troponin T and troponin I for prediction of mortality in unstable coronary artery disease. Eur Heart J 1997;18(suppl):123.
54. Meo A, Quaranta G, Rebuzzi AG, *et al.* Comparison of sensibility and specificity of troponin I and troponin T for cardiac events in unstable angina. Eur Heart J 1997;(suppl):393
55. Olatidoye AG, Feng Y, Wu A. Do troponin T and I have equal prognostic significance in the same population of unstable angina patients ? Circulation 1997;96(suppl I):333.
56. Hamm CW, Goldmann BU, Heeschen C, Kreymann G, Berger J, Meinertz T. Emergency room triage of patients with acute chest pain by means of rapid testing for cardiac troponin T or troponin I. N Engl J Med 1997;337:1648-53.
57. Lindahl B, Venge P, Wallentin L, for the FRISC Study Group. Troponin T identifies patients with unstable coronary artery disease who benefit from long-term antithrombotic protection. J Am Coll Cardiol 1997;29:43-8.
58. Van de Werf F. Cardiac troponins in acute coronary syndromes. N Engl J Med 1996;335:1388-9.
59. Koning R, Lavoine A, Andres H, *et al.* Cardiac troponin I, a marker of severe unstable angina: a clinical, electrocardiographic and quantitative angiographic study. J Am Coll Cardiol 1997;29(suppl A):210-1.
60. Benamer H, Aubry P, Himbert D, *et al.* Increased cardiac troponin I is associated with complex coronary artery disease in unstable angina. Eur Heart J 1997(suppl):393.
61. Myler RK, Shaw RE, Stertzer SH, *et al.* Lesion morphology and coronary angioplasty: current experience and analysis. J Am Coll Cardiol 1992;19:1641-52.

Chapter 5

MYOGLOBIN

Johannes Mair

Although myoglobin was demonstrated to appear in blood and urine in the course of acute myocardial infarction (AMI) as early as 1956 by Kiss and Reinhard [1], for the next two decades enzyme determinations were preferred for the biochemical diagnosis of myocardial necrosis for methodological reasons. The first radioimmunoassay for the quantification of serum myoglobin was reported in 1975 [2]. This new technology made possible for the first time the accurate measurement of myoglobin concentrations in blood, and plasma myoglobin elevation after AMI was re-emphasised. However, many authors were not able to detect myoglobin in urine despite relatively high plasma concentrations [3], and it appears that myoglobin is only detectable in urine after extensive AMIs which release high amounts of myoglobin. It soon turned out, and was repeatedly demonstrated, that myoglobin concentrations rise several hours earlier after the onset of chest pain than creatine kinase (CK) and CKMB enzyme activities [2-5]. Since then myoglobin has become an established early marker for myocardial infarction and is a standard against which all new markers proposed for early AMI diagnosis have to be compared. However, in the 1970's myoglobin determinations, were mainly of academic interest because these radioimmunoassays were too time consuming (the results were only available on the next day or even later), and additionally the earlier diagnosis of AMI had at this time no real clinical consequences for patient care and management. During recent years the rapid early diagnosis of AMI using biochemical markers has received increasing attention, in particular for early exclusion of AMI in patients with non-diagnostic electrocardiogram (ECG) because of the increasing economic need for cost-effective utilisation of expensive coronary care units. This has forced assay manufacturers to overcome analytical limitations, and now-a-days rapid, automated immunoassays for plasma and whole blood myoglobin measurement are available.

Biochemistry

Myoglobin is an oxygen-binding heme protein of low molecular mass (MW: 17.8 kD) found in striated muscles, where it accounts for 5-10% of all cytoplasmic proteins (myoglobin concentration: approximately 4-5 mg/g wet weight). It is not found in smooth muscle. Myoglobin is an oxygen transport protein and is located in the striated muscle fibres close to the sarcolemma, to the contractile apparatus, and to intracellular membranous or fibrillar structures. In skeletal muscle myoglobin is mainly found in the slow-twitch ("red") fibres. Its most striking characteristic is its ability to bind oxygen reversibly. It binds oxygen with a higher affinity than haemoglobin, and myoglobin is likely to facilitate oxygen diffusion in striated muscle fibres and also to serve as an oxygen store within the muscle fibre [6,7]. However, its precise physiological role is still controversial.

Some investigators have reported that heart muscle contains multiple forms of myoglobin [8]. Whether there is a cardiac-specific myoglobin isoform in humans remains to be demonstrated. The current myoglobin assays cannot discriminate between myoglobin released from the human heart or from human skeletal muscle, and, therefore, myoglobin measurement is not heart-specific.

The turnover in striated muscles, in particular skeletal muscle, causes a continuous release of myoglobin into the circulation, which leads to detectable serum myoglobin concentrations in the reference populations. A dependence of reference values on age, sex, and race has been reported by some but not by all investigators [9-12]. Myoglobin is rapidly cleared from the serum by the kidneys, and its biological half-life time is only about 10 minutes [13]. Therefore, all diseases associated with impaired renal function lead to increased serum myoglobin concentrations.

Clinical Results

Myoglobin as an early marker for myocardial infarction

Myoglobin is markedly more sensitive than CK and CKMB activities during the first hours after the onset of chest pain (see Table 1) [4,11,14,15]. However, the differences are much less pronounced when newer technologies for CK and CKMB measurements, i.e. CK isoforms and CKMB mass, are used [16,17].

Table 1. Average sensitivities of cardiac markers in relation to the time of onset of chest pain.

Marker	0 - 2 hours	2 - 4 hours	4 - 6 hours
Myoglobin	0.35	0.80	0.95
CK isoforms	0.30	0.70	0.90
CKMB mass	0.30	0.70	0.90
Troponin	0.25	0.60	0.85
CK activity	0.15	0.35	0.70
CKMB activity	0.10	0.25	0.50

In a study (unpublished results) for which we collected blood every 15 minutes the sensitivities of myoglobin and CKMB mass differed significantly only in the baseline sample, which was taken before the start of thrombolytic treatment (see Figure 1). Thereafter, there were no significant differences between either marker (Figure 1). Myoglobin usually starts to rise within 2-4 hours after the onset of chest pain and myoglobin is detectable in all AMI patients between 6-10 hours from the onset of chest pain [14-17]. Peak values are found in patients with early reperfusion of the infarct-related coronary artery within 7 hours from AMI onset (see Figure 2). In all other AMI patients peaks are observed thereafter, and myoglobin returns into the reference interval

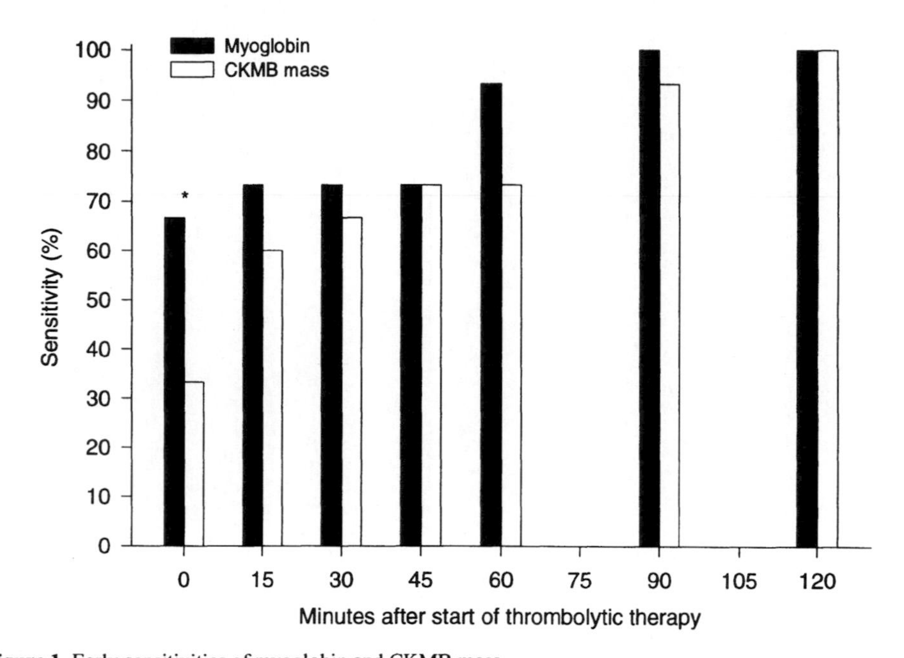

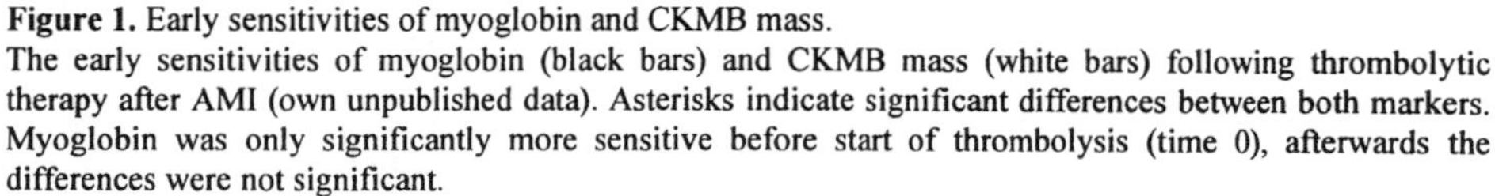

Figure 1. Early sensitivities of myoglobin and CKMB mass.
The early sensitivities of myoglobin (black bars) and CKMB mass (white bars) following thrombolytic therapy after AMI (own unpublished data). Asterisks indicate significant differences between both markers. Myoglobin was only significantly more sensitive before start of thrombolysis (time 0), afterwards the differences were not significant.

usually within 24-36 hours after the onset of AMI [4,11,14-17]. Myoglobin release correlates with infarct size [18,19]. Non-Q wave infarctions usually have smaller myoglobin peaks as compared to Q wave infarctions. The rapid disappearance of myoglobin in uncomplicated AMI makes this analyte very suitable to detect a reinfarction in patients in whom chest pain reoccurs [19].

Myoglobin is an excellent marker for ruling out an AMI in a patient admitted with chest pain [4,11,14-17]. Ideally, determinations should be carried out upon patient admission, then after 2,4, and 6 hours. If myoglobin is still within the reference interval 8 hours after the onset of chest pain, an AMI can be ruled out with near certainty. Chest pain patients without myoglobin increase rarely develop clinical complications [14]. Myoglobin was found to be a strong independent predictor of AMI in patients with symptoms of short duration, which was particularly useful in patients with an equivocal ECG at hospital admission [15,16]. After 12 hours the myoglobin peak concentration may already have passed, so its efficiency as a marker drops considerably and is markedly lower than that of CKMB after this time point. CKMB determination is the preferred option in patients admitted later than 10 hours from chest pain onset, because myoglobin may already have normalised at the time of hospital admission.

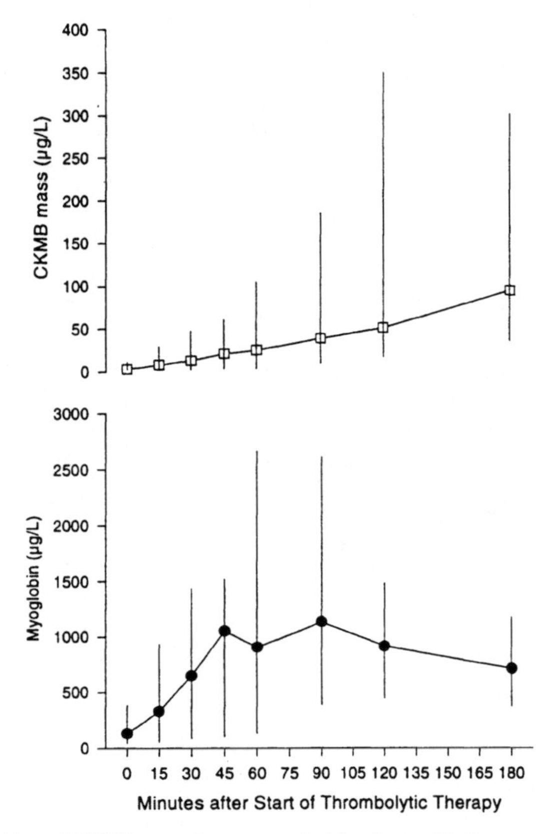

Figure 2. Myoglobin and CKMB mass time courses in 16 patients with Q wave myocardial infarction. Data given as median and interquartile range (bars); own unpublished data.

Despite its lack of heart specificity, myoglobin is also very useful for ruling in AMI early after the onset of chest pain [11,14-16]. In the non-traumatised chest pain patient admitted to the emergency department skeletal muscle damage is rare enough not to markedly influence the diagnostic efficiency of myoglobin in this patient population and missing the diagnosis AMI is the real problem for the physicians in the atypical patient. Myoglobin specificity in non-traumatised chest pain patients was reported to be in the range of that of CKMB [11]. But myoglobin should not be used for AMI testing in patients after resuscitation or in patients with renal failure.

Monitoring of thrombolytic therapy

Early reperfusion of the infarct-related coronary artery, either spontaneous or therapeutically induced by thrombolytic agents or acute percutaneous transluminal coronary angioplasty (PTCA), leads to a markedly more rapid and higher rate of increase and earlier peak values. Patients with and without early reperfusion may be reliably differentiated by their myoglobin rate of increase, if concomitant skeletal muscle injury can be excluded clinically. In all studies which compared different cardiac markers for non-invasive reperfusion monitoring myoglobin was always, tendentiously, the best cardiac marker to assess reperfusion non-invasively. This is because its diagnostic performance was less susceptible to changes in the threshold value used than that of other markers [20,21]. In the largest study published so far [21] Laperche *et al.* reported that a relative increase in myoglobin of ≥4 in 90 minutes after the start of thrombolytic therapy could reliably distinguish between reperfused and non-reperfused patients when thrombolytic therapy was started between 3-6 hours after the onset of chest pain. However, this myoglobin threshold remains to be tested in a prospective clinical trial.

Diagnosis of perioperative myocardial infarction in coronary artery bypass grafting (CABG)

Myoglobin is an early marker of perioperative myocardial infarction in patients undergoing elective CABG [22-24]. The myoglobin time course allows the differentiation of patients with and without myocardial infarction several hours earlier than other markers, such as CKMB or troponins. Myoglobin concentrations in patients without perioperative myocardial infarction increase with aortic unclamping after reperfusion of the heart after cardioplegic cardiac arrest, usually peak after one hour, and decrease to almost baseline values within four hours. In contrast, myoglobin concentrations in patients with perioperative myocardial infarction further increase after one hour post aortic unclamping. Patients with and without infarction can be discriminated as early as 3 hours after aortic unclamping [22]. Because of its short half-life in the circulation myoglobin also allows the recognition of the time point of infarction onset more accurately than other markers.

Concomitant skeletal muscle damage in CABG patients, due to the preparation of peripheral veins, median sternotomy, or the preparation of the internal mammary artery, is usually negligible as a source of false positive myoglobin results. Myoglobin measurements in coronary sinus blood samples identified the myocardium as the major source of myoglobin release in CABG patients [23]. However, when using alternative bypass vessels or techniques, such as the right gastroepiploic artery or the inferior epigastric artery or minimally invasive CABG, the surgical skeletal muscle injury is greater and myoglobin results must be assessed with caution. Clearly, in these patients, heart-specific markers are to be preferred for diagnostic purposes.

Myoglobin in patients with unstable angina pectoris

Elevated myoglobin concentrations have been reported in patients with unstable angina, perhaps reflecting small areas of myocardial muscle fibre death undetected by CKMB activity measurements [14,25,26]. Information on the prognostic significance of myoglobin release in these patients is limited and restricted to the development of in-hospital complications [14] which are more frequent in the patients with myoglobin increase. However, from a practical point of view it seems to be somewhat problematic to assess borderline or slight increases in myoglobin, a non heart-specific marker, in critically ill patients in whom it is not always possible to rule out skeletal muscle damage with near certainty. Therefore it is advisable to use more cardiac-specific markers for risk stratification in patients with unstable angina.

Specificity of myoglobin

Damage to skeletal muscles, either via trauma, surgery, or diseases, such as degenerative or inflammatory conditions, hyperthermia and hypothermia, hypoxia, or alcohol abuse, also elevate myoglobin in serum [5]. Only extraordinary, unaccustomed, physical exercise leads to an increase in myoglobin serum concentrations above the upper limit of the reference interval, particularly in untrained individuals. Uncomplicated intramuscular injections rarely cause false positive myoglobin results [5]. Myoglobin is also rarely elevated after cardiac catheterisation. Because myoglobin is cleared from the circulation by the kidneys, patients with a decreased glomerular filtration rate from low perfusion or renal failure can also have elevated myoglobin levels [27].

Summary

Myoglobin is a very suitable marker for the early diagnosis of AMI, and it contributes to the diagnosis in the subgroup of patients presenting early with non-diagnostic ECG. Myoglobin is useful for the non-invasive assessment of reperfusion in AMI patients. Fortunately, the problem of false positive myoglobin results is not as great as it may seem at first glance, particularly when assessing myoglobin concentrations in the non-traumatised chest pain patient. The causes of such false positive results can usually be ruled out or identified easily from the clinical history, and serum urea nitrogen (BUN) and creatinine allow the assessment of renal function in a patient. However, there are several clinical settings in which myoglobin should not be used for diagnosing myocardial damage, such as patients presenting with chest pain after heavy physical exercise, patients with neuromuscular disorders, or patients with obvious muscular trauma or renal failure.

References

1. Kiss A, Reinhard W. Über den Nachweis des Myoglobins im Serum und im Harn nach Herzinfarkt. Wien Klein Wochenschr 1956;58:154-8.
2. Stone MJ, Willerson JT, Gomez-Sanchez CE, Waterman MR. Radioimmunoassay of myoglobin in human serum: results in patients with myocardial infarction. J Clin Invest 1975;56:1334-9.
3. Reichlin M, Visco JP, Klocke FJ. Radioimmunoassay for human myoglobin: initial experience in patients with coronary artery disease. Circulation 1978;57:52-9.
4. Drexel H, Dworzak E, Kirchmair W, Milz MM, Puschendorf B, Dienstl F. Myoglobinemia in the early phase of acute myocardial infarction. Am Heart J 1983;105:642-51.
5. Roxin LE, Cullhed I, Groth T, *et al.* The value of serum myoglobin determinations in the early diagnosis of acute myocardial infarction. Acta Med Scand 1984;215:417-21.
6. Wittenberg JB. Myoglobin-facilitated oxygen diffusion: role of myoglobin in oxygen entry into muscle. Physiol Rev 1970;50:559-636.
7. Braunlin EA, Wahler GM, Swayze CR *et al.* Myoglobin facilitated oxygen diffusion maintains mechanical function of mammalian cardiac muscle. Cardiovasc Res 1986;20:627-36.
8. Wu JT, Pieper RK, Wu LH, Peters JL. Isolation and characterization of myoglobin and its two major isoforms from sheep heart. Clin Chem 1989;35:778-82.

9. Chen IW, David R, Maxon HR, Sperling M, Stein EA. Age-, sex-, and race-related differences in myoglobin concentration in serum of healthy persons. Clin Chem 1980:26:1864-8.
10. Delanghe J, Chapelle JP, Vanderschueren S. Quantitative nephelometric assay for determining myoglobin evaluated. Clin Chem 1990;36:1675-8.
11. Mair J, Artner-Dworzak E, Lechleitner P, *et al.* Early diagnosis of acute myocardial infarction by a newly developed rapid immunoturbidimetric assay for myoglobin. Br Heart J 1992;68:462-8.
12. Wu AHB, Laios I, Green S, *et al.* Immunoassays for serum and urine myoglobin: myoglobin clearance assessed as a risk factor for acute renal failure. Clin Chem 1994;40:796-802.
13. Klocke FJ, Copley DP, Krawczyk JA, Reichlin M. Rapid renal clearance of immunoreactive canine plasma myoglobin. Circulation 1982;65:1522-8.
14. Isakov A, Shapira I, Burke M, Almog C. Serum myoglobin levels in patients with ischemic myocardial insult. Arch Intern Med 1988;148:1762-5.
15. Ohman EM, Casey C, Bengtson JR, Pryor D, Tormey W, Horgan JH. Early detection of acute myocardial infarction: additional diagnostic information from serum concentrations of myoglobin in patients without ST elevation. Br Heart J 1990;63:335-8.
16. de Winter RJ, Koster RW, Sturk A, Sanders GT. Value of myoglobin, troponin T, and CKMB mass in ruling out an acute myocardial infarction in the emergency room. Circulation 1995;92:3401-7.
17. Mair J, Morandell D, Genser N, Lechleitner P, Dienstl F, Puschendorf B. Equivalent early sensitivities of myoglobin, creatine kinase MB mass, creatine kinase isoform ratios, and cardiac troponin I and T for acute myocardial infarction. Clin Chem 1995;41:1266-72.
18. Yamashita T, Abe S, Arima S, *et al.* Myocardial infarct size can be estimated from serial plasma myoglobin measurements within 4 hours of reperfusion. Circulation 1993;87:1840-9.
19. Honda Y, Katayama T. Detection of myocardial infarction extension or reattack by serum myoglobin radioimmunoassay. Int J Cardiol 1984;6:325-35.
20. Zabel M, Hohnloser SH, Köster W, Prinz M, Kasper W, Just HJ. Analysis of creatine kinase, CK-MB, myoglobin, and troponin T time-activity curves for early assessment of coronary artery reperfusion after intravenous thrombolysis. Circulation 1993;87:1542-50.
21. Laperche T, Steg PG, Dehoux M, *et al.* A study of biochemical markers of reperfusion early after thrombolysis for acute myocardial infarction. Circulation 1995;92:2079-86.
22. Mair P, Mair J, Seibt I, Balogh D, Puschendorf B. Early and rapid diagnosis of perioperative myocardial infarction in aortocoronary bypass surgery by immunoturbidimetric myoglobin measurements. Chest 1993;103:1508-11.
23. Seguin J, Saussine M, Ferriere M, *et al.* Comparison of myoglobin and creatine kinase MB levels in the evaluation of myocardial injury after cardiac operations. J Thorac Cardiovasc Surg 1988;95:294-7.
24. Kinoshita K, Tsuruhara Y, Tokunaga K. Delayed time to peak serum myoglobin level as an indicator of cardiac dysfunction following open heart surgery. Chest 1991;99:1398-402.
25. Hoberg E, Katus HA, Diederich KW, Kübler W. Myoglobin, creatine kinase-B isoenzyme, and myosin light chain release in patients with unstable angina pectoris. Eur Heart J 1987;8:989-94.
26. Mainard F, Massoubre B, LeMarec H, Madec Y. Study of a myoglobin test in patients hospitalized for suspected myocardial infarction. Clin Chim Acta 1985;153:1-8.
27. Hart PM, Feinfeld DA, Briscoe AM, *et al.* The effect of renal failure and hemodialysis on serum and urine myoglobin. Clin Nephrology 1982;18:141-3.

Chapter 6

GLYCOGEN PHOSPHORYLASE ISOENZYME BB

Johannes Mair, Bernd Puschendorf, and Ernst-Georg Krause

In the final analysis all currently available routine markers are not sufficiently sensitive within the first 3-4 hours after the onset of acute myocardial infarction (AMI), and the diagnostic performance of the electrocardiogram (ECG) is still clearly superior to that of cardiac markers during this time interval [1,2]. Because the diagnostic accuracy of the ECG on hospital admission is also limited [3] the search for more sensitive markers of ischaemia and myocardial damage is still ongoing. Based on its metabolic function, its pathophysiology, and supported by clinical results, glycogen phosphorylase isoenzyme BB (GPBB) is a promising enzyme for the early laboratory detection of ischaemic myocardial injury.

Biochemistry

Glycogen phosphorylase (GP) is an enzyme which plays an essential role in the regulation of carbohydrate metabolism by the mobilization of glycogen [4]. It catalyses the first step in glycogenolysis in which glycogen is converted to glucose-1-phosphate, utilizing inorganic phosphate. The physiological role of muscle phosphorylase is to provide the fuel for the energy supply required for muscle contraction. Its activity is allosterically regulated by the binding of adenosine monophosphate (AMP) and phosphorylation. Phosphorylase kinase converts GP *b* into its more active form GP *a*. Phosphorylase exists in the cardiomyocyte in association with glycogen and the sarcoplasmatic reticulum and forms a macromolecular complex (sarcoplasmatic reticulum glycogenolysis complex) [5,6]. The degree of association of GP with this complex depends, essentially, on the metabolic state of the myocardium. With the onset of tissue hypoxia, when glycogen is broken down and disappears, glycogen phosphorylase is converted from a particulate into a soluble form, and the enzyme becomes free to move around in the cytoplasma. [5-7].

GP exists as a dimer under normal physiological conditions. The dimer is composed of two identical subunits. Three GP isoenzymes are found in human tissues that are named after the tissue in which they are preferentially expressed, GPLL (liver), GPMM (muscle), and GPBB (brain) [4]. The three isoenzymes can be distinguished by functional and immunological properties. They are encoded by three distinct genes. The genes of the three human GP isoenzymes have been cloned and sequenced [8]. The proteins predicted by the cDNA sequences are 846 (LL), 842 (MM) and 862 (BB) amino acids long. Amino acids 1-830 match and differences are mainly found at the C-terminus, which is the catalytic domain of the protein. In pairwise sequence comparison the brain type protein is 80% identical to the liver-type and 83% identical to the muscle-type. GPBB has 21 and 16 additional amino acid residues on its C-terminal portion that are not present on the MM and LL isoenzymes, respectively.

Adult human skeletal muscle contains only one isoenzyme, GPMM. GPLL is the predominant isoenzyme in human liver and all other human tissues except for heart, skeletal muscle and brain. GPBB is the predominant isoenzyme in human brain. Its molecular weight as a monomer is approximately 94 kD. In the human heart the isoenzymes BB and MM are found, but GPBB is the predominant isoenzyme in myocardium as well. By far the highest concentrations of GPBB were found in human brain and heart. The tissue concentrations of GPBB in heart and brain are comparable [9]. Although immunoblot, electrophoresis, and northern blot data are partly conflicting [4,8-11], there is evidence that GPBB isoenzyme might not be restricted to brain and heart in humans. Much lower GPBB concentrations have been reported for example in leukocytes, spleen, kidney, bladder, testis, digestive tract and aorta. However, in all these tissues GPLL is by far the predominant GP isoenzyme.

Pathophysiology: Myocardial Oxygen Deficiency and GPBB Release

As mentioned above the sarcoplasmic reticulum-glycogenolysis complex represents a functionally coupled association. Using isolated glycogen particles a burst in glycogenolysis could be initiated by either calcium ions or cAMP accompanied by a breakdown of glycogen [7]. These data are evidence for a close relationship between glycogenolysis and excitation and contraction coupling and/or beta-adrenergic stimulation of the myocardium. There is no doubt that during acute ischaemia a sympathetic activation of the myocardium is followed by a transient rise in cardiac cyclic AMP levels and by an activation of GP due to a conversion of the non-phosphorylated *b* form into phosphorylase *a* by phosphorylase kinase [12-15]. Concomitantly the rate of glycogenolysis was found to be accelerated. Kinetic properties of GPBB allow furthermore a glycogen breakdown catalyzed by the *b* form. Compared with GPMM this isoenzyme is characterized by low values of K_m for the substrate orthophosphate as well as of $K_{0.5}$ for the activator AMP [10]. An ischaemia-induced rise in the levels of intracellular orthophosphate and AMP in myocardium may, therefore, induce a second, long lasting, acceleration of glycogenolysis under these conditions. Indeed cardiac glycogen breakdown was found to be continued during post-ischaemic reperfusion when the *a* form of GP had declined to pre-ischaemic control levels but the orthophosphate level was still high [16]. In experimental studies, as well as in patients with acute myocardial infarction, the released GPBB was exclusively found in the *b* form [17,18]. Thus, it is suggested that the activity of GPBB (form *b*) catalyses the prolonged degradation of glycogen in the sarcoplasmatic reticulum-glycogenolysis complex in the ischaemic area of the myocardium [13].

In conscious dogs a rapid release of GPBB was measured in the cardiac lymph after a transient ligation of a coronary artery for no longer than 10 min, which did not lead to

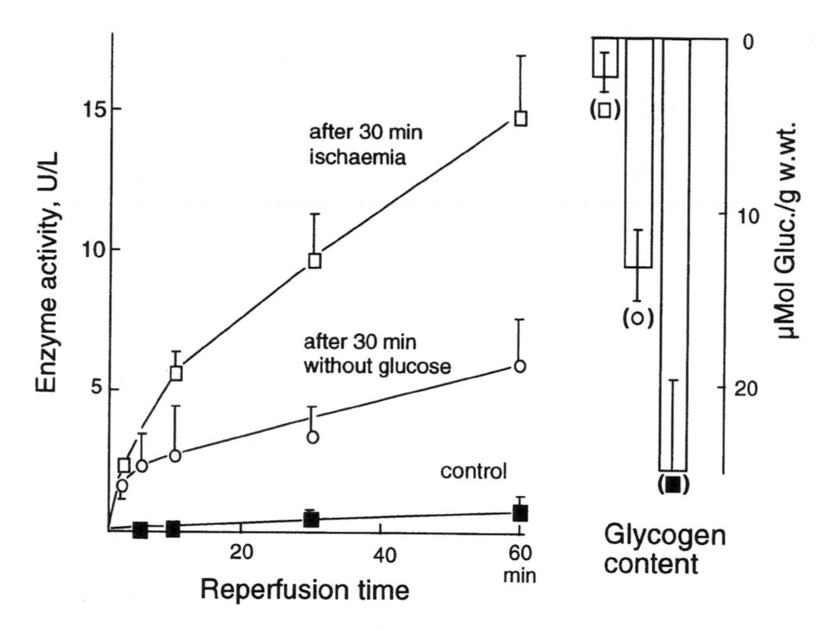

Figure 1. Release of glycogen phosphorylase and glycogen content of rabbit hearts perfused in the Langendorff mode.
Enzyme efflux was assayed during reperfusion after global ischaemia (□), after substrate depletion (O) and under aerobic control perfusion conditions (■). Finally the perfused hearts were freeze-clamped and glycogen content was determined. Values are means plus 1 standard error of the mean. Reprinted with permission from Krause *et al.* [20].
Abbreviations: Gluc. = glucosyl residue; w. wt. = wet weight.

histological signs of myocardial necrosis [17]. An efflux of GP from the myocardium after hypoxia or substrate depletion has been observed earlier in the isolated perfused rat and rabbit heart (see Figure 1) [19-21]. The GP release in these experiments correlated with the remaining myocardial glycogen content [21]. In this Langendorff model of the isolated perfused rabbit heart the addition of the drug imipramine under aerobic conditions to a cardioplegic perfusion solution only caused a release of CK, but not of GP (see Figure 2) [20,21]. The myocardial glycogen content remained unaffected as well. Imipramine causes in a certain concentration range a selective increase in the plasma membrane permeability without myocardial hypoxia. On the other hand, the stimulation of glycogenolysis by high doses of epinephrine did not cause a decrease in myocardial GP activity, although the glycogen content of the tissue was greatly diminished by the added epinephrine (see Table 1) [19]. These experimental results suggest that the release of GPBB requires both a burst in glycogenolysis and a concomitantly increased plasma membrane permeability as known for ischaemically injured cardiomyocytes [15].

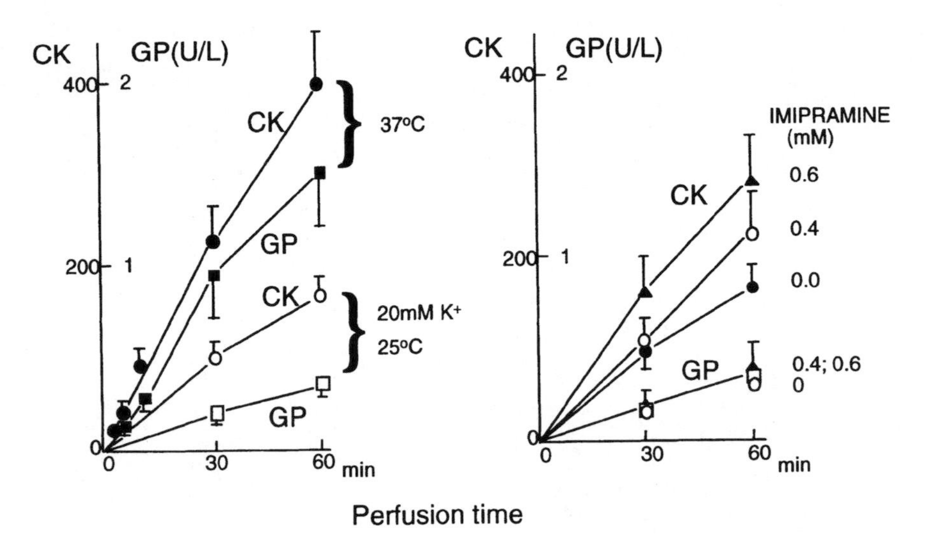

Figure 2. Release of glycogen phosphorylase (GP) and creatine kinase (CK) from normothermic beating and hypothermic, K^+-arrested, perfused rabbit hearts.
The effects of 0.4 and 0.6 mM imipramine were only studied with non-beating hearts at 24°C. Reprinted with permission from Krause *et al.* [20].

Table 1. Activity of glycogen phosphorylase (GP) and glycogen concentration in perfused rat hearts after addition of epinephrine.

Treatment	GP activity, *a* and *b* form (U/mg protein)	Glycogen (μmol glucose/ g wet weight)
Controls	0.41 ± 0.05	13.4 ± 2.4
Epinephrine	0.50 ± 0.09	3.9 ± 0.8

Adapted with permission from Krause *et al* [20].

Given its molecular mass the early release of GPBB raises questions regarding the mechanisms of its release from ischaemic myocardium. GPBB plays a key role in the energy metabolism of ischaemic myocardium. When glycogen is broken down and disappears, GPBB becomes free to move from the peri-sarcoplasmatic reticulum compartment directly into the extracellular fluid, if cell membrane permeability is simultaneously increased, which is usually the case in ischaemia. A high GPBB concentration gradient, which immediately is formed in the compartment of the sarcoplasmatic reticulum glycogenolytic complex, may be the reason for the high efflux rate of this enzyme. In contrast to other cytosolic proteins, this gradient may be at least partly also be realized via T-tubuli and may contribute to the efflux of GPBB (see Figure 3).

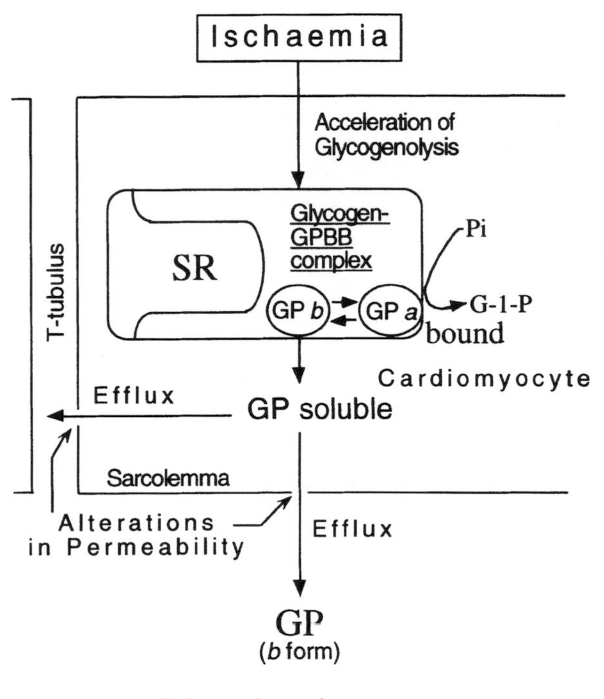

Figure 3. Scheme of GPBB release from myocardium in ischaemia.
Glycogen phosphorylase (GP) together with glycogen are tightly associated with the vesicles of sarcoplasmic reticulum (SR) under normal conditions. A release of GPBB, the main isoenzyme in the myocardium, essentially depends on the degradation of glycogen, which is catalyzed by GP *a* (the phosphorylated, active form of the isoenzyme) and by GP *b* (nonphosphorylated, AMP-dependent form). Ischaemia is known to favor the conversion of bound GP *b* into GP *a*, thereby accelerating glycogen breakdown, which seems to be the ultimate prerequisite for getting GP into a soluble form. An efflux of GPBB into the extracellular fluid may only follow if ischaemia-induced structural alterations in the cell membrane are manifested. For more details see text (Adapted with permission from Krause *et al.* [20]).
Abbreviations: Pi: inorganic phosphate; G-1-P: glucose-1-phosphate.

In summary, the ischaemia-sensitive glycogen degradation, which is regulated by Ca_{2+}, metabolic intermediates and catecholamines, seems to be a crucial prerequisite for the efflux of GPBB. This outlines the specific sensitivity of this enzyme marker to indicate transient imbalances in heart energy metabolism, as is the case during angina pectoris attacks and/or in the infarcting myocardium. Therefore, this enzyme is a promising analyte for the detection of ischaemic myocardial injury.

Methodology

The first hints that blood GP increases above its upper reference limit early after the onset of AMI, before CK increases, were obtained approximately two decades ago, measuring total GP activity with a relatively insensitive enzymatic assays [18]. However, the breakthrough was the later development of a sensitive and specific immunoenzymometric assay for the measurement of the isoenzyme GPBB [23]. We used this assay in all our clinical studies. The upper limit of this research assay was 7μg/L [23]. The challenge still remains to develop a rapid assay which is suitable for bedside or "stat" use in the routine laboratory. Currently two diagnostic companies are working on the development of a GPBB assay for routine use. It has to be stressed that the upper reference limit of GPBB is strongly dependent on the assay used, because the available GPBB assays have not been calibrated against a common GPPB standard.

First Clinical Results

Acute myocardial infarction

Differences in the sensitivities of GPBB in comparison to myoglobin, CKMB mass, CK and cardiac troponin T were reported within the first 2-3 hours after the onset of AMI [22,23]. GPBB was the most sensitive parameter during the first four hours after AMI onset (see Table 2). In the majority of AMI patients GPBB increased between one and four hours after the onset of chest pain. Therefore, GPBB may be a very important marker for the early diagnosis of AMI. GPBB usually peaks before CK, CKMB or troponin T and returns within the reference interval within 1-2 days after the onset of AMI (see Figure 4).

Table 2. Early sensitivities of GPBB and other biochemical markers before start of thrombolytic treatment in patients with acute myocardial infarction who were admitted within 4 hours after the onset of chest pain

	GPBB	CK activity	CKMB mass	Myoglobin	Troponin T
Sensitivity	0.77 (0.55-0.92)	0.20 (0.04-0.48)	0.47 (0.21-0.73)	0.47 (0.21-0.73)	0.40 (0.16-0.68)

The 95% confidence interval is given in parentheses.
Adapted with permission from Krause *et al.* [20].

Similar to soluble markers, such as myoglobin and CKMB, we demonstrated that the time course of GPBB in patients with AMI was markedly influenced by early reperfusion of the infarct-related coronary artery occurs [23]. The well established "wash out" phenomenon after successful thrombolysis lead to a more rapid increase in GPBB with earlier and higher peak values (see Figure 5). Therefore, GPBB may be useful, alongside other soluble myocardial proteins, to assess the effectiveness of thrombolytic therapy non-invasively. However, decision limits to detect successful and what is clinically more important, failed reperfusion, remain to be established in a controlled study involving coronary angiography.

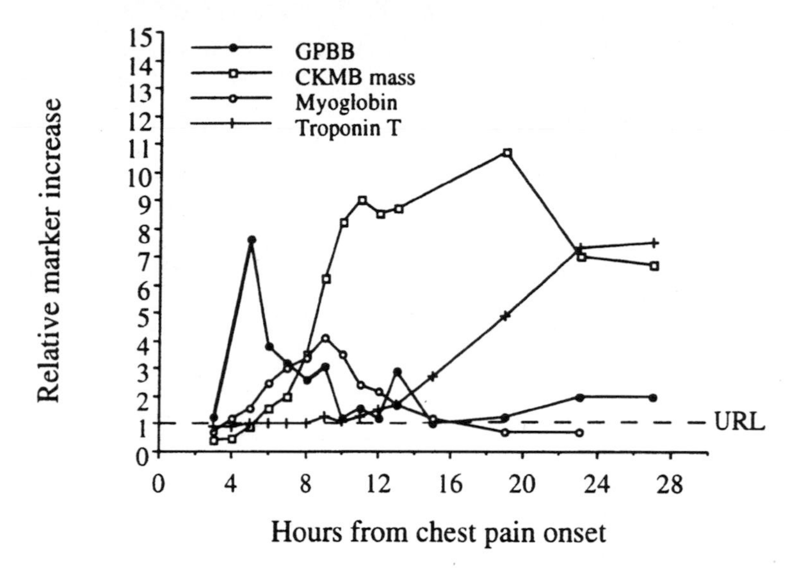

Figure 4. Glycogen phosphorylase BB, CKMB mass, myoglobin, and cardiac troponin T time courses in a patient with a small non-Q wave myocardial infarction.
Data are given as x-fold increase of the upper reference limt (URL). Reprinted with permission from Krause *et al* [20].

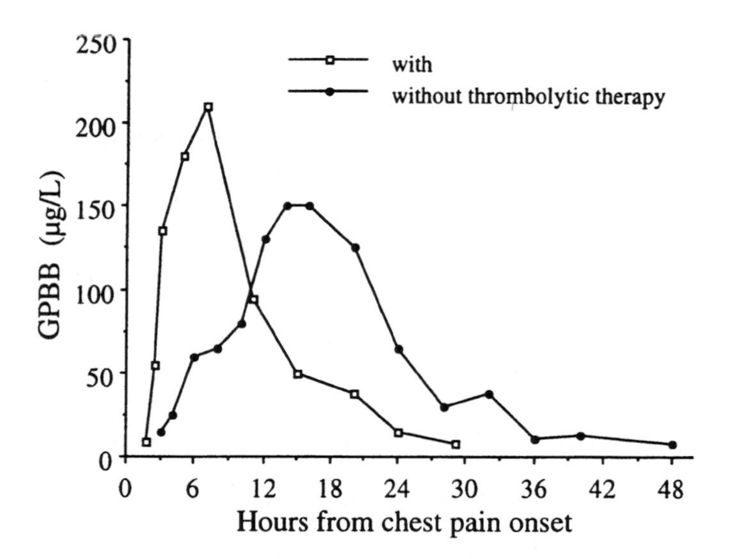

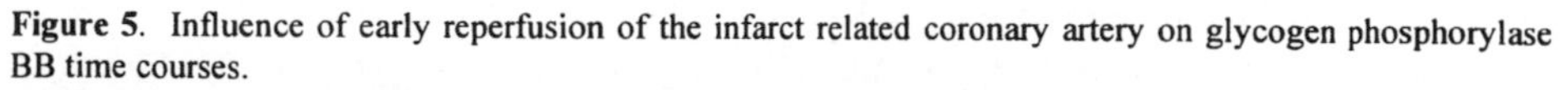

Figure 5. Influence of early reperfusion of the infarct related coronary artery on glycogen phosphorylase BB time courses.
GPBB time course in a patient who was successfully reperfused by thrombolytic therapy and in a patient who did not receive fibrinolytic treatment because of contraindications. The GPBB upper reference limit of the assay was 7µg/L. Reprinted with permission from Krause *et al* [20].

Unstable angina pectoris

The application of GPBB is not restricted to conventional myocardial infarction. An early release of GPBB was demonstrated in patients with Braunwald class III unstable angina who showed ST-T alterations at rest. GPBB was increased above the upper reference limit in the majority of these patients at hospital admission [24]. Whether the early GPBB release in these patients was due to minimal necrosis of myocardial tissue or severe reversible ischaemic injury is not known. GPBB showed the best diagnostic performance of all tested markers (see Figure 6) to detect acute ischaemic coronary syndromes (AMI or severe unstable angina at rest with transient ST-T alterations) on hospital admission [23]. GPBB plasma concentrations in patients with stable angina resembled those of healthy individuals or patients without angina [23].

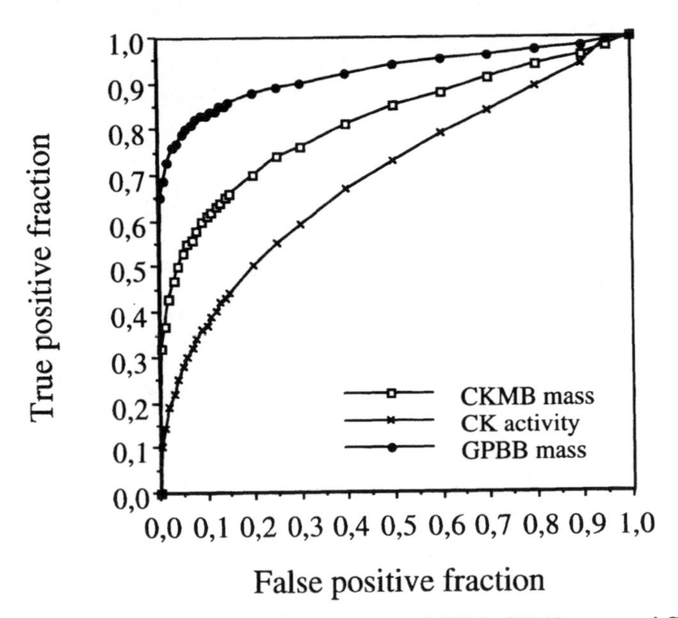

Figure 6. Receiver operating characteristic (ROC) curves of GPBB, CKMB mass, and CK activity for the identification of acute coronary syndromes in non-traumatized chest pain patients at hospital admission. A cohort of 107 emergency room patients with chest pain was studied. An acute coronary syndrome was defined as either acute myocardial infarction or unstable angina with reversible ST-T alteration in the admission ECG recording. The larger the area under the ROC curve the better is the discriminatory power of the marker. The area under the GPBB ROC curve is significantly greater than those of the other markers. Reprinted with permission from Krause *et al* [20].

Coronary artery bypass grafting (CABG)

GPBB is also a sensitive marker for the detection of perioperative myocardial ischaemia and infarction in patients undergoing CABG [25]. In uncomplicated patients GPBB peaked within 4 hours after aortic unclamping and returned to baseline values within 20 hours. GPBB release correlated with aortic crossclamping time, which reflects the duration of myocardial hypoxia during cardioplegic cardiac arrest. GPBB time courses of patients with perioperative myocardial infarction (PMI) differed markedly from those of uneventful patients in time to peak values (peaks occurred later) and peak concentrations (>50 µg/L). In addition, patients with severe episodes of perioperative myocardial ischaemia which did not fulfill standard PMI criteria showed markedly elevated GPBB concentrations compared with uncomplicated patients. In patients with emergency CABG GPBB, but not CKMB, correlated with clinical evidence of myocardial ischaemia [25]. In summary, GPBB is a very sensitive marker of perioperative ischaemic myocardial injury in CABG patients.

Diagnostic specificity

GPBB is not a heart-specific marker and its specificity is limited. However, increases in GPBB are specific for ischaemic myocardial injury when damage to the brain and consequent disturbance of the blood-brain barrier can be excluded. According to experimental studies and clinical observations, increases in GPBB do not occur in response to therapeutic circumstances in which cardiac work is increased and glycogen might be mobilized, such as after administration of catecholamines and glucagon [19,20,25]. The diagnostic specificity of GPBB for myocardial injury in non-traumatized chest pain patients was in the range of CKMB [23], which suggests sufficient specificity in clinical practice. However, further studies on the diagnostic specificity of GPBB, will also have to address this issue in an unselected cohort of patients which includes severely traumatized patients with and without head injuries, patients with liver damage or renal failure. As long as the diagnostic specificity of GPBB for myocardial damage is not fully delineated, a positive GPBB result should be confirmed later by cardiac troponin I or cardiac troponin T measurement.

Summary and Conclusions

Clearly, our first clinical results have to be confirmed in a larger number of patients, but they allow us to conclude that GPBB is a promising marker for the detection of ischaemic myocardial injury. This is probably explained by its function as a key enzyme of glycogenolysis. GPBB was a very sensitive marker for the diagnosis of AMI within four hours after the onset of chest pain because this marker increased in a considerable proportion of AMI patients within 2-3 hours after the onset of chest pain [22,23]. The application of GPBB is not restricted to the diagnosis of conventional AMI. GPBB is also increased early in patients with unstable angina and reversible ST-T alterations in the resting ECG at hospital admission [22,23]. Therefore, GPBB could be useful for early risk stratification in these patients. GPBB is also a sensitive marker for the detection of perioperative myocardial ischaemia and infarction in CABG patients [25]. The diagnostic specificity of GPBB appears to be sufficient for clinical practice, in non-traumatized chest pain patients it was in the range observed for CKMB [23]. Thus, if these first clinical results on GPBB can be confirmed and an assay which is suitable for routine use is available, a future scenario for laboratory testing for myocardial injury could be the combination of, for example, cardiac troponin I and GPBB measurement, which combines cardiac-specificity with high early sensitivity for ischaemic myocardial damage.

References

1. Mair J, Morandell D, Genser N, Lechleitner P, Dienstl F, Puschendorf B. Equivalent early sensitivities of myoglobin, creatine kinase MB mass, creatine kinase isoform ratios, and cardiac troponin I and T for acute myocardial infarction. Clin Chem 1995;41:1266-72.
2. Mair J, Smidt J, Lechleitner P, Dienstl F, Puschendorf B. A decision tree for the early diagnosis of acute myocardial infarction in non-traumatic chest pain patients at hospital admission. Chest 1995;108:1502-9.
3. Rozenman Y, Gotsman MS. The earliest diagnosis of acute myocardial infarction. Annu Rev Med 1994;45:31-44.
4. Newgard CB, Hwang PK, Fletterick RJ. The family of glycogen phosphorylases: structure and function. Crit Rev Biochem Molec Biol 1989;24:69-99.
5. Meyer F, Heilmeyer LMG Jr, Haschke RH, Fischer EH. Control of phosphorylase activity in a muscle glycogen particle: isolation and characterization of the protein-glycogen complex. J Biol Chem 1970;245:6642-8.
6. Entman ML, Kaniike K, Goldstein MA, *et al.* Association of glycogenolysis with cardiac sarcoplasmic reticulum. J Biol Chem 1976;251:3140-6.
7. Entman ML, Bornet EP, van Winkle WB, Goldstein MA, Schwartz A. Association of glycogenolysis with cardiac sarcoplasmic reticulum: II. Effect of glycogen depletion, deoxycholate solubilization and cardiac ischemia: evidence for a phosphorylase kinase membrane complex. J Mol Cell Cardiol 1977;9:515-28.
8. Newgard CB, Littmann DR, Genderen C, Smith M, Fletterick RJ. Human brain glycogen phosphorylase. Cloning sequence analysis, chromosomal mapping, tissue expression and comparison with the human liver and muscle isozymes. J Biol Chem 1988;263:3850-7.
9. Kato A, Shimizu A, Kurobe N, Takashi M, Koshikawa K. Human brain-type glycogen phosphorylase: quantitative localization in human tissues determined with an immunoassay system. J Neurochem 1989;52:1425-32.
10. Will H, Krause E-G, Böhm M, Guski H, Wollenberger A. Kinetische Eigenschaften der Isoenzyme der Glykogenphosphorylase b aus Herz- und Skelettmuskulatur des Menschen. Acta Biol Med Germ 1974;33:149-60.
11. Proux D, Dreyfus J-C. Phosphorylase isoenzymes in tissues: prevalence of the liver type in man. Clin Chim Acta 1973;48:167-72.
12. Wollenberger A, Krause E-G, Heier G. Stimulation of 3', 5'-cyclic AMP formation in dog myocardium following arrest of blood flow. Biochem Biophys Res Commun 1969;36:664-70.
13. Dobson JG, Mayer SE. Mechanism of activation of cardiac glycogen phosphorylase in ischemia and anoxia. Circ Res 1973;33:412-20.
14. Krause E-G, Wollenberger A. Cyclic nucleotides in heart in acute myocardial ischemia and hypoxia. Adv Cyclic Nucl Res 1980;12:49-61.
15. Reimer KA, Jennings RB. Myocardial ischemia, hypoxia and infarction. In: Fozzard HA, Haber E, Jennings RB, Katz AM, Morgan HE, editors. The Heart and Cardiovascular System. 2nd ed., New York: Raven Press, 1991:1875-1973.
16. Kalil-Filho R, Gersdtenblith G, Hansford RG, Chacko VP, Vandegaer K, Weiss RG. Regulation of myocardial glycogenolysis during post-ischemic reperfusion. J Mol Cell Cardiol 1991;23:1467-79.
17. Michael LH, Hunt JR, Weilbaecher D, *et al.* Creatine kinase and phosphorylase in cardiac lymph: coronary occlusion and reperfusion. Am J Physiol 1985;248:H350-9.
18. Krause E-G, Will H, Böhm M, Wollenberger A. The assay of glycogen phosphorylase in human blood serum and its application to the diagnosis of myocardial infarction. Clin Chim Acta 1975;58:145-54.
19. Schulze W, Krause E-G, Wollenberger A. On the fate of glycogen phosphorylase in the ischemic and infarcting myocardium. J Mol Cell Cardiol 1971;2:241-51.
20. Krause E-G, Rabitzsch G, Noll F, Mair J, Puschendorf B. Glycogen phosphorylase isoenzyme BB in diagnosis of myocardial ischaemic injury and infarction. Mol Cell Biochem 1996;160/161:289-95.
21. Krause E-G, Härtwig A, Rabitzsch G. On the release of glycogen phosphorylase from heart muscle: effect of substrate depletion, ischemia and of imipramine. Biomed Biochim Acta 1989;48:S77-82.
22. Rabitzsch G, Mair J, Lechleitner P, *et al.* Isoenzyme BB of glycogen phosphorylase b and myocardial infarction. Lancet 1993;341:1032-3.
23. Rabitzsch G, Mair J, Lechleitner P, *et al.* Immunoenzymometric assay of human glycogen phosphorylase isoenzyme BB in diagnosis of ischemic myocardial injury. Clin Chem 1995;41:966-78.

24. Mair J, Puschendorf B, Smidt J, *et al.* Early release of glycogen phosphorylase in patients with unstable angina and transient ST-T alterations. Brit Heart J 1994;72:125-7.
25. Mair P, Mair J, Krause E-G, Balogh D, Puschendorf B, Rabitzsch G. Glycogen phosphorylase isoenzyme BB mass release after coronary artery bypass grafting. Eur J Clin Chem Clin Biochem 1994;32:543-7.

Chapter 7

FATTY ACID-BINDING PROTEIN AS A PLASMA MARKER FOR THE EARLY DETECTION OF MYOCARDIAL INJURY

Jan F. C. Glatz

Fatty acid-binding protein (FABP) is a relatively small (14-15 kD) protein that is abundantly present in the soluble cytoplasm of almost all tissue cells, including those from cardiac and skeletal muscles. Upon muscle injury FABP is released from the damaged cells into blood plasma, similar to the well-established release under such circumstances of other soluble proteins, such as creatine kinase, lactate dehydrogenase and myoglobin. This finding indicates the possibility of using FABP as a plasma marker to monitor the occurrence and extent of muscle injury. Research performed by several groups during the last decade has revealed that FABP is a useful marker for the early detection of muscle injury in general, and for acute myocardial infarction (AMI) in particular. In this chapter, the biochemistry of FABP will be reviewed, followed by a description of its release characteristics from injured muscle in relation to that of other cardiac marker proteins, and a discussion of the application of FABP in routine clinical practice.

Biochemistry of FABP

The presence in tissue cytosol preparations of proteins which can reversibly and non-covalently bind long-chain fatty acids was first described by Ockner and co-workers in 1972 [1]. In subsequent years it was established that, of these, distinct cytoplasmic fatty acid-binding proteins (FABPs) types occur, each showing a unique pattern of tissue distribution [reviewed in 2]. These FABP types are generally named after the tissue in which they were first identified and/or mainly occur, e.g. L(iver)-FABP, I(ntestinal)-FABP, H(eart)-FABP. However, L-FABP is also found in small intestine and kidney, and H-FABP shows a rather widespread tissue distribution because this FABP type occurs not only in heart but also in skeletal muscle, smooth muscle, specific parts of the brain, distal tubule cells of the kidney, stomach parietal cells, lactating mammary gland, lung, placenta, and ovaries [2]. Currently, nine distinct types of FABP have been identified, with some tissues containing more than one type. The FABPs are relatively abundant in tissues with active fatty acid metabolism such as liver, adipose tissue and heart, which show a tissue content of 0.5 - 1 mg FABP per g wet weight of tissue [2].

The cytoplasmic FABPs belong to a multigene family of intracellular lipid-binding proteins, which also includes the cellular retinoid-binding proteins [2,3]. These non-enzymatic proteins each contain 126-137 amino acid residues (molecular mass 14-15 kD) and show a similar tertiary structure which resembles that of a clam shell [3]. The lipid ligand is bound in between the two halves of the clam by interaction with specific amino acid residues within the binding pocket, the so-called b-barrel, of the protein [3]. In general, the cytoplasmic FABPs show an affinity for long-chain fatty acid binding which is comparable to that of plasma albumin, the K_d ranging from 2 to 500 nM depending on the type of FABP and type of fatty acid studied [2,4]. Most FABP types bind exclusively long-chain fatty acids, but L-FABP also binds heme, bilirubin and various other amphiphilic ligands [2].

H-FABP has been fully characterised in man, cattle, rat and mouse. It contains 132 amino acid residues (14.5 kD), is post-translationally modified by acylation of its N-terminus, and is an acidic protein (p*I* 5) [5]. Recombinant H-FABP preparations generally show some isoelectric heterogeneity due to the existence of N-terminal variants [5], but for human H-FABP it has been shown that these differences do not affect the immunoreactivity of the protein [6].

The FABPs appear to be stable proteins, exhibiting an intracellular turnover with a half-life time of about 2-3 days [7]. Their cellular expression is regulated primarily at the transcriptional level. In general, the FABP expression is responsive to changes in lipid metabolic activity as induced by various (patho)physiological and pharmacological manipulations [2]. For instance, the H-FABP content of heart and skeletal muscles increases by endurance training [8], and is also higher in the diabetic state [9], but is slightly decreased in the hypertrophied heart [10,11].

A number of biological functions have been established or are tentatively attributed to the FABPs. The primary function is their facilitation of the cytoplasmic translocation of long-chain fatty acids, which normally is hampered by a very low solubility of these compounds in aqueous solutions (Figure 1). FABP can thus be regarded as an intracellular counterpart of plasma albumin. Additional related functions include (*i*) the intracellular trapping of fatty acids to maintain a low intracellular concentration of non-protein bound fatty acids, (*ii*) a direct involvement in fatty acid metabolism by serving as a stimulatory or inhibitory cofactor for reactions in which fatty acids are substrates or regulators, and (*iii*) the selective binding of specific fatty acid types to influence their metabolic fate in the cell. Interestingly, the FABPs were found to function also in cell growth and differentiation [12]. Finally, FABP has been postulated to participate in signal transduction pathways and in fatty acid regulation of gene expression, and to protect cells against the adverse (detergent-like) effects of long-chain fatty acids [2]. The latter function would be of special importance for the ischaemic heart, because the tissue accumulation of fatty acids and their derivatives occurring in this condition has been associated with arrhythmias, increased myocardial infarct size and depressed myocardial contractility [13]. H-FABP may then be crucial to sequester accumulating fatty acids and thus prevent tissue injury. However, at present, the available evidence for such a role for H-FABP remains inconclusive [2].

For the sake of completeness it should be noted that fatty acid-binding proteins are also found in the plasma membrane of most parenchymal cells. These membrane-associated FABPs comprise the 40-kD plasmalemmal FABP ($FABP_{pm}$), 60-kD fatty acid-transport protein (FATP), and 88-kD fatty acid translocase (FAT, also known as CD36), and they presumably function in the transmembrane transport (cellular uptake) of fatty acids (Figure 1) [2,14,15]. However, these proteins are clearly distinct from the cytoplasmic FABPs.

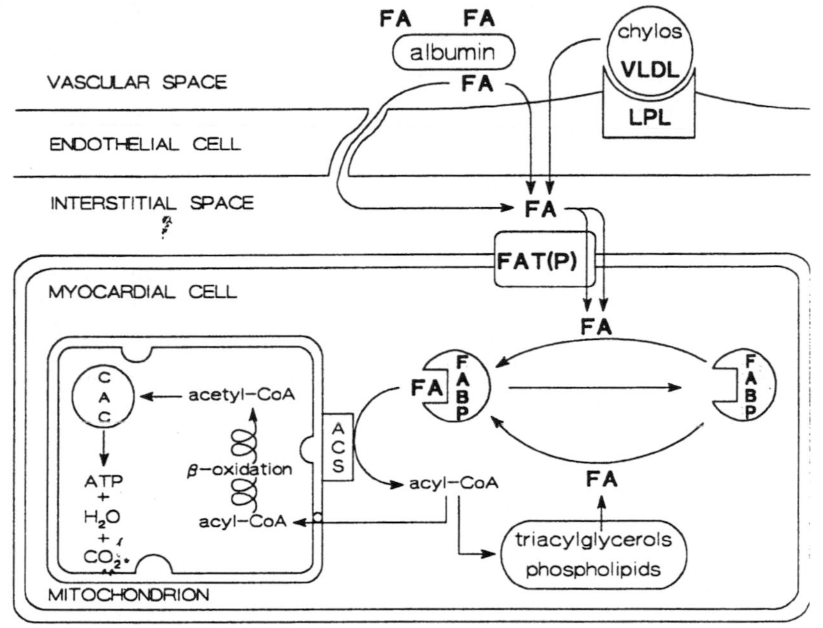

Figure 1. Schematic overview of the uptake and metabolism of long-chain fatty acids (FA) in cardiac tissue, illustrating the role of cytoplasmic fatty acid-binding protein (FABP) in the trans-cytoplasmic transport of FA from the sarcolemma to their sites of esterification or oxidation.
Abbreviations: FA, long-chain fatty acid; chylos, chylomicrons; VLDL, very low-density lipoproteins; LPL, lipoprotein lipase; FAT, fatty acid translocase; FATP, fatty acid-transport protein; FABP, fatty acid-binding protein; ACS, acylcoenzyme A synthethase; CAC, citric acid cycle.

Release of FABP upon muscle injury

The release of (H-)FABP from injured muscle was first demonstrated with isolated working rat hearts [16]. Cellular injury was induced by subjecting the hearts to 60 min of global no-flow ischemia, and was followed by reperfusion. During post-ischaemic reperfusion FABP was released into the coronary perfusate at a rate and to an extent similar to that of lactate dehydrogenase (LDH) [16]. In subsequent studies it was shown that following experimental myocardial infarction in the dog [17] or the rat [18,19], FABP is released into plasma and also partly excreted into urine. These data indicated the potential use of FABP as a plasma or urine marker of myocardial injury in experimental animals and triggered studies on the possible application of FABP for the detection of cardiac and skeletal muscle injury, especially acute myocardial infarction, in humans.

Shortly after the observations in experimental animals were made, several groups reported the release of FABP into plasma of patients with acute myocardial infarction (AMI) [20-23]. As an example, Figure 2 shows mean plasma release curves of FABP and of several other plasma marker proteins for 15 AMI patients (treated with thrombolytic therapy) from whom blood samples were obtained frequently during the first 72 hours of hospitalisation [24]. The FABP release curve appears markedly different from that of most marker proteins, but closely resembles that of myoglobin (Figure 2). Thus, for both FABP and myoglobin peak plasma concentrations are reached within 4 hours of AMI, while for creatine kinase (CK) this takes about 12 hours and for lactate dehydrogenase (LDH) about 20 hours. Furthermore, plasma FABP and myoglobin return to their respective reference values within 24 hours of AMI (Figure 2), a finding which indicates the usefulness of both markers, particularly for the assessment of reinfarction [25]. However, for AMI patients not treated with thrombolytics, peak levels are reached about 8 hours after AMI and elevated plasma FABP and myoglobin concentrations are found up to 24-36 hours after onset of chest pain [25]. In comparison, the release of the myofibrillar proteins troponin T (TnT) and troponin I (TnI) from injured myocardium follows a different pattern with elevated plasma concentrations occurring from about 8 hours up to more than one week after infarction [26, 27]. Hence, the so-called diagnostic window of the various marker proteins differs considerably.

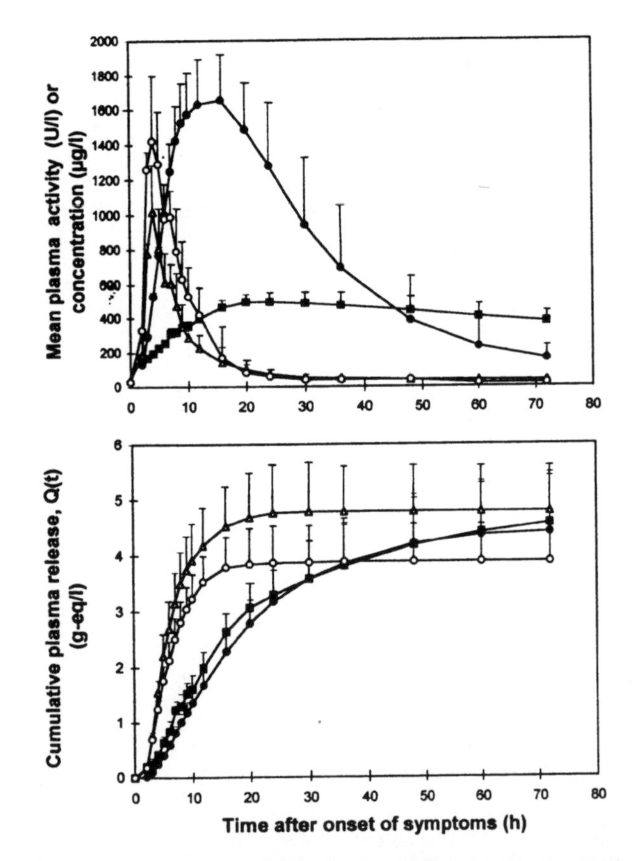

Figure 2. Mean plasma concentration or activities (upper panel) and mean cumulative release expressed in gramequivalents (g-eq) of healthy myocardium per litre of plasma (lower panel) of FABP (multiplied by 10, o), myoglobin (Æ), creatine kinase (o) and LDH isoenzyme-1 (_) as a function of time after onset of symptoms, in 15 patients after acute myocardial infarction. Data are mean ± SEM. Adapted from ref. 24.

The marked differences in the time course of plasma concentrations or activities among the soluble proteins (CK, LDH, FABP and myoglobin) are caused (*i*) by a more rapid washout of the smaller proteins (FABP, myoglobin) from the interstitium to the vascular compartment, and (*ii*) by differences among the proteins in their rate of elimination from plasma [28]. Studies with isolated cardiac myocytes subjected to simulated ischaemia showed that protein release from the damaged myocytes is independent of molecular mass [29]. This indicates that during the protein-release phase the sarcolemma does not act as a selective sieve through which small proteins are preferentially lost. The fact that smaller proteins can be detected in blood plasma earlier after muscle injury than larger proteins therefore relates to a greater permeability of the endothelial barrier for smaller proteins [28].

With respect to protein elimination from plasma, FABP and myoglobin, unlike the cardiac enzymes, are removed from the circulation predominantly by renal clearance [21-23, 28]. This explains not only their rapid clearance from plasma after AMI (Figure 2) but their relatively low plasma concentrations in healthy individuals. The latter concentrations are determined mainly by the release of protein from skeletal muscle, as its total mass far exceeds that of cardiac muscle. Because the skeletal muscle FABP content is relatively low compared to that of myoglobin, the plasma reference concentration of FABP also is relatively low (Table 1). This notion is also reflected in the ratio of the concentrations of myoglobin and FABP in plasma from healthy subjects (myoglobin/FABP ratio ca. 20) which resembles the ratio in which these proteins occur in skeletal muscle (myoglobin/FABP ratio 15-70) (see below).

Table 1. Comparison of selected soluble markers of myocardial injury.

Protein	Molecular mass (kD)	Cardiac muscle content (mg/g or U/g)	Skeletal muscle content * (mg/g or U/g)	Reference plasma concentration (μg/L or U/L)
Fatty acid-binding protein (FABP)	14.5	0.57	0.04 - 0.14	1.7
Myoglobin	17.6	2.7	2.2 - 6.7	32
Creatine kinase (CK)	80	865	2380 - 3110	40
Creatine kinase MB (CK-MB)	80	132	15 - 36	3
Lactate dehydrogenase isoenzyme-1 (LDH-1)	136	123	42 - 69	80

Data are obtained from refs. 11, 24, 25 and 30. Enzyme activities are expressed as μmol/min (U) per g or per litre, measured at 25°C.
* Range given for muscles of different fibre type composition.

The role of the kidney in the clearance of FABP and myoglobin from plasma also indicates that increased plasma concentrations of these proteins are likely to be found in case of renal insufficiency. Indeed, it was recently reported that patients with chronic renal failure and normal heart function show a several-fold increase in plasma concentrations of both FABP and myoglobin [31]. In addition, Kleine *et al* [22] reported a patient with AMI and severe renal insufficiency in whom the plasma FABP concentration remained markedly elevated for at least 25 hours after infarction.

Early diagnosis of acute myocardial infarction

The rapid release of FABP into plasma after myocardial injury, and its relatively low plasma reference concentration, favour the use of FABP especially, for the early diagnosis of AMI. The higher sensitivity of FABP for confirmation of AMI in the early hours after chest pain onset has now been firmly documented [21-23]. FABP appears at least as sensitive as myoglobin [32]. Only in the last few years have studies been performed that allow the proper assessment of both the sensitivity and specificity of FABP for AMI diagnosis. Preliminary data from a prospective multicentre study, comprising four European hospitals and including over 300 patients admitted with chest pain suggestive of AMI, revealed a superior performance of FABP over myoglobin, both in terms of sensitivity and specificity of AMI diagnosis [33]. For instance, specificities >90% were reached for FABP at 10 μg/L and for myoglobin at 90 μg/L. Using these upper reference concentrations, in the subgroup of patients admitted within 3 hours after onset of symptoms (n=90) the diagnostic sensitivity of the first blood sample taken was for FABP 48% and for myoglobin 37%, while for patients admitted 3-6 hours after AMI (n=47) the sensitivity was for FABP 83% and for myoglobin 74% (*EUROCARDI* Study Team, unpublished observations). Similar data were obtained in two other recent studies [34,35].

FABP was also found to be useful for the early detection of postoperative myocardial tissue loss in patients undergoing coronary bypass surgery [27,36]. In these patients myocardial injury may be caused by global ischaemia/reperfusion and, additionally, by postoperative myocardial infarction. In a recent study we found that in such patients plasma CK, myoglobin and FABP are already significantly elevated 0.5 hours after reperfusion. In the patients who developed postoperative myocardial infarction, a second increase was observed for each plasma marker protein, while a significant increase was recorded earlier for FABP (4 hours after reperfusion) than for CK or myoglobin (8 hours after reperfusion) [36]. These data suggest that FABP would allow a more early exclusion of postoperative myocardial infarction, thus permitting the earlier transfer of these patients from the intensive care unit to the ward.

Finally, FABP is a useful marker for early detection of successful coronary reperfusion [37] and for the immunohistochemical detection of very recent myocardial infarctions [38,39].

Estimation of myocardial infarct size

Myocardial infarct size is usually estimated from the serial measurement of cardiac proteins in plasma and calculation of the cumulative release over time (plasma curve area) taking into account the elimination rate of the protein from plasma [40]. This approach requires that the proteins are completely released from the heart after AMI and recovered quantitatively in plasma. Complete recovery is well documented for CK, LDH and myoglobin (but does not apply for the structural proteins TnT and TnI), and can also be shown for FABP [24,41]. Figure 2 (lower panel) presents the cumulative release patterns of these four proteins, expressed in gram-equivalents (g-eq) of healthy myocardium per litre of plasma (infarct size). The release of FABP and that of myoglobin are completed much earlier than that of either CK or LDH, but despite this kinetic difference for each of the proteins, the total quantities released yield comparable estimates of the mean extent of myocardial injury when evaluated at 72 hours after the onset of AMI (Figure 2). It can be concluded that, provided frequent blood samples are taken and there is no renal failure, FABP gives a clinically useful estimate of myocardial infarct size.

Discrimination of cardiac from skeletal muscle injury

A classic problem with several marker proteins is their presence in significant quantities not only in heart muscle but also in skeletal muscle cells. The use of these markers for the diagnosis of AMI may then be hampered in case of extensive skeletal muscle injury, such as multi-organ failure, post-operative states, or vigorous exercise.

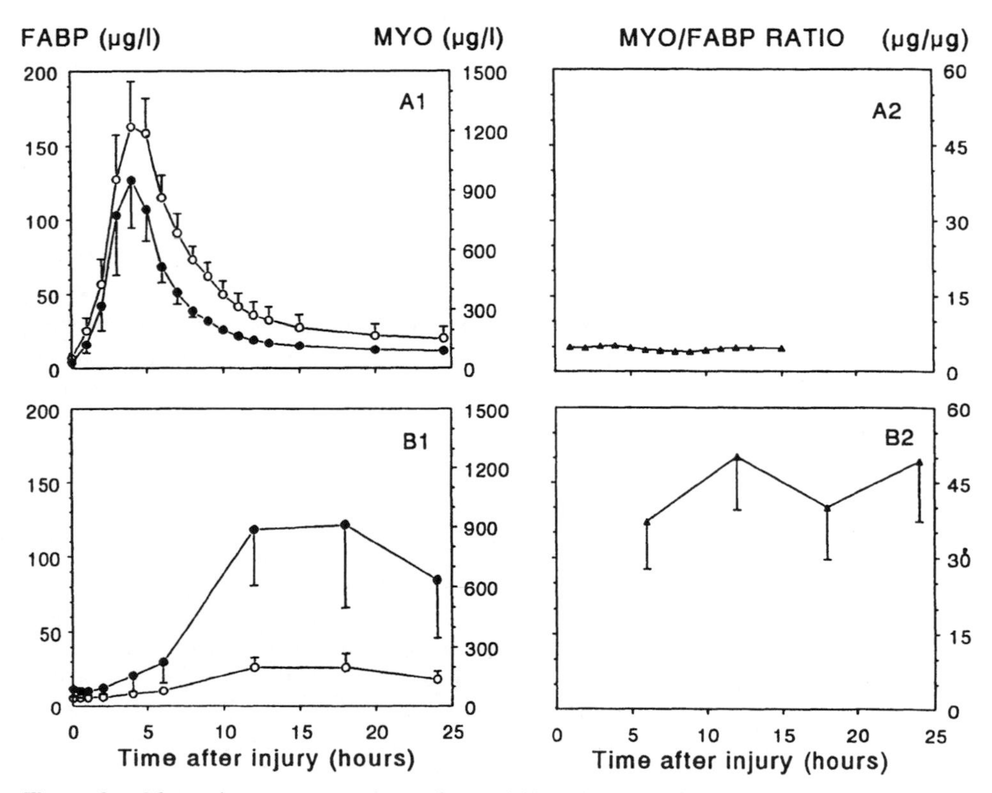

Figure 3. Mean plasma concentrations of myoglobin (o) and FABP (o) (left panels) and the myoglobin/FABP ratio (Æ) (right panels) in 9 patients after AMI (and receiving thrombolytic therapy) (A), and in 9 patients after aortic surgery (B). Data are mean ± SEM. Adapted from ref. 25.

The presence of (the same type of) FABP in both heart and skeletal muscle also complicates the discrimination between myocardial and skeletal muscle injury. However, this problem can be overcome by the combined measurement of myoglobin and FABP concentrations in plasma and expressing their ratio, as this plasma ratio is a reflection of the ratio in which these proteins occur in the affected tissue cells and differs between heart muscle (myoglobin/FABP ratio 4-5) and skeletal muscles (myoglobin/FABP ratio 20-70, depending on the type of muscle) (Table 1) [25, 27, 42]. In Figure 3 this finding is illustrated for patients after AMI in whom the plasma myoglobin/FABP ratio was ca. 5 during the entire period of elevated plasma concentrations (upper panels), and for patients who underwent aortic surgery, which causes no-flow ischemia of the lower extremities, in whom the plasma myoglobin/FABP ratio was ca. 45 (lower panels). In addition, van Nieuwenhoven *et al* [25] described a patient who was defibrillated shortly after AMI, a treatment that most likely results in injury of intercostal pectoral muscles, and in whom the plasma myoglobin/FABP ratio increased from 8 to 60 during the first 24 hours after AMI. Finally, in case AMI patients show a second increase of plasma concentrations of marker proteins, the ratio may be of help to delineate whether this second increase was caused either by a recurrent infarction or by the occurrence of additional skeletal muscle injury, as in the former case the ratio will remain unchanged [25].

The diagnostic performance of FABP as an early plasma marker of myocardial injury may thus further increase when the criterion of a plasma myoglobin/FABP ratio <10 is taken as an additional parameter [32]. However, the myoglobin/FABP ratio cannot provide absolute cardiac specificity [27].

Assay of FABP in plasma

To date, a large number of immunochemical assays for (H-)FABP have been described, mostly enzyme-linked immunosorbent assays (ELISA) of the antigen capture type [6, 21, 22, 43], but also a competitive immunoassay [20] and an immunofluorometric assay [44]. In most cases monoclonal antibodies are used in the assays showing virtually no cross-reactivity with other FABP types. These assays have been used for retrospective analyses of plasma FABP in patient samples. However, the fastest immunoassay reported [6] still takes 45 min to complete, a time generally regarded as too long to rely on the test in the clinical decision making process in case of suspected AMI.

More rapid FABP immunoassays include a microparticle-enhanced turbidimetric assay to be performed on a conventional clinical chemistry analyser (performance time 10 min) [45], and an electrochemical immunosensor (performance time 20 min) [46, 47]. This latter sensor is based on screen printed graphite electrodes and uses an immunosandwich procedure and an amperometric detection system [47]. Measurements of plasma samples from a patient with AMI with this immunosensor and with an ELISA show an excellent correlation (Figure 4). These developments indicate that the application of FABP as a plasma marker in the early phase diagnosis of AMI could soon enter clinical practice.

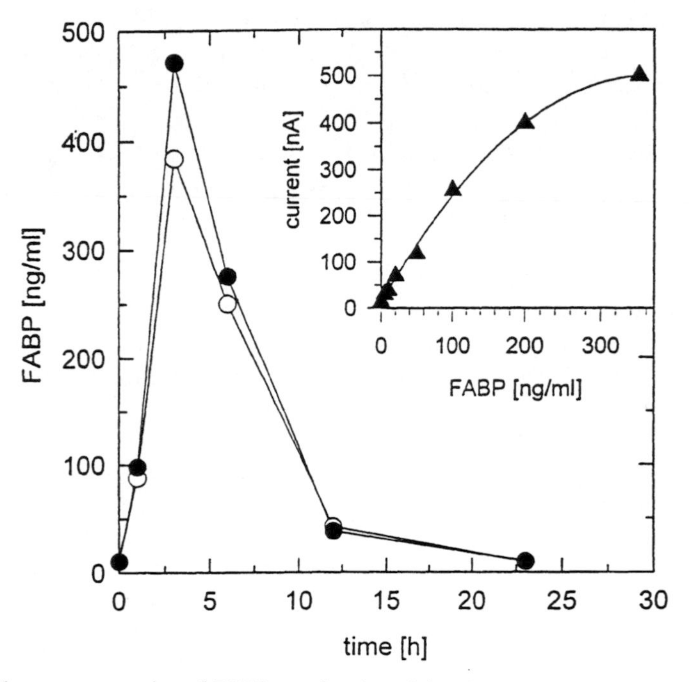

Figure 4. Plasma concentration of FABP as a function of time in a patient with AMI, measured with a sandwich ELISA (o) and with an immuosensor device (o). Inset shows the calibration curve for FABP in plasma measured with the immunosensor (Æ). Reproduced from ref. 47, with permission.

Conclusions

Biochemical markers of myocardial cell damage continue to be important tools for differentiating patients with and those without AMI, because specific ST segment changes on the admission ECG remain absent in a large number of patients with AMI [27, 48-51]. At present, attention is focused on defining those cardiac marker proteins, or combinations of marker proteins, that show a high sensitivity as well as specificity for AMI detection, especially in the early hours after admission. Theoretically, the perfect early marker of cardiac myocyte necrosis would be a soluble protein of small size and showing absolute cardiospecificity. The soluble nature and small size would result in a rapid release from the damaged cells and appearance in plasma, while the cardiospecificity is needed to exclude injury of tissues other than heart muscle. However, the existence of such a protein has not (yet) been reported, as the known small soluble proteins do not appear to be cardiospecific (e.g. myoglobin), while identified cardiospecific proteins are all structural (e.g. cardiac TnT and cardiac TnI) [27,48].

Fatty acid-binding protein (FABP), which has been only recently introduced as a plasma marker of AMI, is a small (14.5 kD) soluble protein that is not cardiospecific but is expressed in skeletal muscle in relatively low amounts. Therefore, FABP appears suitable for the early diagnosis of AMI. The release and plasma kinetics of FABP closely resemble those of myoglobin, which to date is generally referred to as the earliest plasma marker of muscle necrosis [27,48,49]. Experimental studies indicate that this resemblance relates to the similar molecular masses of FABP (14.5 kD) and myoglobin (17.6 kD). Interestingly, recent clinical studies with patients suspected of AMI even reveal a superior performance of FABP over myoglobin (as well as other marker proteins) for the early detection of AMI. This finding most likely relates to marked differences in the tissue contents of FABP and myoglobin in cardiac and skeletal muscles, which result in relatively low upper reference concentrations in plasma for FABP compared to myoglobin. These differences in tissue content are also reflected in the plasma concentrations of these proteins following either cardiac or skeletal muscle injury, such that the ratio of the plasma concentrations of myoglobin and FABP can be applied to discriminate myocardial from skeletal muscle injury.

The use of FABP as a diagnostic plasma marker in the clinical setting is hampered by two drawbacks. First, the elimination of FABP from plasma is mainly by renal clearance, and caution should be exercised when using FABP in cases of renal insufficiency, since the pre-infarct plasma concentration then is already likely to be elevated. Assay of plasma creatinine or urea would be helpful in identifying patients with renal insufficiency. Second, the so-called diagnostic window of FABP extends to 12-24 hours after the onset of AMI, limiting its use to those patients who are admitted to hospital early after chst pain. Thus, in patients presenting to hospital more than 12 hours after the onset of chest pain, or in cases where the time of onset of symptoms is unsure, the combined measurement in plasma of an early marker (FABP) and a late marker (e.g. cardiac TnT or cardiac TnI) should be recommended [cf. 52]

Acknowledgements

The author wishes to thank Prof. Dr. W.T. Hermens for critical reading of the manuscript. Work in the authors laboratory was supported by grants from the Netherlands Heart Foundation (D90.003 and 95.189) and the European Community (BMH1-CT93.1692 and CIPD-CT94.0273).

References

1. Ockner RK, Manning JA, Poppenhausen RB, Ho WKL. A binding protein for fatty acids in cytosol of intestinal mucosa, liver, myocardium, and other tissues. Science 1972; 177: 56-8.
2. Glatz JFC, Van der Vusse GJ. Cellular fatty acid-binding proteins. Their function and physiological significance. Prog Lipid Res 1996; 35: 243-82.
3. Banaszak L, Winter N, Xu Z, Bernlohr DA, Cowan S, Jones TA. Lipid binding proteins: a family of fatty acid and retinoid transport proteins. Adv Protein Chem 1994; 45: 89-151.
4. Richieri G V, Ogata RT, Kleinfeld AM. Equilibrium constants for the binding of fatty acids with fatty acid-binding proteins from adipocyte, intestine, heart, and liver measured with the fluorescent probe ADIFAB. J Biol Chem 1994; 269: 23918-30.
5. Schaap FG, Specht B, Van der Vusse GJ, Börchers T, Glatz JFC. One-step purification of rat heart-type fatty acid-binding protein expressed in Escherichia coli. J Chromatogr 1996; B 679: 61-7.
6. Wodzig KWH, Pelsers MMAL, Van der Vusse GJ, Roos W, Glatz JFC. One-step enzyme-linked immunosorbent assay (ELISA) for plasma fatty acid-binding protein. Ann Clin Biochem 1997; 34: 263-8.
7. Bass NM. The cellular fatty acid binding proteins: Aspects of structure, regulation, and function. Int Rev Cytol 1988; 111: 143-84.

8. Van Breda E, Keizer HA, Vork MM, *et al.* Modulation of fatty acid-binding protein content of rat heart and skeletal muscle by endurance training and testosterone treatment. Eur J Physiol 1992; 421: 274-9.
9. Glatz JFC, Van Breda E, Keizer HA, *et al.* Rat heart fatty acid-binding protein content is increased in experimental diabetes. Biochem Biophys Res Commun 1994; 199: 639-46.
10. Vork MM, Trigault N, Snoeckx LHEH, Glatz JFC, Van der Vusse GJ. Heterogeneous distribution of fatty acid-binding protein in the hearts of Wistar Kyoto and Spontaneously Hypertensive rats. J Mol Cell Cardiol 1992; 24: 317-21.
11. Kragten JA, Van Nieuwenhoven FA, Van Dieijen-Visser MP, Theunissen PHMH, Hermens WT, Glatz JFC. Distribution of myoglobin and fatty acid-binding protein in human cardiac autopsies. Clin Chem 1996; 42: 337-8.
12. Sorof S. Modulation of mitogenesis by liver fatty acid binding protein. Cancer Metastasis Rev 1994; 13: 317-36.
13. Van der Vusse GJ, Glatz JFC, Stam HCG, Reneman RS. Fatty acid homeostasis in the normoxic and ischemic heart. Physiol Rev 1992; 72: 881-940.
14. Schaffer JE, Lodish HF. Molecular mechanism of long-chain fatty acid uptake. Trends Cardiovasc Med 1995; 5: 218-24.
15. Van Nieuwenhoven FA, Van der Vusse GJ, Glatz JFC. Membrane-associated and cytoplasmic fatty acid-binding proteins. Lipids 1996; 31: S-223-7.
16. Glatz JFC, Van Bilsen M, Paulussen RJA, Veerkamp JH, Van der Vusse GJ, Reneman RS. Release of fatty acid-binding protein from isolated rat heart subjected to ischemia and reperfusion or to the calcium paradox. Biochim Biophys Acta 1988; 961: 148-52.
17. Sohmiya K, Tanaka T, Tsuji R, *et al.* Plasma and urinary heart-type cytoplasmic fatty acid-binding protein in coronary occlusion and reperfusion induced myocardial injury model. J Mol Cell Cardiol 1993; 25: 1413-26.
18. Knowlton AA, Apstein CS, Saouf R, Brecher P. Leakage of heart fatty acid binding protein with ischemia and reperfusion in the rat. J Mol Cell Cardiol 1989; 21: 577-83.
19. Volders PGA, Vork MM, Glatz JFC, Smits JFM. Fatty acid-binding proteinuria diagnosis myocardial infarction in the rat. Mol Cell Biochem 1993; 123: 185-90.
20. Knowlton AA, Burrier RE, Brecher P. Rabbit heart fatty acid-binding protein. Isolation, characterization, and application of a monoclonal antibody. Circ Res 1989; 165: 981-8.
21. Tanaka T, Hirota Y, Sohmiya K, Nishimura S, Kawamura K. Serum and urinary human heart fatty acid-binding protein in acute myocardial infarction. Clin Biochem 1991; 24: 195-201.
22. Kleine AH, Glatz JFC, Van Nieuwenhoven FA, Van der Vusse GJ. Release of heart fatty acid-binding protein into plasma after acute myocardial infarction in man. Mol Cell Biochem 1992; 116: 155-62.
23. Tsuji R, Tanaka T, Sohmiya K, *et al.* Human heart-type cytoplasmic fatty acid-binding protein in serum and urine during hyperacute myocardial infarction. Int J Cardiol 1993; 41: 209-17.
24. Wodzig KWH, Kragten JA, Hermens WT, Glatz JFC, Van Dieijen-Visser MP. Estimation of myocardial infarct size from plasma myoglobin or fatty acid-binding protein. Influence of renal function. Eur J Clin Chem Clin Biochem 1997; 35: 191-98.
25. Van Nieuwenhoven FA, Kleine AH, Wodzig KWH, *et al.* Discrimination between myocardial and skeletal muscle injury by assessment of the plasma ratio of myoglobin over fatty acid-binding protein. Circulation 1995; 92: 1848-54.
26. Kragten JA, Hermens WT, Van Dieijen-Visser MP. Cardiac troponin T release into plasma after acute myocardial infarction: Only fractional recovery compared with enzymes. Ann Clin Biochem 1996; 33: 314-23.
27. Mair J. Progress in myocardial damage detection: New biochemical markers for clinicians. Crit Rev Clin Lab Sci 1997; 34: 1-66.
28. Hermens WT. Mechanisms of release of proteins from injured muscular tissue. In: Kaski JC, Holt DW, editors. Myocardial damage: Early detection by novel biochemical markers. Dordrecht, Kluwer Academic Publishers, 1997: xxx-xx.
29. Van Nieuwenhoven FA, Musters RJP, Post JA, Verkleij AJ, Van der Vusse GJ, Glatz JFC. Release of proteins from isolated neonatal rat cardiac myocytes subjected to simulated ischemia or metabolic inhibition is independent of molecular mass. J Mol Cell Cardiol 1996; 28: 1429-34.
30. Willems GM, Van der Veen FH, Huysmans HA, *et al.* Enzymatic assessment of myocardial necrosis after cardiac surgery: Differentiation from skeletal muscle damage, hemolysis, and liver injury. Am Heart J 1984; 109: 1243-52.

31. Górski J, Hermens WT, Borawski J, Mysliwiec M, Glatz JFC. Increased fatty acid-binding protein concentration in plasma of patients with chronic renal failure. Clin Chem 1997; 43: 193-5.
32. Abe S, Saigo M, Yamashita T, *et al.* Heart fatty acid-binding protein is useful in early and myocardial-specific diagnosis of acute myocardial infarction. Circulation 1996; 94: I-323 (abstr.).
33. Kristensen SR, Haastrup B, Hørder M, *et al.* Fatty acid-binding protein: a new early marker of AMI. Scand J Clin Lab Invest 1996; 56 suppl. 225: 36-7 (abstr.).
34. Okamoto F, Tanaka T, Sohmiya K, *et al.* Heart type fatty acid-binding protein as a new biochemical marker for acute myocardial infarction. J Mol Cell Cardiol 1997; 29: A306 (abstr.).
35. Panteghini M, Bonora R, Pagani F, *et al.* Heart fatty acid-binding protein in comparison with myoglobin for the early detection of acute myocardial infarction. Clin Chem 1997; 43: S157 (abstr.).
36. Fransen EJ, Maessen JG, Hermens WT, Glatz JFC. Demonstration of ischaemia-reperfusion injury separate from postoperative infarction in CABG patients. Ann Thoracic Surg, in press.
37. Ishii J, Nagamura Y, Nomura M, *et al.* Eraly detection of successful coronary repefusion based on serum concentration of human heart-type cytoplasmic fatty acid-binding protein. Clin Chim Acta 1997; 262: 13-27.
38. Kleine AH, Glatz JFC, Havenith MG, Van Nieuwenhoven FA, Van der Vusse GJ, Bosman FT. Immunohistochemical detection of very recent myocardial infarctions in man with antibodies against heart type fatty acid-binding protein. Cardiovasc Pathol 1993; 2: 63-9.
39. Watanabe K, Wakabayashi H, Veerkamp JH, Ono T, Suzuki T. Immunohistochemical distribution of heart-type fatty acid-binding protein immunoreactivity in normal human tissues and in acute myocardial infarct. J Pathol 1993; 170: 59-65.
40. Hermens WT, Van der Veen FH, Willems GM, Mullers-Boumans ML, Schrijvers-Van Schendel A, Reneman RS. Complete recovery in plasma of enzymes lost from the heart after permanent coronary occlusion in the dog. Circulation 1990; 81: 649-59.
41. Glatz JFC, Kleine AH, Van Nieuwenhoven FA, Hermens WT, Van Dieijen-Visser MP, Van der Vusse GJ. Fatty acid-binding protein as a plasma marker for the estimation of myocardial infarct size in humans. Br Heart J 1994; 71: 135-40.
42. Yoshimoto K, Tanaka T, Somiya K, *et al.* Human heart-type cytoplasmic fatty acid-binding protein as an indicator of acute myocardial infarction. Heart Vessels 1995; 10: 304-9.
43. Ohkaru Y, Asayama K, Ishii H, *et al.* Development of a sandwich enzyme-linked immunosorbent assay for the determination of human heart type fatty acid-binding protein in plasma and urine by using two different monoclonal antibodies specific for human heart fatty acid-binding protein. J Immunol Meth 1995; 178: 99-111.
44. Katrukha A, Bereznikova A, Filatov V, *et al.* Development of sandwich time-resolved immunofluorometric assay for the quantitative determination of fatty acid-binding protein (FABP). Clin Chem 1997; 43: S106 (abstr.).
45. Robers M, Van der Hulst FF, Pelsers MMAL, Roos W, Eisenwiener HG, Glatz JFC. Development of a rapid microparticle-enhanced turbidimetric immunoassay for fatty acid-binding protein in plasma. Proceedings of XVI International Congress of Clinical Chemistry, London, U.K., 1996, 458-9.
46. Siegmann-Thoss C, Renneberg R, Glatz JFC, Spener F. Enzyme immunosensor for diagnosis of myocardial infarction. Sensors Actuators 1996; B30: 71-6.
47. Schreiber A, Feldbrügge R, Key G, Glatz JFC, Spener F. An immunosensor based on disposable electrodes for rapid estimation of fatty acid-binding protein, an early marker of myocardial infarction. Biosens Bioelectr, in press.
48. Adams JE, Abendschein DR, Jaffe AS. Biochemical markers of myocardial injury. Is MB creatine kinase the choice for the 1990s? Circulation 1993; 88: 750-63.
49. Bhayana V, Henderson AR. Biochemical markers of myocardial damage. Clin Biochem 1995; 28: 1-29.
50. Tormey WP. The diagnosis and management of acute myocardial infarction - a role for biochemical markers? Ann Clin Biochem 1996; 33: 477-81.
51. Keffer JH. Myocardial markers of injury. Evolution and insights. Am J Clin Pathol 1996; 105: 305-20.
52. Lindahl B, Venge P, Wallentin L, and the BIOMACS Study Group. Early diagnosis and exclusion of acute myocardial infarction using biochemical monitoring. Coron Artery Dis 1995; 6: 321-8.

Chapter 8

MECHANISMS OF PROTEIN RELEASE FROM INJURED HEART MUSCLE

Wim Th. Hermens

Cardiac marker proteins are now routinely determined in serum or plasma of patients suspected of myocardial injury, and serial measurements of such proteins are used for estimation of total myocardial injury, or 'infarct size', from cumulative protein release [1-8]. Also, protein release rates are used for the detection of coronary reperfusion.

In spite of the general clinical application of these methods some basic underlying concepts are subject to controversy. It is not generally accepted, for instance, that release of intracellular proteins indicates necrosis, that is, irreversible myocyte injury [9,10]. Higher protein release rates, as observed after coronary reperfusion, are sometimes interpreted as enhanced washout [11] and sometimes as accelerated necrosis or 'reperfusion injury' [12]. Also, partial degradation of marker proteins, preventing such proteins from reaching the circulation, has been claimed by some authors [11,13] and denied by others [14]. Because such degradation could be influenced by therapy, for instance by coronary reperfusion, this could invalidate the comparison of infarct size between treated and untreated patients [15-18].

Apart from these fundamental issues, a number of practical aspects remain to be settled. For instance, different values are used in the literature for the elimination rates of circulating proteins, which has resulted in widely differing values for calculated infarct size and has frustrated the comparison of such estimates from different studies [14]. Also, some authors account for extravasated amounts of protein [19,20], while others assume that the proteins remain confined to plasma [21,22].

These issues are discussed below with emphasis on the quantitative aspects. An attempt is made to explain some seeming discrepancies and to indicate some remaining aspects that need clarification.

Loss of cardiomyocytes as a unique measure for irreversible injury

Protein concentrations in plasma are often only used for global classification of the extent of tissue injury, for instance as 'large', 'medium-sized' or 'small'. For tissues with a high capacity of cellular regeneration this seems appropriate, because the extent of necrosis will then probably only be relevant in as much as it exceeds the formation of new cells. It has been estimated, for instance, that patients with acute hepatic injury can lose more than 300g of liver tissue in 48 hours and still may recover uneventfully [23]. This is not true for the heart. Although some mitosis may occur in the ageing rat heart

[24], as well as in *in-vitro* culture of adult human cardiomyocytes [25], it is generally assumed that the adult human cardiomyocyte has no *in-vivo* regenerative capacity [26]. Accordingly, any loss of cardiomyocytes implies a true, irreversible, injury.

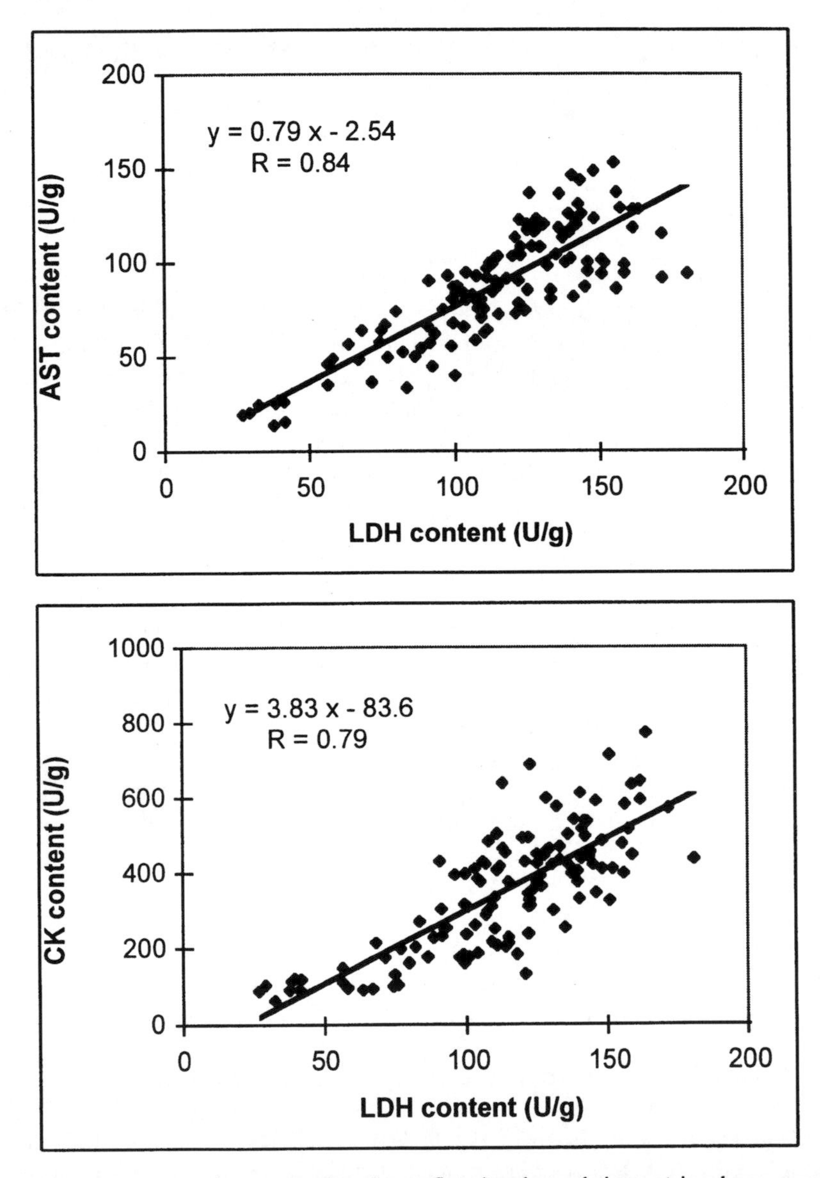

Figure 1. Mean plasma protein concentrations (upper figure) and cumulative protein release, expressed in gram-equivalents of healthy myocardium (lower figure), in 14 patients with acute myocardial infarction, who were treated with intravenous streptokinase. Because myoglobin and FABP returned to normal within 24 hours, measurements were not continued. Much earlier entrance into plasma of myoglobin and FABP, compared to CK and HBDH, is shown in the lower figure but the total release of the four proteins is comparable. For AST, gram-equivalents were calculated from the tissue content of the cytosolic isoform, and continuing release of mitochondrial AST is apparent. It is also shown that only a minimal fraction of total tissue TnT is released.

This is an important aspect of the use of marker proteins in estimation of heart injury. It is often stated that functional parameters, like left ventricular ejection fraction or wall motion score, are clinically more relevant than the extent of muscle loss. This is obviously true in the acute phase of disease and may even be true with respect to prognosis. However, myocardial stunning may cause temporary depression of cardiac function in the acute phase, while cardiac remodelling and hypertrophy will compensate for muscle loss in a later phase. This implies that cardiac contractility cannot simply be related to the extent of muscle loss, while the reduction of such loss is usually the aim of therapy. Thus, in the evaluation of such therapy, for instance of recanalisation therapy after acute myocardial infarction, marker proteins offer unique possibilities. Similar concerns can be expressed about the techniques for morphometric assessment of necrosis, often considered as the 'gold standard'. Necrotic areas, as defined by such techniques as staining with nitroblue-tetrazolium (NBT), are quite heterogenous and may contain considerable fractions of surviving muscle cells. Overall, such areas have been reported to retain more than 50% of muscle cell proteins [21] and, therefore, morphometry may grossly overestimate the extent of necrosis.

Another important point is that total protein release, between the onset of myocardial injury up to its completion, can be measured. Plasma levels of these proteins are usually still in the normal range when the patient is admitted to hospital, especially for the larger proteins (see Figure 1), and only negligible release has occurred in the period before admission. In contrast, only endpoints can be obtained for parameters such as ventricular ejection fraction or histological injury scores, and one has to assume that these variables had normal values before the acute event occurred, which often will not be true. This adds to an inherent insensitivity of these parameters and it is hard to imagine, for instance, how measurement of cardiac contractility could have detected the minimal degree of myocardial necrosis that may have proved important in the prognosis of cardiovascular disease (see below).

Variability due to pathological changes in myocardial protein content

In order to translate myocardial protein release into grams of lost muscle, the myocardium should have a well-defined protein content. Tissue protein content should preferably be measured in biopsies, rather than in autopsies, because of the risk of post-mortem protein degradation. Using human donor hearts as a reference, normal tissue content and 10-15% variability (SD) was reported for the activities of creatine kinase (CK) and lactate dehydrogenase (LDH) in biopsies taken during valve replacement [27]. About 10% variability was found for CK, a-hydroxybutyrate dehydrogenase (HBDH) and aspartate aminotransferase (AST) in biopsies obtained during various forms of heart surgery [28].

In cardiac autopsies, variabilities of about 15% were found for LDH, HBDH, and fatty acid-binding protein (FABP) [29]. These cytosolic proteins are released from the tissue by simple homogenisation, and the larger variations of about 30% reported for the tissue content of a structural protein such as troponin T (TnT), could either reflect true increased variability or could be caused by the more elaborate techniques required to extract such proteins from tissue [30,31]. The latter is also true for proteins from specific intracellular compartments, such as lysosomal and mitochondrial enzymes [32].

In considering these data, sample size should also be taken into account. Due to uncertain fractional catabolic rate constant (FCR) values and normal plasma concentrations, the lower limit for infarct size as calculated from cytosolic cardiac enzymes such as CK, HBDH or AST, is 0.1-0.5 gram wet weight [33], whereas biopsies are typically about 10 times smaller. Apart from increasing experimental scatter, the use of such small samples will tend to increase observed variability due to the presence of structural inhomogeneities and mixed cell populations. Summarising these data, it is estimated that, at least for cytosolic markers, protein release can be expressed in grams of myocardium with an accuracy of 10-15%.

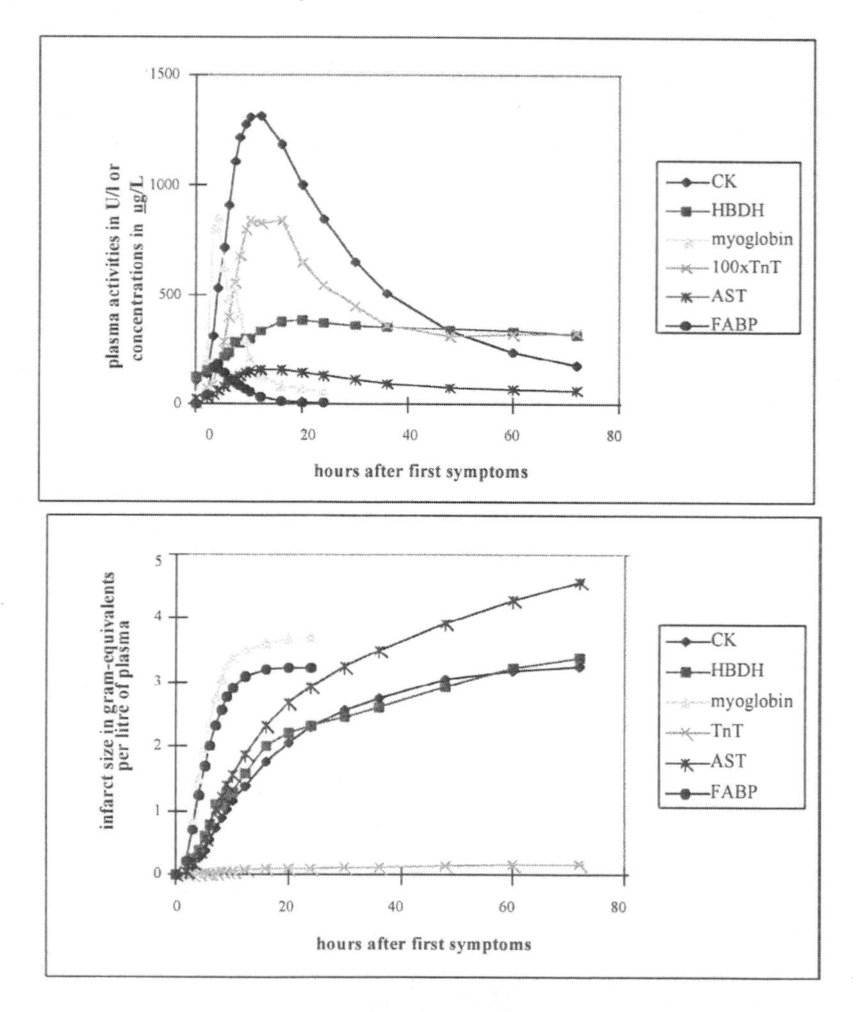

Figure 2. Correlated myocardial LDH, CK and AST content in 122 tissue samples of 0.5-2.5 g from 8 hearts, obtained from patients who died within 6 hours after the onset of acute myocardial infarction. Values were expressed in catalytic units (U) per gram wet weight of tissue. Enzyme activities are generally much lower than in healthy hearts but, although LDH seems to be better preserved than the other enzymes, the changes can be globally explained by dilution of healthy myocardium with variable fractions of connective tissue.

The data discussed so far were based on biopsies from relatively healthy hearts or autopsies from patients who died from non-cardiac causes. However, pathological changes, as found in diseased hearts, will cause altered myocardial protein content. In such hearts, hypertrophy is a common finding and is accompanied by fibrosis [34]. The healthy muscle is invaded by fibroblasts, collagen, and fat, and it has been shown that severely diseased hearts may contain large fractional volumes of connective tissue [35]. Due to these changes, average reductions of 30-40%, and an overall variability of 25%, have been reported for myocardial CK content in biopsies from failing human hearts [36]. Similar reductions and variabilities of 30-50% were found for the tissue content of CK, HBDH and AST in non-infarcted areas - supposedly representing the pre-infarction situation - of hearts from patients who died after myocardial infarction [37].

Variation in myocardial protein content of this magnitude could obviously invalidate the expression of cumulative protein release in grams of myocardium. In the quoted study [37], however, the *ratios* of enzyme showed only 15-20% variation. This is also shown in Figure 2 for left ventricular samples taken from 8 human hearts obtained within 6 hours of acute cardiac malfunction, that is, before a significant release of LDH, CK and AST from the heart could occur (see below). Although LDH was less reduced than CK and AST, probably because of LDH activity in fibroblasts [34], it was concluded that the major part of the pathological changes observed in these hearts could be described as dilution of normal myocardium with variable volume fractions of connective tissue [37]. This implied that release of muscle enzymes could still be expressed in grams of healthy myocardium, although one such gram could correspond to several grams of diseased muscle. This was indicated by expression of infarct size in gram-*equivalents*.

Little is known quantitatively of the pathological changes in myocardial content of structural proteins, but reductions up to 80% were reported for TnT and troponin I (TnI) in failing porcine hearts [38]. This indicates that pathological changes in the myocardial content of contractile proteins could be larger than changes in cytosolic proteins and this may invalidate expression of the release of contractile proteins in gram-equivalents of myocardium.

The relation between cellular release of proteins and cardiomyocyte death

On the basis of clinical experience, the release of myocardial marker proteins into plasma has traditionally been considered as a sign of necrosis. This interpretation has recently been confirmed when it was shown that even minor release of proteins such as TnT [39] or CKMB [40] indicates a worse prognosis. This traditional interpretation has not remained unchallenged, however. In the normal dog heart, lymph may contain higher concentrations of muscle proteins than plasma, while, as shown in Table 1, the reverse is normally true. This was interpreted as cellular release of proteins under normal physiological conditions [41]. In cultures of adult rat heart cells, hypoxia induced sarcolemmal protrusions separating from the intact cell ('blebbing'), and this was considered as proof for protein release from living cells [10].

The quantitative relevance of these findings is difficult to assess. Assuming a cardiac lymph flow in dog heart of 10 ml/kg/h (see Table 1), a heart weight of 100 g and a CK concentration in lymph exceeding the plasma concentration by 100 U/L, total

CK release into the lymphatic system during 10 years of dog life would be about 9000 U, that is the CK content of about 6 grams of dog myocardium. This is well within the range of an estimated total 15-20% loss of heart muscle due to ageing in man [26], and the increased lymphatic concentrations could, thus, reflect physiological loss of muscle cells due to ageing, rather than protein release from living cells.

Table 1. Parameter values for protein distribution in the dog

	plasma	heart	skeletal muscle	skin	viscera
weight (kg)	1*	0.1	8.5	3	2.8
interstitial fluid volume (ml/kg)	-	140	100	400	250
endothelial permeability (ml/kg/h)	-	30-300	0.75	4	6
lymph flow, mlk/kg/h	-	6-30	3	16	3
lymph-to-plasma concentration ratio	-	0.6-0.8	0.2	0.2	0.5
protein pool (fraction of plasma pool)	1	-	0.17	0.24	0.35

* Values are for a standard dog of about 21 kg and 1 kg of plasma (see ref. [62]).

With respect to blebbing by still viable cells, it should be realised that total bleb volume was only a small fraction of total cell volume. Moreover it was mentioned that the blebs did not release encapsulated proteins [10]. Recent interest in this blebbing phenomenon during apoptotic cell death [42] has also attracted attention to the rapid elimination of microvesicles from the circulation [43]. Together these data indicate that blebbing is probably not a mechanism for the release of significant amounts of free protein molecules into plasma.

Although compelling evidence for protein release from living cardiomyocytes thus seems to be lacking, little can de said about the possibility that such cells may have died (long) before any protein is released. As long as a strictly scientific definition of 'life' is lacking, this issue will probably remain unresolved. As discussed below, however, there is much evidence suggesting that the release of marker proteins by the cardiomyocyte is a sudden, all-or-none, phenomenon that can profitably be used as an indicator of cell death.

Sarcolemmal disruption and protein transport from heart to plasma

In-vivo, proteins liberated by cardiomyocytes still have to be transported from interstitial space to the circulation and this complicates the exact timing of the release process. This is not so in cultures of neonatal rat cardiomyocytes, where cells are under similar external conditions and protein release can be instantaneously assessed. Under these circumstances, the time interval between sudden induction of hypoxia and protein release may differ by hours for different cells [9]. Apparently, the resistance of individual cells to hypoxic injury has a large variability. It was shown in this study, however, that at each separate time point the amount of released HBDH was proportional to the fraction of cells that could be stained with Trypan Blue, a small molecular dye that cannot penetrate into the living cell.

After hypoxia or metabolic inhibition of such cultured cells, completely simultaneous release was observed for proteins differing in molecular weight between 15 kDa (FABP) and 140 kDa (LDH) [44]. Also, simultaneous appearance of different proteins was observed in human right duct lymph during cardiac surgery [25]. As discussed below, cardiac interstitium is drained relatively fast by lymph and, together, these results suggest that for each individual cell the sacolemma becomes rather suddenly and totally permeable. Cellular variability in resistance to such sarcolemmal rupture causes gradual release during hours, but this process can be synchronised by re-oxygenation, causing sudden and massive protein release, *in vitro* [9], as well as *in vivo* [12].

In spite of simultaneous release of proteins by individual cells, entrance of marker proteins into the bloodstream is not simultaneous and depends on molecular size (see Figure 1). This is due to the fact that 70-90% of total protein transport to the plasma is affected by transendothelial diffusion of proteins into the microvessels and, thereby, depends on molecular size [41,46,47]. Proteins reach the circulation because the vessels are flushed, even in ischaemic myocardium. This may seem surprising, but it is explained by some residual collateral flow [48]. Normal myocardial blood flow is of the order of 1 ml/g/min and the muscle contains only about 0.06 ml/g of blood [49]. So, even for only 1% residual flow, the blood vessels in the ischaemic area will be flushed every few minutes. Local retrograde venous flow may also add to this washout process.

Because the transport of proteins from an ischaemic area is not flow-limited, but dependent on diffusion, small molecules will reach the plasma faster than large ones. In man it has been estimated that the half-life for myocardial washout is about 1 hour for myoglobin and about 7 hours for CK [12]. In contrast, buffer-perfused isolated hearts often release their small and large proteins together, due to artificially increased interstitial washout.

The remaining 10-30% of protein transport is effected by lymph. This transport is relatively fast and proteins pass from the cardiac interstitial space to the right lymphatic duct, which empties into the external jugular vein, in about 20 minutes[41]. This implies that in the modelling of circulating cardiac proteins, the efflux of proteins from the heart can be assumed to be a direct input into plasma.

The recovery of myocardial marker proteins in plasma

Local degradation of cardiac proteins in the myocardium, either before cellular release or during transport from the heart to plasma, could cause incomplete protein recovery in plasma and underestimation of injury. Indeed, inactivation of CK in dog lymph has been found [50], and in necrotic myocardium CK proved to be much more labile than LDH or myoglobin [51]. This suggests that CK is a well-suited probe for testing the risk of protein degradation.

In animal experiments, this issue has produced conflicting results. A report that only about 15% of myocardial CK was recovered after permanent coronary occlusion in the dog [13] was contradicted by a similar study, reporting complete recovery of CK as well as of HBDH [14]. In the latter study it was suggested that myocardial CK depletion may be easily overestimated due to edema and or to processing of large tissue samples. After temporary coronary occlusion in the dog, reperfusion has been claimed to reduce CK recovery [11], but was also found to increase it [52]. Conversely, for myoglobin, quantitative recovery was found [22]. Apart from being controversial, it is also uncertain whether such experimental results can be extrapolated to man. It was estimated in the dog, for instance, that the half-life for washout of CK from ischemic myocardium was 1-2 hours [12], instead of 7 hours. Also, collateral circulation is more developed in the dog than in man [48].

Some data indicate that local protein degradation in man is probably limited. In autolysing human heart, kept at 19 °C, only 5-6% of CK activity disappeared in the first 10 hours post-mortem, versus 4%, 3% and 1% for LDH, HBDH and AST [28]. Considering the quoted half-life time for CK washout of about 7 hours, this indicates that probably only a minor fraction of myocardial CK will be lost due to autolysis. Also, infarct size as calculated from CK, LDH and myoglobin are approximately equal (see Figure 1) which again indicates that no preferential denaturation of CK occurs.

A possibility that presently cannot be excluded, and still would be compatible with equal estimates from different proteins, is that some myocardial regions could become totally isolated from the circulation. This situation could arise, for instance, due to plugging of the microvessels by leukocytes [53] and myocardial proteins from such areas could be totally prevented from reaching the circulation. Existence of such isolated areas is suggested by the 'no-reflow' phenomenon, traditionally observed in the dog after coronary reperfusion [54], and this would be especially alarming, because reduced release of myocardial proteins after reperfusion therapy, as observed in many studies [1-8], could then be an artefact.

There is convincing evidence, however, that this is not true. In the first place, the reduction of myocardial protein release, observed in the quoted studies, was matched by favorable trends in other parameters such as reduced mortality and preservation of myocardial function. More directly it was shown that the relation between cumulative release of HBDH and the reduction of left ventricular ejection fraction after acute myocardial infarction was not altered by thrombolytic therapy [18]. It has been claimed that the release of CK and CKMB was, indeed, influenced by thrombolytic therapy [15-17], but in these studies protein release was found to be enhanced, rather than reduced, by reperfusion therapy.

Circulatory models for the calculation of cumulative protein release

In order to calculate cumulative protein release from protein concentrations in plasma, one needs a circulatory model that accounts for elimination of protein from the circulation, and for distribution of protein into compartments other than plasma.

As mentioned before, myocardial proteins reach the circulation mainly by direct entry into plasma. From plasma, protein molecules then enter into extravascular compartments, mainly located in the interstitial space of skeletal muscle, skin and viscera [55]. Some parameters of this protein distribution are shown in Table 1. After circulating for some time, the protein molecules are eliminated from plasma, mainly by receptor-mediated endocytosis in the liver [56] or by renal clearance [57]. The FCR for this elimination is the dominant parameter in the calculation of cumulative protein release from time-curves of protein concentrations in plasma.

Values of FCR cannot be simply obtained from the downslopes of protein curves because, due to tailing protein release, such disappearance constants (k_d) are typically 3 to 5 times lower than the FCR [20]. In animal experiments, values of FCR can be obtained by intravenous infusion of protein preparations, but this technique is usually not considered applicable in man. If it is assumed, however, that two proteins with different FCR values are released in a fixed quantitative ratio, r, the values of r, FCR_1 and FCR_2 can be obtained from release curves as observed in patients with myocardial infarction [20,58]. An independent check of this method is provided by the requirement that obtained values of r should be equal to the ratio of protein contents in myocardium.

As explained before, fixed release ratios, or strictly simultaneous release, is indeed found for cellular release into the interstitium. For simultaneous release into plasma, however, the proteins should also be of comparable molecular size. This requirement is not too critical. Studies with radiolabelled proteins and high molecular dextrans have shown that the lymph/blood barrier does not discriminate in the passage of molecules with a molecular weight exceeding 40-60 kDa [59,60], and values of FCR for CK (83 kDa), HBDH and LDH (140 kDa), or AST (100 kDa) could thus be obtained, as shown in Table 2. For the small proteins in this Table, such as myoglobin and FABP, values of FCR were estimated from glomerular filtration rates, corrected for age, sex and plasma creatinine [61], and exchange parameters were obtained from the literature [19,60].

Extravascular distribution spaces in skeletal muscle, skin and the viscera have quite different magnitudes and exchange rates, as shown in Table 1. Due to their much higher endothelial permeability, extravasation in skin and viscera will proceed much faster than in skeletal muscle, but the high muscle mass contains a much larger interstitial volume. For accurate calculations of protein distribution, for instance in measurements of albumin turnover from frequently sampled radiolabelled albumin, all three extravascular compartments have to be included in the model. Because of limited sample rates, however, and sometimes also because of a limited accuracy of the protein assay, the precision of calculation for myocardial markers is usually much lower. It should also be realised that individual FCR values cannot usually be determined and average values, as shown in Table 2, have to be used. Due to 10-20% biological variation in FCR, this also introduces a considerable error [62]. As a result, it has been noted that changing from three extravascular compartments to a single one only marginally influenced the calculations [62]. Therefore, only the parameter values of an equivalent two-compartment model, with a single extravascular compartment, are presented in Table.2, and the cumulative release as shown in Figure 1 was calculated from these parameters.

Because it is assumed that protein elimination by the liver or kidney occurs directly from plasma, and that at each moment the rate of elimination equals the product of FCR and the plasma concentration, total eliminated quantities of protein can be calculated from integrated plasma curves, multiplied by FCR. This implies that, if the concentrations can be measured up to the time of complete re-normalisation, cumulative protein release will be equal to cumulative elimination of protein and can simply be calculated as the area under the plasma curve, multiplicated by FCR. In that case, temporary extravasation of protein molecules is irrelevant because they will have returned to plasma in the end, and one effectively uses a simple one-compartment (plasma) model. This method can only be used for proteins that are rapidly eliminated from plasma, such as myoglobin and FABP; for most other proteins the plasma concentrations will not yet have returned to normal during the period of observation.

Table 2. Mean parameter values for the two-compartment model

Protein	mol. weight kDa	myocardial content**	FCR* ($t_{1/2}$) h^{-1}(h)	TER* h^{-1}	ERR* h^{-1}
CK	83	865 U/g	0.20 (3.5)	0.014	0.018
CKMB	83	132 U/g	0.34 (2.0)	0.014	0.018
HBDH***	140	123 U/g	0.015 (46)	0.014	0.014
LDH	140	155 U/g	0.015 (46)	0.014	0.014
LDH-1[#]	140	140 U/g	0.015 (46)	0.014	0.018
LDH-5[$]	140	< 2 U/g	0.13 (5.3)	0.014	0.018
cAST	100	54 U/g	0.09 (7.7)	0.014	0.014
mAST	100	92 U/g	0.19 (3.6)	0.014	0.018
FABP	15	570 mg/g	1.4 (0.50)^	1.9	0.9
myoglobin	18	2300 mg/g	1.4 (0.50)^	1.9	0.9
TnT	32	234 mg/g	0.11 (6.3)	0.014[&]	0.018

* FCR = fractional catabolic rate constant
*TER = transendothelial escape rate constant
*ERR = extravascular return rate constant
** enzymatic activity units (25°C) or mg of protein per gram wet weight
*** the HBDH assay measures the LDH isoenzymes in heart
[#] dominant isoform in heart
[$] dominant isoform in skeletal muscle and liver
^ depending on age, sex and plasma creatinine
[&] probably as complexes

Conclusions

Loss of cardiomyocytes is irreversible. This makes measurement of the release of myocyte proteins into plasma a unique means of characterising myocardial injury. This technique offers specific advantages, especially in the evaluation of therapy aiming at myocardial salvage, and for detection of minimal necrosis.

Although routine determination of myocardial marker proteins is now generally performed in the clinic, some underlying concepts involved in the interpretation of these data are still subject to controversy. Notably, the assumption that protein release is strictly coupled to cell death has been studied for several decades, without producing generally accepted results. This is probably caused by the lack of scientific criteria for discrimination between living and dead cells.

When myocardial protein release is expressed in gram-equivalents of healthy myocardium, similar estimates of the extent of injury are obtained from different cytosolic marker proteins. These results are obtained in spite of a large variability of the tissue content of these markers in diseased myocardium, which reflects pathological 'dilution' of healthy myocardium with connective tissue. For the cardiospecific contractile proteins, pathological changes in tissue content are probably even larger.

Sarcolemmal disruption, due to hypoxia or metabolic depravation, is an all-or-nothing phenomenon which causes simultaneous release of large and small proteins into cardiac interstitium. Subsequent entry into plasma occurs mainly by transendothelial diffusion of proteins into the microvessels, and this causes relatively rapid appearance of smaller proteins in the plasma.

Local protein degradation, either before cellular release or during transport to plasma, could cause incomplete recovery of myocardial marker proteins from plasma. However, because labile and stable cytosolic markers yield equal estimates of injury, such degradation is improbable and could only occur if a mechanism existed that prevents all proteins from reaching the circulation. In animal experiments, some evidence for such a mechanism, that is, plugging of the microvessels by leukocytes, was found after coronary reperfusion. In man, however, increased rather than decreased protein recovery after reperfusion has been claimed.

The fractional metabolic rate constant FCR is the dominant parameter for calculation of cumulative protein release, while the parameters for the extravascular exchange do not discriminate between proteins with a molecular weight exceeding 40-60 kDa. Sophisticated multi-compartment modelling is not required for such calculations and a simple two-compartment model can be used, which may even be simplified to a single (plasma) compartment if the total time-concentration curve, including the normalisation phase, can be measured.

References

1. Anderson JL, Marshall HW, Askins JC, *et al.* A randomised trial of intravenous and intracoronary streptokinase in patients with acute myocardial infarction. Circulation 1984;70;606-18.
2. The ISAM Study Group. A prospective trial on intravenous streptokinase in acute myocardial infarction. N Engl J Med 1986;314:1465-71.
3. Simoons ML, Serruys PW, Van den Brand M, *et al.* Early thrombolysis in acute myocardial infarction: Limitation of infarct size and improved survival. J Am Coll Cardiol 1986;7:717-28.
4. Van de Werf F, Arnold AER for the E.C.S.G. Intravenous tissue plasminogen activator and size of infarct, left ventricular function, and survival in acute myocardial infarction. Br Med J 1988;297; 1374-9.
5. The TEAHAT Study Group. Very early thrombolytic therapy in suspected acute myocardial infarction. Am J Cardiol 1990;65;401-7.
6. Karagounis L, Sorensen SG, Menlove RL, Moreno F, Anderson JL. Does thrombolysis in myocardial infarction (TIMI) perfusion grade 2 represent a mostly patent artery or a mostly occluded artery ?

Enzymatic and electrocardio-graphic evidence from the TEAM-2 study. J Am Coll Cardiol 1992;19:1-10.

7. De Boer MJ, Suryapranata H, Hoorntje JCA, *et al.* Limitation of infarct size and preservation of left ventricular function after primary coronary angioplasty compared with intravenous streptokinase in acute myocardial infarction. Circulation 1994;90:753-61.
8. Baardman T, Hermens WTh, Lenderink T, *et al.* Differential effects of tissue plasminogen activator and streptokinase on infarct size and on rate of enzyme release: influence of early infarct-related artery patency. The GUSTO Enzyme Substudy. Eur Heart J 1996;17:237-246.
9. Van der Laarse A, Hollaar L, Van der Valk LJM. Release of alphahydroxybutyrate dehydrogenase from neonatal rat heart cell cultures exposed to anoxia and reoxygenation: comparison with impairment of structure and function of damaged cardiac cells. Cardiovasc Res 1979;13;345-53.
10. Piper HM, Schwartz P, Spahr R, *et al.* Early enzyme release from myocardial cells is not due to irreversible cell damage. J Mol Cell Cardiol 1984;16:385-8.
11. Vatner SF, Baig H, Manders WT, Maroko PR. Effects of coronary artery reperfusion on myocardial infarct size calculated from creatine kinase. J Clin Invest 1978;61,1048-56.
12. Cobbaert Ch, Hermens WT, Kint PP, *et al.* Thrombolysis-induced coronary reperfusion causes acute and massive interstitial release of cardiac muscle cell proteins. Cariovasc Res 1997;33:147-155.
13. Roberts R, Henry PD, Sobel BE. An improved basis for enzymatic estimation of infarct size. Circulation 52 743-54,1975.
14. Hermens WTh, Van der Veen FH, Willems GM, *et al.* Complete recovery in plasma of enzymes lost from the heart after permanent coronary artery occlusion in the dog. Circulation 1990;81;649-59.
15. Tamaki S, Murakami T, Kadota K, *et al.* Effects of coronary artery reperfusion on relation between creatine kinase-MB release and infarct size estimated by myocardial emission tomography with thallium-201 in man. J Am Coll Cardiol 1983;2;1031-8.
16. Ong L, Reiser P, Coromilas J, *et al.* Left ventricular function and rapid release of creatine kinase MB in acute myocardial infarction. N Engl J Med 1983;309;1-6.
17. Blanke H, von Hardenberg D, Cohen M, *et al.* Patterns of creatine kinase release during acute myocardial infarction after nonsurgical treatment and correlation with infarct size. J Am Coll Cardiol 1984; 3; 675-80.
18. Van der Laarse A, Kerkhof PLM, Vermeer F, *et al.* Relation between infarct size and left ventricular performance assessed in patients with first acute myocardial infarction randomized to intracoronary thrombolytic therapy or to conventional treatment. Am J Cardiol 1988;61;1-7.
19. Groth T, Sylvén Ch. Myoglobin kinetics in patients suffering from acute myocardial infarction in its early phase - as studied by the single injection method. Scand J clin Lab Invest 1981;41:79-85.
20. Willems GM, Muijtjens AMM, Lambi FHH, Hermens WTh. Estimation of circulatory parameters in patients with acute myocardial infarction. Significance for calculation of enzymatic infarct size. Cardiovasc Res 1979;13;578-87.
21. Grande P, Hansen BF, Christiansen C, Naestoft J. Estimation of acute myocardial infarct size in man by serum CK-MB measurements. Circulation 1982;65;756-64.
22. Ellis AK, Saran BR. Kinetics of myoglobin release and prediction of myocardial depletion after coronary artery reperfusion. Circulation 1989;80:676-683.
23. Peltenburg HG, Hermens WTh, Willems GM, *et al.* Estimation of fractional catabolic rate constants for the elimination of cytosolic liver enzymes from plasma. Hepatology 1989;10:833-9.
24. Anversa P, Palackal Th, Sonnenblick EH, *et al.* Myocyte cell loss and myocyte cellular hyperplasia in the hypertrophied aging rat heart. Circ Res 1990;67:871-85.
25. Li RK, Mickle DAG, Weisel RD, *et al.* Human pediatric and adult ventricular cardiomyocytes in culture: assessment of phenotype change with passaging. Cardiovasc Res 1996;32:362-73.
26. Olivetti G, Melissari M, Capasso JM, Anversa P. Cardiomyopathy of the aging human heart. Myocyte loss and reactive cellular hypertrophy. Circ Res 1991;68:1560-8.
27. Ingwall JS, Kramer MF, Fifer MA, *et al.* The creatine kinase system in normal and diseased human myocardium. N Engl J Med 1985;313;1050-4.
28. Van der Laarse A, Dijkshoorn NJ, Hollaar L, Caspers T. The (iso-)enzyme activities of lactate dehydrogenase, alpha-hydroxybutyrate dehydrogenase, creatine kinase and aspartate aminotransferase in human myocardial biopsies and autopsies. Clin Chim Acta 1980;104;381-91.
29. Kragten JA, van Nieuwenhoven FA, van Dieijen-Visser MP, *et al.* Distribution of myoglobin and fatty acid-binding protein in human cardiac autopsies. Clin Chem 1996;42:337-8.

30. Katus HA, Remppis A, Scheffold Th, *et al.* Intracellular compartmentation of cardiac troponin T and its release kinetics in patients with reperfused and nonreperfused myocardial infarction. Am J Cardiol 1991;67:1360-7.
31. Kragten JA, Hermens WTh, van Dieijen-Visser MP. Cardiac troponin T release into plasma after acute myocardial infarction: only fractional recovery compared with enzymes. Ann Clin Biochem 1996;33:314-23.
32. Altona JC, Van der Laarse J, Bloys van Treslong CHF. Release of compartment-specific enzymes from neonatal rat heart cell cultures during anoxia and reoxygenation. Cardiovasc Res 1984;18:99-106.
33. Hermens WTh, Willems GM, Visser MP. Quantification of circulating proteins. Martinus Nijhoff Publishers, Boston 1982; Ch.5.
34. Revis NW, Thomson RY, Cameron AJV. Lactate dehydrogenase isoenzymes in the human hypertrophic heart. Cardiovasc Res 1977;11:172-6.
35. Hoyt RH, Ericksen E, Collins SM, Skorton DJ. Computer-assisted quantitation of myocardial fibrosis in histological sections. Arch Pathol Lab Med 1984;108:280-3.
36. Nascimben L, Ingwall JS, Pauletto P, *et al.* Creatine kinase system in failing and nonfailing human myocardium. Circulation 1996;94:1894-1901.
37. Van der Veen FH, Visser R, Willems GM, *et al.* Myocardial enzyme depletion in infarcted human hearts: infarct size and equivalent tissue mass. Cardiovasc Res 1988;22:611-9.
38. Ricchiuti V, Zhang J, Apple FS. Cardiac troponin I and T alterations in hearts with severe left ventricular modeling. Clin Chem 1997;43:990-5.
39. Hamm ChW, Ravkilde J, Gerhardt W, *et al.* The prognostic value of serum troponin T in unstable angina. N Engl J Med 1992;327:146-50.
40. Abdelmeguid AE, Topol EJ, Whitlow PL, *et al.* Significance of mild transient release of creatine kinase-MB fraction after percutaneous coronary interventions. Circulation 1996;94:1528-36.
41. Spieckermann PG, Nordbeck H, Preusse CJ. From heart to plasma. In: Enzymes in Cardiology. DJ Hearse, J de Leiris (eds). John Wiley & Sons, New York 1979;81-96.
42. Majno G, Joris I. Apoptosis, oncosis and necrosis. An overview of cell death. Am J Pathol. 1995;146: 3-15.
43. Lee Y, Horstman LL, Jania J, *et al.* Elevated platelet microparticles in transient ischemic attacs, lacunar infarcts and multiinfarct dementias.Thromb Res 1996;72:295-304.
44. Van Nieuwenhoven FA, Musters RJP, Post JA, *et al.* Release of proteins from isolated neonatal rat cardiomyocytes subjected to simulated ischemia or metabolic inhibition is independent of molecular mass. J Mol Cell Cardiol 1996;28:1429-34.
45. Hansson HE. Efflux of enzymes in right duct lymph and serum after coronary perfusion and ischemic arrest. Scand J Thor Cardiovasc Surg 1976;10;157-66.
46. Szabo G. Movement of proteins into the blood capillaries. In: Ergebnisse der Angiologie vol. 12: Basic Lymphology. M. Földi (ed). Schattauer Verlag, Stuttgart 1976;11;31-50.
47. Van der Veen FH, Hermens WTh, Willems GM, *et al.* Time course of cellular enzyme release in dog heart injury. Circ Res 1990;67:1257-66.
48. Schaper W. The collateral circulation of the heart. North Holland Publishing Cy, Amsterdam 1971.
49. Feigl EO. Coronary physiology. Physiol Rev 1983;63;1-161.
50. Clark GL, Robison AK, Gnepp DR, *et al.* Effects of lymphatic transport of enzyme on plasma creatine kinase time-activity curves after myocardial infarction in dogs. Circ Res 1978;43:162-69.
51. Johnson RN, Sammel NL, Norris RM. Depletion of myocardial creatine kinase, lactate dehydrogenase, myoglobin and K+ after coronary artery ligation in dogs. Cardiovasc Res 1981;15:529-537.
52. Jarmakani JM, Limbird L, Graham ThC, Marks RA. Effect of reperfusion on myocardial infarct, and the accuracy of estimating infarct size from serum creatinine phosphokinase in the dog. Cardiovasc Res 1976;10:245-53.
53. Schmid-Schönbein GW, Engler RL. Granulocytes as active participants in acute myocardial ischemia and infarction. Am J Cardiovasc Path 1986;1;15-30.
54. Kloner RA, Ganote CE, Jennings RB. The no-reflow phenomenon after temporary coronary occlusion in the dog. J Clin Invest 1974;54;1496-1508.
55. Auckland K, Nicolaysen G. Interstitial fluid volume. Local regulatory mechanisms. Physiol Rev 1981;61;556-643.
56. Smit MJ, Duursma AM, Bouma JMW. Receptor-mediated endocytosis of lactate dhydrogenase M4 by liver macrophages: a mechanism for elimination of enzymes from plasma. J Biol Chem 1987;262:13020-6.

57. Klocke FC, Copley DP, Krawczic JA, Reichlin M. Rapid renal clearance of immunoreactive canine plasma myoglobin. Circulation 1982;65:1522-8.
58. Willems GM, Visser MP, Krill MTA, Hermens WTh. Quantitive analysis of plasma enzyme levels based upon simultaneous determination of different enzymes. Cardiovasc Res 1982;16;120-31.
59. Garlick DG, Renkin EM. Transport of large molecules from plasma to interstitial fluid and lymph in dogs. Am J Pysiol 1970;219:1595-1605.
60. Renkin EM. Multiple pathways of capillary permeability. Circ Res 1977;41;735-43.
61. Gault MH, Longerich LL, Harnett JD, Wesolowski C. Predicting glomerular function from adjusted serum creatinine. Nephron 1992;62:249-56.
62. Van Kreel BK, Van der Veen FH, Willems GM, Hermens WTh. Circulatory models in assessment of cardiac enzyme release in dogs. Am J Physiol 1993;33:H747-54.

Chapter 9

PROLONGED CHEST PAIN AND THE EARLY DETECTION OF MYOCARDIAL DAMAGE BY NOVEL BIOCHEMICAL MARKERS - PRACTICAL CONSIDERATIONS

Robbert J. de Winter

In recent years rapid analytical techniques have been developed which allow accurate measurement of biochemical serum markers useful for the detection of early myocardial damage and acute myocardial infarction (AMI). Early diagnosis of AMI is likely to improve patient management and reduce complications. The availability of sensitive biochemical markers may result in the identification of low-risk patients and allow for early triage and rational use of Coronary Care Unit (CCU) beds. The potential clinical implications of rapid and accurate new assays are obvious and have triggered great interest. This is reflected by the large number of reports on the early diagnostic value of biochemical markers such as myoglobin [1 - 3], CK-MB mass [4,5],CK-isoforms [6], troponin T (TnT) [7 - 9], troponin I (TnI) [10,11] and fatty acid binding protein (FABP) [12]. At least partially, this interest may be "industry-driven" and for practically every biochemical marker, assays are currently available from different manufacturers. The interest is also driven by health economics and managed-care programs. For example in the United States health maintenance organizations (HMOs) may refuse the reimbursement of costs for patients "wrongly" admitted to CCU [13]. Growing interest in the novel biochemical markers is also driven by the obvious implications of the early and accurate diagnosis of myocardial damage. However, the advent of a large number of these markers has raised a series of practical issues. With the large choice of novel markers available, it is difficult for the treating physician to decide what marker(s) to use or even what strategy is appropriate for risk stratification and management of acute chest pain patients. The focus of this chapter will be on the role of biochemical markers in the detection or exclusion of myocardial damage in chest pain patients and the possible implications for triage and patient management in the first 24 hours from the onset of chest pain.

The diagnosis of acute myocardial infarction and minor myocardial necrosis

The WHO defines AMI in the presence of two-out-of-three criteria [14], i.e. ECG changes, typical chest pain and elevated cardiac enzymes. These criteria were established as an epidemiological tool to enable the creation of coronary registers [15], and in light of present knowledge these criteria may need updating. One of the reasons for this is that a substantial proportion of patients who present with acute chest pain but do not fulfil WHO criteria for AMI have evidence of myocardial damage as assessed by CK-MB mass, TnT or TnI [16 -20]. WHO criteria are used as the gold standard for AMI in many studies evaluating sensitivities and specificities of serum markers and in clinical practice. However, a novel cardiac specific marker TnT, is gradually emerging as a "gold standard" for the diagnosis of myocardial damage. The absence of detectable levels of TnT in plasma is considered as evidence for the absence of myocardial necrosis [21]. It is also argued that elevated TnT in patients with chronic renal disease may represent low-grade ongoing myocardial damage [22]. Whether cardiac marker release can occur in the absence of myocyte necrosis is still controversial [23-25]. In addition, studies have shown that CK-MB content may differ in normal hearts compared to hearts with evidence of coronary artery disease [26,27]. However, for practical purposes it can be assumed that with the new assays, even mild myocardial *necrosis* may be detected with remarkable accuracy. This is important as the majority of patients who attend emergency departments due to acute chest pain, have normal or non-diagnostic electrocardiograms (ECG) and therefore the final diagnosis will rely on the biochemical reliability of the markers. The prognosis for patients with minor myocardial damage who do not fulfil the WHO criteria for AMI is comparable to the prognosis of patients with non-Q-wave AMI who do fulfil the WHO criteria, making the distinction according to the WHO classification less relevant.

Early recognition and management of acute myocardial infarction - any need for biochemical markers?

When a patient is admitted to hospital with prolonged typical chest pain and nitrate resistant ST segment-elevation on the ECG, serum markers are routinely measured, but do they influence patient management to any great extent? Probably not, as such patients are usually treated promptly with primary coronary angioplasty (PTCA) or fibrinolytic agents. In most hospitals with facilities for primary PTCA, coronary angiography will be performed to document the presence of an occlusive stenosis of the large epicardial coronary artery and the extent of coronary disease. Successful reperfusion will be achieved in the large majority of cases following this approach. Only later, the extent of myocardial damage, residual left ventricular function and the need for oral anticoagulants, anti-arrhythmic drugs or ACE inhibitors will be assessed by measuring cardiac enzymes, observing the clinical course, performing ECG monitoring, chest X-rays and two-dimensional echocardiography. In hospitals without primary PTCA, the decision to initiate thrombolytic therapy is made without awaiting cardiac enzyme results. The larger trials that investigated the effects of thrombolytic therapy in patients with AMI did not incorporate myocardial damage marker measurements among their inclusion criteria [28 - 31]. By the time cardiac markers become elevated in the circulation, the “golden hour” of reperfusion therapy has passed [32]. Therefore in patients with AMI and patients with prolonged chest pain and ST segment elevation (or left bundle branch block not known to have been present previously) who are considered candidates for reperfusion therapy, there is no rationale for the diagnostic use of novel biochemical markers of damage.

Role of the 12 lead electrocardiogram

In a study by Gray and Hampton [33] comprising 15,832 patients, the clinical "impression" and the 12 lead ECG at admission had a sensitivity of 55% and a specifity of 88% for the diagnosis of AMI. In the MILIS study [34], including 3,697 patients, pre-specified ECG criteria had a sensitivity of 81% and a specificity of 69% for the diagnosis of AMI. In this study, 45% of patients (n=1,652) did not have ST-segment elevation, ST-segment depression, LBBB or new Q-waves; of these, 21% developed AMI. In a Swedish study including 7,157 consecutive patients, 14% of patients with an abnormal ECG but without typical features of acute ischemia developed AMI, whereas only 6% of patients with a normal baseline ECG developed AMI. However, when patients were prospectively classified on the basis of the clinical history, physical examination and characteristics of the admission ECG into 4 categories (obvious AMI, strong suspicion, vague suspicion and no suspicion), AMI was present in 88%, 34%, 8% and 1% respectively [35]. These data show that although the diagnostic value of the 12 lead ECG at admission may be limited, if the ECG results are combined with clinical information available on admission, the majority of AMI patients will be identified using simple inexpensive triage. Of importance, the admission 12 lead ECG may represent a valuable predictor of major in-hospital complications. In fact, it has been shown that the prognostic value of the admission ECG may be higher than its diagnostic value [36 - 39]. In a study by Goldman *et al.* [39], evaluating the need for intensive care in patients with acute chest pain, the risk of major complications could be estimated on the basis of clinical presentation (e.g. ECG, symptoms, physical examination) and additional clinical observations made during a 12-hours observation period. In the very low risk patient group (no signs of infarction or myocardial ischemia on the ECG, absence of "risk factors" and no clinical complications after 12 hours observation), the incidence of major events within 72 hours was 0.4% in a validation study of 4,676 patients. Unfortunately, in this study biochemical marker measurement was not part of the triage protocol.

The "exclusion" of myocardial necrosis does not exclude trouble

Patients with a diagnosis of unstable angina are at risk of progression to myocardial infarction or death. Before unstable angina patients were routinely treated with aspirin and/or heparin, in-hospital event rate was high (2% mortality, 10% myocardial infarction, 20% refractory angina) [40]. The use of aspirin and/or heparin has reduced the event rate by 40-60% [40,41]. Novel therapeutic strategies may improve mortality and morbidity even further; the FRIC investigators [42] compared the effects of low molecular weight heparin given subcutaneously with those of unfractionated intravenous heparin, in addition to aspirin, for the treatment of unstable coronary artery disease. They reported a six-day incidence of the combined death, myocardial infarction or recurrence of angina of 7.6 % and 9.3% respectively [42]. In a study by Lindahl *et al.* [43] for the FRISC study group including 976 patients with unstable coronary artery disease, there was a 4.4% incidence of cardiac death and a 12.3%

incidence of AMI for the whole study group during a follow-up of 5 months [43]. In 36.7% of patients there was a need for coronary revascularization. All patients were treated with aspirin 75 mg daily. In this study, an elevated TnT identified patients at increased risk. However, even in the TnT-negative patients there was still an incidence of cardiac death and myocardial infarction of 4.3%. In a single center study including unstable angina patients with a follow-up of three years, Stubbs *et al.* [44] reported an incidence of death or non-fatal MI of 17% in the TnT negative patients, and a need for revascularization of 21%. Although the first two studies included both patients with unstable angina and non-Q-wave myocardial infarction, event rates for the two groups were similar. These data show that patients with unstable coronary artery disease are at a similar risk for subsequent cardiac events as non Q-wave infarct patients despite treatment. Of importance, even patients who do not have myocardial damage as detected by troponin T may be at increased risk. How to identify these patients is of critical importance.

The importance of ruling-out myocardial damage

From what I have mentioned before, it can be suggested that the detection of myocardial damage with biochemical markers and new rapid assays does not contribute significantly to the triage or management of patients with a clear clinical indication for reperfusion therapy. Moreover, the exclusion of minor myocardial damage by markers such as TnT in patients with unstable coronary artery disease does not necessarily indicate that these patients are truly at low risk. The decision to admit patients with unstable coronary artery disease to the CCU can be safely based on clinical physical examination and the admission 12-lead ECG [39]. Treatment in patients with unstable angina and non-Q-wave infarction is similar particularly in the first days after admission, and may not be affected by the presence or absence of minor myocardiac damage.

However, there is a group of patients in whom markers of myocardial damage may play a major role. These are patients who present with chest pain, have non-diagnostic electrocardiograms and no events develop during a several hour observation period. These patients constitute a very low risk group and can be discharged safely, and quickly, provided the results of biochemical markers of myocardial damage show no alteration. To rule out myocardial damage in this subgroup of patients, it is important to know the sensitivity and specificity of the biochemical markers used.

The new biochemical markers

To accurately rule out myocardial damage based on the measurement of biochemical marker levels, a high <u>negative predictive value</u> of a normal test result is needed. In other words, for the test to be useful in a population of low-risk patients, with a low incidence of myocardial damage, the test in question has to have high sensitivity.

Several reviews on the usefulness of biochemical markers for the diagnosis of AMI have been published recently [45 - 47]. Markers such as myoglobin, CK-MB mass, CK-MB-isoforms, troponin T and troponin I have all been shown to have good sensitivity for the diagnosis of AMI [1 - 11, 46]. However, in the early hours after the onset of symptoms, the sensitivity of the markers may change rapidly due to their time-dependent appearance in the peripheral blood. It is important in this respect to distinguish between early sensitivity assessed relative to the time of admission and relative to the time of onset of symptoms.

Myoglobin, CK-MB mass, troponin T. In the first six hours after the onset of symptoms, myoglobin has better sensitivity than CK-MBmass or troponin T for the diagnosis of AMI [5]. All three markers however have been shown to have limited sensitivity in the first six hours [1, 5, 48]. Moreover, it has been shown that the sensitivity of myoglobin decreased after 7 hours, whereas the sensitivities of CK-MB mass, and troponin T reached nearly 100% 10-12 hours after the onset of symptoms. Early sensitivities of CK-MB mass and troponin T for the detection of AMI are comparable and, in addition, in patients in whom AMI has been excluded, most troponin T positive patients can also be identified with serial samples of CK-MB mass [49].

CK-MB isoforms. Puleo *et al.* reported a sensitivity of 95.7% at six hours after the onset of symptoms for the CK-MB-isoforms using a high-voltage electrophoresis assay, compared with 48% for CK-MB activity [6]. However, Laurino *et al* [1] found similar sensitivities and specificities for CK-MB2 isoform and CK-MB mass.

Troponin I. Cardiac troponin I was reported to have a similar sensitivity for the detection of AMI to CK-MB mass, with blood sampling every 12 hours [50] and may have comparable sensitivity for early detection of AMI [51]. Cardiac troponin I appears to have high specificity for cardiac injury [52], a wide time-window [53] and prognostic value in unstable angina [20]. Based on these findings, some authors believe troponin I could represent a suitable candidate to replace CK-MB, total LD and LD isoenzyme analysis for the routine assessment of AMI [54].

Differences between the sensitivities of these novel markers for the detection of myocardial damage are small and level out within the time-window of 12 hours usually used for observation of patients with chest pain as proposed by the Chest Pain Study Group [55]. Moreover, no studies have been carried out to establish if these novel markers are useful in the clinical setting, regarding both the decision to discharge a patient from the emergency room or the selection of the appropriate level of care after the patient has been admitted to hospital. This problem was addressed by Selker *et al.* [56] for the US National Heart Attack Alert Program, in a report published 1997. In their manuscript 'Technologies for Identifying Acute Cardiac Ischemia in the Emergency Department' [56] they conclude: "*A prospective intervention study, with follow-up of all (including non-admitted) patients, of the effect of serial CK and CK-MB on patient outcomes is needed before a strategy incorporating CK-MB into medical decision making can be fully evaluated, is recommended*"... The evaluation of the use of myoglobin and cardiac troponin T and I was summarized as ... "*the use of new biochemical markers in the emergency department as a routine measure to improve either the initial triage or therapy of patients with AMI is currently unproven*" [56].

Acute Chest Pain Units

A report from the multicenter Chest Pain Study [57] in the late 1980s showed that 4% of patients attending hospital for assessment of acute chest pain and who had documented AMI at presentation, or within three days, were sent home with the diagnosis of non ischaemic chest pain. In this study, the short term mortality was significantly higher among patients in whom the diagnosis of AMI [57] was missed compared to those AMI patients admitted to hospital. Both this finding and perhaps fear of malpractice claims, of which the wrong diagnosis of AMI and management is a leading cause [58], may explain why only 15-30% of patients currently admitted to CCU actually have AMI [57,58]. With increasing costs of critical care beds, it is mandatory to develop strategies to limit unnecessary hospital admissions [38]. Triage protocols for short term observation periods [55], and the development of "short-stay units", "coronary observation units" or "Heart ER" units perhaps adjacent to the emergency room and equipped with ECG rhythm-monitoring and with a low nurse-to-patient ratio may be necessary [13, 59 - 61]. Previous reports have found that patients at low risk of infarction can be accurately identified within 24 hours after admission [62,63]. Lee *et al.* [55] reported that this observation time may be shortened safely to a 12-hour strategy in low-risk patients. In their study [55], myocardial damage was assessed using total CK and CK-MB(activity). Data from the same group suggest that assessment of low-risk chest pain patients in a coronary observation unit is safe and adequate for ruling-out AMI, with low complication rates and excellent survival for those patients sent home directly from such a unit [55]. Mean length of stay in this study was 1.2 days.

Gaspoz *et al* [60] demonstrated that this type of units may be cost-saving. In another study including 1010 patients, Gibler *et al.* [61] assessed a 9-hour follow up protocol in the emergency department that included serial CK-MBmass testing, serial 12-lead ECG, two-dimensional echocardiography and graded exercise testing. They found that the protocol was effective. However, of the 1010 patients, only 12 were diagnosed as having AMI (1.2%) and 113 patients (11,2%) had cocaine-induced chest pain. Moreover, a 12-lead ECG and CK-MB determination were scheduled for the next day but only 25% of patients returned for testing. Zalenski *et al.* [64] found that only 63 of 599 screened patients (14,1%) were eligible for a 12-hour observation protocol in a short-stay unit according to the recursive partition function of the chest pain study group [55]. These observations leave many questions unanswered considering both safety and cost-effectiveness. Despite these concerns and in light of current financial constraints, properly designed chest pain centers associated with emergency departments may prove useful [65]. Their popularity is increasing in different countries. How the information discussed earlier applies to emergency departments in European hospitals still remains to be determined.

Early ruling-out of Acute Myocardial Infarction, which marker?

Most hospitals continue to use total creatine kinase (CK) measurements to rule-out AMI in patients presenting with acute chest pain, despite the limitations of this marker related to the influence of muscle mass, effects of exercise, gender, age, and race [45]. The sensitivity and specificity of CK are relatively low; myocardial damage may be present in a substantial number of patients even when total CK is within the "normal" range [66 - 70]. Ad-hoc protocols to rule-out AMI (short stay ≤12 hours) currently involve the use of a rapid assay for one of the new biochemical markers. One has to bear in mind that these protocol/marker strategies are still largely untested, and that safety and cost-effectiveness of this approach will vary according to the characteristics of the patient population that is served by a particular institution. Figure 1 illustrates one of such strategies.

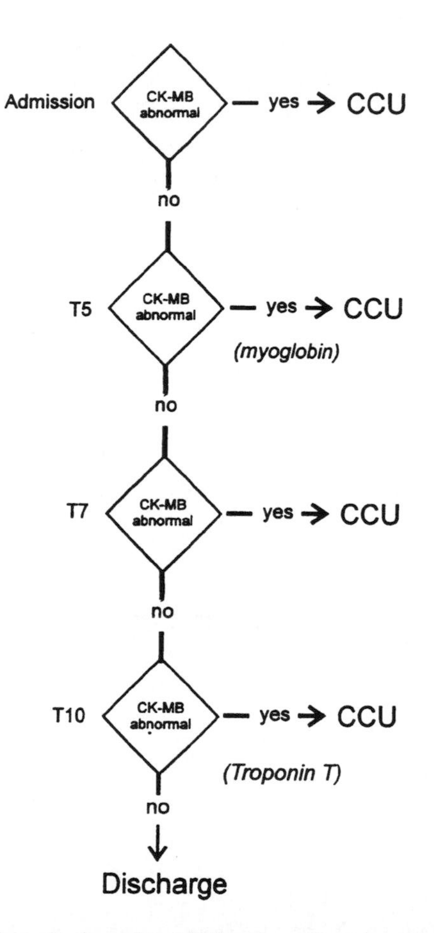

Figure 1. Decision tree for early diagnosis of AMI at the cardiac emergency room. A single CK-MB > 7.0 μg/L or a difference between two serial samples > 2.0 μg/L was defined as abnormal. A myoglobin measurement at T5 (time point of maximal diagnostic accuracy) is optional. A troponin T measurement at T10 is optional. In case of uncertainty concerning the onset of symptoms, or concomitant skeletal muscle damage, a troponin measurement is useful due to its cardiac specificity and presence in plasma for a longer period of time. Patients admitted to the CCU will have additional CK-MB measurements later that T10, but for the ruling-out of AMI, T10 is last sampling point. In addition, history, admission ECG or clinical observations, such as recurrent chest pain with accompanying electrocardiographic changes may necessitate CCU admission irrespective of CK-MB results. T5, T7, T10: five, seven and ten hours after the onset of symptoms. CCU: coronary care unit.

Serial CK-MB mass testing has been evaluated for ruling-out AMI in a CCU setting [71,72] and in the cardiac emergency room [5]. These are feasible and safe when included as part of a 24-hour or 12-hour protocol. There are now several new assays available with good precision at the lower concentration ranges around the cut-off level which offer good possibilities for this marker. Test results with these assays are available within 20 minutes and large automated analyzers make it feasible to have them available round the clock. At our institution, a University Hospital of the University of Amsterdam, a cardiac emergency room facility was originally established by the late Professor D. Durrer in 1978 [73,74]. For many years, a CK-MB activity assay was used for the diagnosis of myocardial damage. As from 1995, we are evaluating a 12-hour rule-out AMI protocol which includes serial sampling of CK-MB mass and a strict follow-up of patients discharged from hospital.

Myoglobin is a very early marker for myocardial damage but is not recommended as a single test for AMI, due to its lack of cardiac specificity. Recently it has been suggested that measurements combining myoglobin with measurements of fatty acid binding protein (FABP) may solve the problem of low cardiac specificity [12]. However, there are other problems with myoglobin such as a limited time-window and an inconsistent definition of the levels required for a positive test. Current cut-off levels vary between 50 and 110mg/L. Troponin T and troponin I have similar release profiles, and have comparable early sensitivity and specificity as CK-MB mass for the detection of AMI [5,7,8,50], reaching approximately 100% sensitivity within 12 hours after the onset of symptoms. In addition, a positive troponin T or I appears to provide prognostic information [19,20]. However, both markers remain largely untested in the setting of the emergency room. Recently, the quality of evidence and demonstrated impact of the troponins on diagnosis, triage, treatment, and patient outcome when used by clinicians in actual practice was rated as *"NK"=not known* by the National Heart Attack Alert Program Working Group [56]. However, although it is likely that serial testing of the combined application of an "early marker" and a "cardiac-specific" marker (e.g. CK-MB mass or troponin) will yield useful information, the clinical impact of this approach and its cost-effectiveness will have to be carefully assessed.

Acknowledgements

I thank Dr R.W.Koster for his critical comments and help while preparing the manuscript.

References

1. Laurino JP, Bender EW, Kessimian N, Chang J, Pelletier T, Usategui M. Comparative sensitivities and specificities of the mass measurements of CK-MB2, CK-MB, and myoglobin for diagnosing acute myocardial infarction. Clin Chem 1996;42:1454-1459.
2. Ohman EM, Casey C, Bengtson JR, Pryor D, Tormey W, Horgan JH. Early detection of acute myocardial infarction: additional diagnostic information from serum concentrations of myoglobin in patients without ST elevations. Br Heart J 1990;63:335-338.
3. Mair J, Artner Dworzak E, Lechleitner P, Morass B, Smidt J, Wagner I, Dienstl F, Puschendorf B. Early diagnosis of acute myocardial infarction by a newly developed rapid immunoturbidimetric assay for myoglobin. Br Heart J 1992;68:462-468.
4. Collinson PO, Rosalki SB, Kuwana T, Garrat HM, Ramhamadamy EM, Baird IM, Greenwood TW.

Early diagnosis of acute myocardial infarction by CK-MB mass measurements. Ann Clin Biochem 1992;29:43-47.

5. De Winter RJ, Koster RW, Sturk A, Sanders GT. The Value of Myoglobin, Troponin T and CK-MB Mass in Ruling-out an Acute Myocardial Infarction in the Emergency Room. Circulation 1995;92:3401-3407.
6. Puleo PR, Meyer D, Wathen C, Tawa CB, Wheeler S, Hamburg RJ, Ali N, Obermueller SD, Triana JF, Zimmerman JL, Perryman B, Roberts R. Use of a rapid assay of subforms of creatine kinase MD to diagnose or rule out acute myocardial infarction. N Engl J Med 1994;331:561-166.
7. Katus HA, Remppis A, Neumann FJ, Scheffold T, Diederich KW, Vinar G, Noe A, Matern G, Keubler W. Diagnostic efficiency of Troponin T measurements in acute myocardial infarction. Circulation 1991;83:902-912.
8. Bakker AJ, Koelemay MJW, Gorgels JPMC, Van Vlies B, Smits R, Tijssen JGP, Haagen FDM. Troponin T and myoglobin at admission: value of early diagnosis of acute myocardial infarction. Eur Heart J 1994;15:45-53.
9. Mair J, Morandell D, Genser N, Lechleitner P, Dienstl A, Puschendorf B. Equivalent Early Sensitivities of Myoglobin, Creatine Kinase MB Mass, Creatine Kinase Isoform Ratios, and Cardiac Troponins I and T for Acute Myocardial Infarction. Clin Chem 1995;41:1266-1272.
10. 10.Larue C, Calzolari C, Bertinchant JP, Leclercq F, Grolleau R, Pau B. Cardiac-Specific Immunoenzymometric Assay of Troponin I in the Early Phase of Acute Myocardial Infarction. Clin Chem 1993;39:972-979.
11. Pervaiz S, Anderson FP, Lohmann TP, Lawson CJ, Feng YJ, Waskiewicz D, Contois JH, Wu AH. Comparative analysis of cardiac troponin I and creating kinase-MB as markers of acute myocardial infarction. Clin Cardiol 1997;20:269-271.
12. Van Nieuwenhoven FA, Kleine AH, Wodzig KW, Hermens WT, Kragten HA, Maessen JG, Punt CD, Van Dieijen MP, Van Der Vusse GJ, Glatz JF. Discrimination between myocardial and skeletal muscle injury by assessment of the plasma ratio of myoglobin over fatty acid-binding protein. Circulation 1995;92:2848-2854.
13. Gibler WB. Chest pain units: Do they make sense now? Ann Emerg Med 1997;29:168-171.
14. World Health Organization criteria for the diagnosis of acute myocardial infarction. Proposal for the multinational monitoring of trends and determinants in cardiovascular disease. Geneva: Cardiovascular Disease Unit of WHO, 1981. jjj 1981.
15. Rowley JM, Hampton JR. Diagnostic critera for myocardial infarction. Br J Hosp Med 1981;26:253-258.
16. De Winter RJ, Koster RW, Schotveld JH, Sturk A, van Straalen JP, Sanders GT. Prognostic value of troponin T, myoglobin, and Ck-MB mass in patients presenting with chest pain without acute myocardial infarction. Heart 1996;75:235-239.
17. Ravkilde J, Nissen H, Horder M, Thygesen K. Independent Prognostic Value of Serum Creatine Kinase Isoenzyme MB Mass, Cardiac Troponin T and Myosin Light Chain Levels in Suspected Acute Myocardial Infarction. J Am Coll Cardiol 1995;25:574-581.
18. Hamm CW, Ravkilde J, Gerhardt W, Jorgensen P, Peheim E, Ljungdahl L, Goldmann B, Katus HA. The prognostic value of serum Troponin T in unstable angina. N Engl J Med 1994;327:146-150.
19. Ohman EM, Armstrong PW, Christenson RH, Granger CB, Katus HA, Hamm CW, O'Hanesian MA, Wagner GS, Kleiman NS, Harrell FE, Califf RM, Topol EJ. Cardiac troponin T levels for risk stratification in acute myocardial infarction. New Engl J Med 1996;335:1333-1341.
20. Antman EM, Tanasijevic MJ, Thompson B, Schactman M, McCabe CH, Cannon CP, Fischer G, Fung AY, Thompson C, Wybenga D, Braunwald E. Cardiac specific troponin I levels to predict the risk of mortality in patients with acute coronary syndromes. New Engl J Med 1996;335:1342-1349.
21. Biasucci LM, Vitelli A, Liuzzo G, Altamura S, Caliguri G, Monaco C, Rebuzzi AG, Ciliberto G, Maseri A. Elevated Levels of Interleukin-6 in Unstable Angina. Circulation 1996;94:874-877.
22. Katus HA, Haller C, Müller-Bardorff M, Scheffold T, Remppis A. Cardiac Troponin T in End-Stage Renal Disease Patients Undergoing Chronic Maintenance Hemodialysis. Clin Chem 1995;41:1201-1202.
23. Heyndrickx GR, Amano J, Kenna R, Fallon JT, Patrick TA, Manders WT, Rogers GG, Rosendorff C, Vatner SF. Creatine Kinase Release Not Associated With Myocardial Necrosis After Short Periods of Coronary Artery Occlusion in Conscious Boboons. JACC 1985;6:1299-1303.
24. Piper HM, Schwarz P, Spahr R, Hutter JF, Spieckermann PG. Early enzyme release from myocardial cells is not due to irreversible cell damage. J Moll Cell Cardiol 1984;16:385-388.
25. Ishikawa Y, Saffitz JE, Mealman TL, Grace AM, Roberts R. Reversible myocardial ischemic injury is

not associated with increased creatine kinase activity in plasma. Clin Chem 1997;43:467-475.
26. Ingwall JS, Kramer MF, Fifer MA, Lorell BH, Shemin R, Grossman W, Allen PD. The creatine kinase system in normal and diseased human myocardium. N Eng J Med 1985;313:1050-1054.
27. Van der Laarse A, Hollaar L, Vliegen HW, Egas JM, Dijkshoorn NJ, Cornelisse CJ, Bogers AJ, Quaegebeur JM. Myocardial (iso)enzyme activities, DNA concentration and nuclear polyploidy in hearts of patients operated upon for congenital heart disease, and in normal and hypertrophic adult human hearts at autopsy. Eur J Clin Invest 1989;19:192-200.
28. Anonymous. Randomised trial of intravenous streptokinase, oral aspirin, both, or neither among 17,187 cases of suspected acute myocardial infarction: ISIS-2. ISIS-2 (Second International Study of Infarct Survival) Collaborative Group. Lancet 1988;2(8607):349-360.
29. ISIS-3 , Collaborative group. ISIS-3: a randomised comparison of streptokinase vs tissue plasminogen activator vs anistreplase and of aspirin plus heparin vs aspirin alone among 41 299 cases of suspected acute myocardial infarction. Lancet 1992;339:753-770.
30. Ooi CE, Weiss J, Doerfler ME, Elsbach P. Endotoxin-neutralizing poperties of the 25 kD N-terminal fragment and a newly isolated 30 kD C-terminal fragment of the 55-6- kD bactericidal/permeability-increasing protein of human neutrophils. J Exp Med 1991;174:649-655.
31. Hampton J, Wilcox R, Armstrong P, Aylward P, Bett N, Charbonnier B, Gulba D, Heikkila J, Jensen G, Lopez-Bescos L, *et al.* Late assessment of thrombolytic efficacy (LATE) study with alteplase 6-24 hours after onset of acute myocardial infarction. Lancet 1993;342:759-766.
32. Boersma E, Maas AC, Deckers JW, Simoons ML. Early thrombolytic treatment in acute myocardial infarction: reappraisal of the golden hour. Lancet 1996;348:771-775.
33. Gray D, Hampton JR. Sensitivity and specificity of the initial working diagnosis in acute myocardial infarction: implications for thrombolysis. Int J Epidemiol 1993;22:222-227.
34. Rude RE, Poole WK, Muller JE, Turi ZG, Rutherford J, Parker C, Roberts R, Raabe DS, Gold HK, Stone PH, Willerson JT, Braunwald E. Electrocardiographic and Clinical Criteria for Recognition of Acute Myocardial Infarction Based on Analysis of 3,697 Patients. Am J Cardiol 1983;52:936-942.
35. Karlson BW, Herlitz J, Wiklund O, Richter A, Hjalmarson A. Early Prediction of Acute Myocardial Infarction from Clinical History, Examination and Electrocardiogram in the Emergency Room. Am J Cardiol 1991;68:171-175.
36. Brush JE, Brand DA, Acampora D, Chalmer B, Wackers FJ. Use of the initial electrocardiogram to predict in-hospital complications of acute myocardial infarction. New Engl J Med 1985;312:1137-1141
37. Stark ME, Vacek JL. The Initial Electrocardiogram During Admission for Myocardial Infarction. Arch Intern Med 1987;147:843-846.
38. Lee TH, Cook EF, Weisberg MC, Sargent RK, Wilson C, Goldman L. Acute Chest Pain in the Emergency Room. Arch Intern Med 1985;145:65-69.
39. Goldman L, Cook EF, Johnson PA, Brand DA, Rouan GW, Lee TH. Prediction of the need for intensive care in patients who come to emergency departments with acute chest pain. N Eng J Med 1996;334:1498-1504.
40. Theroux P, Quimet H, McCans J, Latour JG, Joly P, Levy G, Pelletier E, Juneau M, Stasiak J, de Guise P, Pelletier GB, Rinzler D, Waters D. Aspirin, Heparin, or both to treat acute unstable angina. New Engl J Med 1988;319:1105-1111.
41. The RISC group , Wallentin L. Risk of myocardial infarction and death during treatment with low dose aspirin and intravenous heparin in men with unstable coronary artery disease. Lancet 1990;336:827-830.
42. Klein W, Buchwald A, Hillis SE, Monrad S, Sanz G, Turpie GG, van der Meer J, Olaisson E, Undeland S, Ludwig K, for the FRISC Investigators. Comparison of Low-Molecular Weight Heparin With Unfractionated Heparin Acutely and With Placebo for 6 Weeks in the Management of Unstable Coronary Artery Disease. Circulation 1997;96:61-68.
43. Lindahl B, Venge P, Wallentin L. Relation Between Troponin T and the Risk of Subsequent Cardiac Events in Unstable Coronary Artery Disease. Circulation 1996;93:1651-1657.
44. Stubbs P, Collinson PO, Moseley D, Greenwood T, Noble M. Prospective study of the role of cardiac troponin T in patients admitted with unstable angina. Br Med J 1997;313:262-264.
45. Bhayana V, Henderson AR. Biochemical markers of myocardial damage. [Review] [208 refs]. Clin Biochem 1995;28:1-29.
46. Adams JE, Abendschein DR, Jaffe AS. Biochemical markers of myocardial injury. Circulation 1993;88:750-763.
47. Apple FS. Acute myocardial infarction and coronary reperfusion. Serum cardiac markers for the 1990s. Am J Clin Pathol 1992;97:217-226.

48. Bakker AJ, Koelemay MJW, Gorgels JPMC, Van Vlies B, Smits R, Tijssen JGP, Haagen FDM. Failure of new biochemical markers to exclude acute myocardial infarction at admission. Lancet 1993;342:1220-1222.
49. De Winter RJ, Koster RW, van Straalen JP, Gorgels JP, Hoek FJ, Sanders GT. Critical difference between serial measurements of CK-MB mass to detect myocardial damage. Clin Chem 1997;43:338-343.
50. Adams JE, Schechtman KB, Landt Y, Ladenson JH, Jaffe AS. Comparable Detection of Acute Myocardial Infarction by Creatine Kinase MB Isoenzyme and Cardiac Troponin I. Clin Chem 1994;40:1291-1295.
51. Anderson FP, Fritz ML, Kontos MC, Jesse RL. Early Diagnosis of Myocardial Infarction with Troponin I, CK-MB, and Myoglobin Assays on the Behring Opus Magnum Analyzer. Clin Chem 1997;43:S129(106).
52. Adams JE, Bodor GS, Dávila-Román VG, Delmez JA, Apple FS, Ladenson JH, Jaffe AS. Cardiac Troponin I; A Marker With High Specificity for Cardiac Injury. Circulation 1993;88:101-106.
53. Bertinchant JP, Larue C, Pernel I, Ledermann B, Fabbro-Peray P, Beck L, Calzolari C, Trinquier S, Nigond J, Pau B. Release kinetics of serum cardiac troponin I in ischemic myocardial injury. Clin Biochem 1996;29:587-594.
54. Apple FS. Measurement of Cardiac Troponin-I in Serum for the Detection of Myocardial Infarction. JIFCC 1996;8:148-150.
55. Lee TH, Juarez G, Cook EF, Weisberg MC, Rouan GW, Brand DA, Goldman L. Ruling Out Acute Myocardial Infarction. A prospective Multicenter Validation of a 12-hour Strategy for Patients at Low Risk. N Engl J Med 1991;324:1239-1246.
56. Selker HP, Zalenski RJ, Antman EM, Aufderheide TP, Bernard SA, Bonow RO, Gibler WB, Hagen MD, Johnson P, Lau J, *et al.* An evaluation of technologies for identifying acute cardiac ischemia in the emergency department: a report from a National Heart Attack Alert Program Working Group [see comments]. Annals of Emergency Medicine 1997;29:13-87.
57. Lee TH, Rouan GW, Weisberg MC, Brand DA, Acampora D, Stasiulewicz C, Walchon J, Terranova G, Gottlieb L, Goldstein-Wayne B, Copen D, Daley K, Brandt AA, Mellors J, Jakubowsky R, Cook EF, Goldman L. Clinical Characteristics and Natural History of Patients with Acute Myocardial Infarction Sent Home from the Emergency Room. Am J Cardiol 1987;60:219-224.
58. Gibler WB, Lewis LM, Erb ER, Makens PK, Kaplan BC, Vaughn RH, Biagini AV, Blanton JD, Campbell WB. Early detection of acute myocardial infarction in patients presenting with chest pain and non-diagnostic ECG's. Ann Emerg Med 1990;19:1359-1366.
59. Gaspoz J, Lee TH, Cook EF, Weisberg MC, Goldman L. Outcome of patients who were admitted to a new short-stay unit to "rule-out" myocardial infarction. Am J Cardiol 1991;68:145-149.
60. Gaspoz J, Lee TH, Weisberg MC, Cook EF, Goldman P, komaroff AL, Goldman L. Cost-Effectiveness of a New Short-Stay Unit to "Rule Out" Acute Myocardial Infarction in Low Risk Patients. JACC 1994;24:1249-1259.
61. Gibler WB, Runyon JP, Levy RC, Sayre MR, Kacich R, Hattemer CR, Hamilton C, Gerlach JW, Walsh RA. A rapid diagnostic and treatment center for patients with chest pain in the emergency department [see comments]. Annals of Emergency Medicine 1995;25:1-8.
62. Lee TH, Rouan GW, Weisberg MC, Brand DA, Cook EF, Acampora D, Goldman L. Sensitivity of routine clinical criteria for diagnosing myocardial infarction within 24 hours of hospitalization. Ann Intern Med 1987;106:181-186..
63. Mulley AG, Thibault GE, Hughes RA, Barnett GO, Reder VA, Sherman EL. The course of patients with suspected myocardial infarction: the identification of low-risk patients for early transfer from intensive care. N Eng J Med 1980;302:943-948.
64. Zalenski RJ, Rydman RJ, McCarren M, Roberts RR, Jovanovic B, Das k, Mensah EK, Kampe LM. Feasibility of a Rapid Diagnostic Protocol for an Emergency Department Chest Pain Unit. Ann Emerg Med 1997;29:99-108.
65. Bahr RD. Growth in chest pain emergency departments throughout the United States: A cardiologist's spin on solging the heart attack problem. Coronary artery disease 1995;25:1-8.
66. Dillon MC, Calbreath DF, Dixon AM, Rivin BE, Roark SF, Ideker RE, Wagner GS. Diagnostic problem in acute myocardial infarction. CK-MB in the absence of abnormally elevated total creatine kinase levels. Arch Intern Med 1982;142:33-38.
67. Heller GV, Blaustein AS, Wei JY. Implications of increased myocardial isoenzyme level in the presence of normal serum creatine kinase activity. Am J Cardiol 1983;51:24-27.
68. White RD, Grande P, Califf L, Palmeri ST, Califf RM, Wagner GS. Diagnostic and Prognostic

Significance of Minimally Elevated Creatine Kinase-MB in Suspected Acute Myocardial Infarction. Am J Cardiol 1985;55:1478-1484.

69. Yusuf S, Collins R, Lin L, Sterry H, Pearson M, Sleight P. Significance of Elevated MB Isoenzyme with Normal Creatine Kinase in Acute Myocardial Infarction. Am J Cardiol 1987;59:245-250.
70. Pettersson T, Ohlsson O, Tryding N. Increased CKMB (mass concentration) in patients without traditional evidence of acute myocardial infarction. A risk indicator of coronary death. Eur Heart J 1992;13:1387-1392.
71. Mair J, Artner-Dworzak E, Dienstl A, Lechleitner P, Morass B, Smidt J, Wagner I, Wettach C, Puschendorf B. Early detection of acute myocardial infarction by measurement of mass concentration of creatine kinase-MB. Am J Cardiol 1991;68:1545-1550.
72. Bakker AJ, Gorgels JPMC, Van Vlies B, Koelemay MJW, Smits R, Tijssen JGP, Haagen FDM. Contribution of creatine kinase MB Mass concentration at admission to early diagnosis of acute myocardial infarction. Br Heart J 1994;72:112-118.
73. Durrer D, Lie KI, van der Wieken LR. Cardiac emergency service; experiences of the initial eight months. Ned Tijdschr Geneeskd 1979;123:1859-1864.
74. Koster RW, Brussen AJSH, Dunning AJ. Acute cardiac care: what happens to patients who are not immediately hopitalized? Ned Tijdschr Geneeskd 1990;134:1454-1457.

Chapter 10

PROGNOSTIC VALUE OF BIOCHEMICAL MARKERS IN ISCHAEMIC HEART DISEASE

Jan Ravkilde

Myocardial infarction (MI), unstable angina pectoris and sudden cardiac death constitute the "acute coronary syndrome" (ACS). Although patients with unstable angina are heterogeneous [1-7] up to 60% of patients who develop acute MI experience a prodrome of unstable angina [8] and approximately 5-20% of patients with unstable angina progress to MI or cardiac death within the first year [3, 9-17]. Non-Q wave MI is also considered an unstable condition associated with a low initial mortality but a higher risk of developing MI or cardiac death at a later time [18]. Allison *et al* [19] reported the presence of "microinfarcts" in patients who died with unstable angina but without clinical evidence of MI. This finding indicates that small areas of myocardial necrosis occur in patients with unstable angina, at least in a high-risk subgroup [19]. Recent post-mortem investigations have further documented that unstable angina leading to MI or sudden cardiac death is frequently preceded by microinfarctions [20, 21]. The release of cardiac enzymes and other myocardial cell constituents is to be expected in the presence of microinfarctions. In order to detect these, markers sensitive to myocardial injury and specific for mycardial tissue, must be employed. The prognostic significance of areas of micronecrosis and its detection by specific markers has been investigated intensively in recent years.

The acute coronary syndrome: Pathogenesis

Coronary atherosclerosis is a progressive disease which usually develops over decades, most often depending on the presence of risk factors. Coronary artery disease may present clinically as stable angina or acute coronary syndrome. The onset of acute coronary syndrome usually has no obvious external triggering factors and the clinical course is often unpredictable [22].

In stable angina pectoris, myocardial ischaemia commonly results from increases in myocardial oxygen demand that outstrip the ability of stenosed coronary arteries to increase oxygen delivery [22, 23]. Although such a mechanism may, on occasions, play a role in patients with unstable angina [24, 25], there is overwhelming evidence that a primary reduction in coronary arterial flow is of greater importance in these patients.

In 1966, Constantinides [26] reported that plaque disruption could lead to coronary thrombosis and acute MI. This mechanism is now an accepted cause of acute MI [27-29]. More recent post-mortem pathoanatomical investigations in patients who died as a result of MI, or unstable angina, or have suffered sudden cardiac death, have added valuable information [20, 21, 30-33]. Falk [20] demonstrated that unstable angina preceding cardiac death is characterized by an ongoing thrombotic process in a major coronary artery. In this process, episodes of mural thrombus formation alternate with intermittent thrombus fragmentation, resulting in peripheral embolization and microembolic occlusion of small intramyocardial arteries which in turn result in microinfarcts. Peripheral microembolization in unstable angina has been also observed in other reports [21]. Coronary angioscopy data has supported the critical pathophysiological role of intracoronary thrombus formation in patients with unstable angina [34-36]. Culprit coronary artery stenoses in unstable syndromes are often eccentric, with overhanging or irregular margins [37]. A hypercoagulable status has been also observed during the acute phase of unstable angina and MI [38-40], which may persist for up to 6 months after the acute phase [40]. The role of coronary vasospasm is also important in the pathogenesis of acute coronary syndromes although intracoronary thrombus formation is recognized as the most important cause of acute coronary syndrome. Vasoconstriction at the site of the culprit lesion may lead to stasis and thrombus formation thus contributing to the development of both unstable angina and acute MI [42]. The most common mechanism underlying the acute coronary syndrome is coronary plaque disruption, which leads to thrombus formation with or without vasospasm, and intermittent or persistent coronary occlusion [23, 42, 43].

Acute coronary syndrome: Diagnosis

Acute MI. The working groups set up by the World Health Organization (WHO) in the 1960's and later in 1971, have published their criteria for the diagnosis of acute MI which are used worldwide. These are based on clinical history, electrocardiographic changes and enzyme levels in serum [44]. These criteria have lately been reviewed in a special report on "Myocardial infarction and coronary deaths in the WHO MONICA project 1985-1987" [45]. Although the conventional criteria for the diagnosis of MI are quite adequate in most cases, these findings are not present or easily discernible in every patient with acute MI. A "typical" clinical history is elicited in approximately 30-70% of patients with acute MI [46, 47] and the so called typical electrocardiographic alterations, carefully described in the report, are not always present and the electrocardiogram may be classified as "non-interpretable" in 12-28% of patients with acute MI [48, 49]. With reference to cardiac enzymes, neither the 1971 report nor the WHO MONICA project describe the significance of these markers in great detail. However, it is generally accepted that measurement of the isoenzymes of CK and LD are useful for the diagnosis of acute MI.

Unstable angina. The syndrome of unstable angina comprises a very heterogenous group of patients [1-7],and it has been difficult to define [50]. Today, unstable angina is defined as a clinical syndrome falling between stable angina and acute MI thus encompassing a variety of clinical presentations of transient episodes of myocardial ischaemia [6]. The diagnosis of unstable angina is mainly based on the patients clinical manifestations and in the absence of diagnostic electrocardiographic changes and/or elevations in cardiac enzymes changes diagnostic of acute MI [51]. The heterogeneity of the patients investigated in diverse series has led to several clinical classifications. Braunwald has proposed a classification of unstable angina which is currently widely used [5]. This classification focuses on two important issues: 1) the severity of angina (i.e., Class I, new onset of severe angina; Class II, angina at rest within the past month but not within the preceding 48 hours and Class III, angina at rest within 48 hours), and 2) the clinical circumstances in which unstable angina occurs (i.e., A, development of angina in the presence of extracardiac conditions ("secondary" unstable angina); B, development of unstable angina in the absence of extracardiac conditions ("primary" unstable angina) and C, development of angina within two weeks after acute MI ("post-infarction" unstable angina)). Although clinically useful, this definition gives at least 9 patient subgroups. These may be even more if patients are further subdivided based on the presence or absence of transient ST-T segment changes and the characteristics of the medical treatment. The classification of patients with unstable angina is not a simple task as unstable angina is a dynamic condition and different pathophysiological mechanisms may operate within patients, at different points in time or simultaneously, resulting in shifts from one clinical subclass to another during hospitalization [3, 5]. The Braunwald classification has, however, gained acceptance and has recently been incorporated into the latest guidelines for "Diagnosing and managing unstable angina" from the National Heart, Lung and Blood Institute in USA [6, 7, 52].

Acute coronary syndrome: Prognosis

The in-hospital prognosis after acute MI has improved from around 25% fatalities in the 1970's to about 10% in the early 1990's. This has been primarily due to effective anti-ischaemic and thrombolytic therapy [15, 53-55]. The one-year mortality after acute MI in hospital survivors decreased from 25% in the 1970's to approximately 15% in the early 1990's, probably due to effective post-MI treatment strategies and improved secondary prevention [14, 53, 56]. The five-year mortality however remained unchanged up to the early 1980's at more than 30% [17, 53]. However, a decline to 20-25% has been seen since the introduction of treatment with thrombolytics and angiotensin-converting enzyme inhibitors [57, 58].

A distinction between Q wave and non-Q wave MI is necessary for prognostic reasons. Years ago, Q-wave MI was synonymous with "transmural" MI and non-Q wave MI with "subendocardial" MI. However, autopsy studies have revealed that a transmural MI may occur even in the absence of Q waves in the ECG, and also subendocardial MI may present as a Q-wave MI [59, 60]. Patients with Q wave MI have a different evolution than patients with non-Q wave MI [51]. The latter tend to have smaller infarcts on presentation and rarely have total occlusion of the infarct-related vessel [37, 60, 61]. Non-Q wave MI is considered to be a relatively unstable condition associated with a lower initial mortality but a higher risk of later MI or cardiac death [60, 61]. The in-hospital mortality in patients with non-Q wave MI is approximately one-third to 50% of that in patients with Q wave MI [18, 61-63]. However, the one-year mortality after hospital discharge is significantly higher in non-Q wave than in Q wave MI [63]. It is noteworthy that subsequent MI in patients with non-Q wave MI tends to occur in the area of the original injury [62, 64]. When long-term (5 years) morbidity and mortality are considered no significant differences are found between Q wave and non-Q wave MI. It is important also to consider the long-term prognosis for "unrecognized" MI is similar to, and as serious as, that of clinically apparent MI [65].

It has been reported that 5-20% of patients with unstable angina have poor prognosis, with progression to acute MI or cardiac death within the first year [3, 9-17].

Biochemical markers of myocardial injury

Isoenzymes of CK and LD are the most widely used diagnostic markers of acute MI. However, they are not cardiospecific and have low sensitivity to detect minor myocardial injury (MMI), albeit retaining a high specificity. Improved immunoassays have been developed in the mid 1980's which allow measurements of the mass concentration of CK-MB in serum instead of its catalytic activity. More recently, new immunoassays have been developed for the measurement of structural proteins of the heart such as troponin T (TnT) and troponin I (TnI), components of the troponin regulatory complex located on the thin filament of the contractile apparatus of the myocyte; and myosin light chains (MLC) a component of the thick filament. An overview of the numerous immunoassays for assessment of CK-MB mass, and MLC has been published elsewhere [66]. Minor ischaemic myocardial injury can be detected by CK-MB mass, TnT, TnI and MLC immunoassays in patients with unstable angina or suspected MI [67-93].

Prognostic implications of abnormal CK-MB Mass, Troponin T, Troponin I and Myosin Light Chains

Several studies have attempted to identify prognostic indicators of clinical value in patients with unstable angina in whom MI has been ruled out based on clinical evaluation, ECG findings or enzyme measurements [9-14, 16, 17, 52, 94-106]. The term "microinfarction" has been used to describe the pattern of increased CK-MB activity despite normal total CK activity in patients with symptoms compatible with MI. Studies have demonstrated prognostic similarities between these patients and those with non-Q wave MI [97, 107, 108]. However, follow-up data have been inconsistent; some patients appear to have an unfavourable outcome [12, 97, 104], whereas others do not [95, 100]. All these studies used electrophoretic determinations of CK and CK-MB activity, and all but two used fraction of CK-MB [13, 100]. A possible explanation for the conflicting results may be the difficulty to electrophoretically discriminate between a slightly increased and a normal CK-MB band [109]. It has been shown that neither fluctuations of CK within the normal reference range, nor CK-MB activity (measured by immunoinhibition) with discrimination limits lower than 24 U/l provide prognostic information [85, 100]. The most likely explanation for this is a poor signal-to-noise ratio when using lower discriminatory values of CK-MB activity, and the presence of intra-individual variability of enzyme levels in the blood. Immunoassays measuring the mass concentration of CK-MB are more sensitive than electrophoresis [110-112] and immunoinhibition [70, 71].

CK-MB mass. It has been shown that approximately one-fourth to one-third of patients with unstable angina, or chest pain patients in whom acute MI has been ruled out, have minor "leakages" of CK-MB mass [70, 71, 74-76, 78, 85] (Table 1). However, the clinical importance of this observation was not clear. Prognostic studies in patients with suspected MI or unstable angina (Braunwald class III) were carried out [74-76, 78, 85-87] and

Table 1. Relationship between events and results of CK-MB mass measurements in patients with unstable angina or patients in whom an AMI was ruled out.

					Events (%)	
	No patients	Type of patients	CK-MB mass positive	CK-MB mass negative	CK-MB mass positive	CK-MB mass negative
Ravkilde 1992 [74]	65	Susp.AMI	24 (37%)[1]	41 (63%)[1]	8 (33%)	4 (10%)
Pettersson 1992 [76]	53	Susp.AMI	14 (26%)[1]	39 (74%)[1]	9 (64%)	2 (5%)
Ravkilde 1995 [85]	124	Susp.AMI	35 (28%)[2]	89 (72%)[2]	8 (23%)	3 (3%)
Markenvard 1992 [75]	101	UAP	29 (29%)[2]	72 (71%)[2]	12 (41%)	7 (10%)
Wu 1995 [86]	131	UAP	8 (6%)[2]	123 (94%)[2]	5 (63%)	6 (5%)
de Winter 1995 [87]	126	UAP	13 (10%)[2]	113 (90%)[2]	1 (8%)	2 (2%)
Bøtker 1991 [70]	21	UAP3	8 (38%)	13 (62%)[1]	2 (25%)	0 (0%)
Cumulative			131 (21%)[1]	490 (79%)	45 (34%)	24 (5%)

[1] Determined based on time course of CK-MB mass, see text for further details.
[2] Determined based on a discrimination limit of CK-MB mass, see text for further details.
UAP = Unstable angina pectoris, UAP3 = Unstable angina pectoris, Braunwald class III,
Susp. AMI = Suspected acute myocardial infarction, Events = Cardiac death or non fatal MI

demonstrated a significantly increased risk of cardiac events (i.e., non-fatal acute MI, cardiac death) associated to fluctuations of CK-MB mass [74], the rate of change in serial samples of CK-MB mass [76], or a fixed discrimination limit [75, 85-87]. In early studies measurement of CK-MB mass were not standardized. Recently (in 1996), a subcommittee of the American Association of Clinical Chemistry (AACC) have completed a standardization of CKMB mass measurements. The advantages of having standard reference levels and discriminatory values are obvious. It appears that both fluctuations of CK-MB and a fixed discrimination limit are suitable tools to identify patients with poor prognosis. An example of individual time courses with relative concentrations of biochemical markers in a patient with unstable angina and subsequent acute MI is illustrated in Figure 1.

Troponin T. The evaluation of TnT was initially carried out in patients with acute MI [72, 113-116]. However, elevated plasma TnT levels were also seen in about one-third of patients with unstable angina or in other chest pain patients in whom an acute MI had been ruled out and a multicentre study suggested that increased TnT levels may be a useful prognostic indicator in patients with unstable angina [73]. The Scandinavian multicentre study assessed the diagnostic performance and prognostic value of TnT in patients admitted to the coronary care unit with the presumptive diagnosis of acute MI. This study concluded that TnT can be employed in the routine diagnosis of acute coronary syndrome, and the prognostic importance of this marker was further confirmed [78].

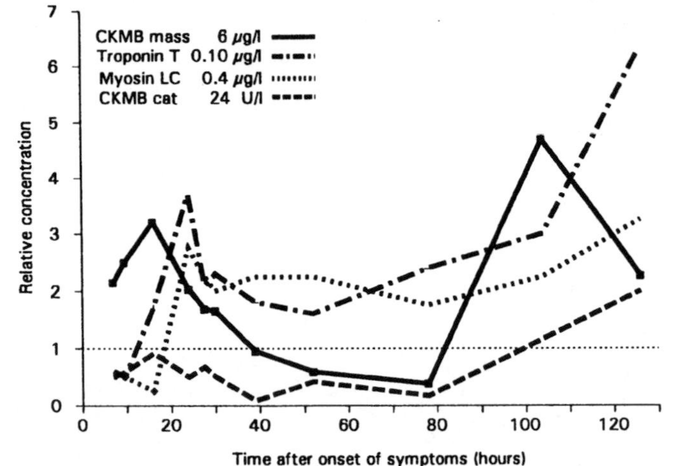

Figure 1. Individual time courses with relative concentrations of biochemical markers in a patient with unstable AP and subsequent acute MI.

Other prognostic studies have corroborated this finding. Patients with elevated TnT have a three to fivefold higher risk of developing acute MI or cardiac death in the months following the acute event compared to those without elevated TnT (Table 2) [72, 79, 80, 82, 86-88, 91, 117]. In contrast to most other studies, the GUSTO IIa trial evaluated the prognostic value of TnT based on the levels found in a single sample obtained at admission [90, 117]. When TnT concentration was above the chosen discriminatory limit, it indicated a risk for early acute MI of 85%. Of interest, even if TnT was negative the risk of acute MI in these patients was found to be as high as 63%. This is most probably due to the fact that little time was allowed to elapse from offset of symptoms to sampling. Thus this study shows that only a "positive" result is useful clinically, as a negative value does not necessarily exclude an evolving MI. Other authors have confirmed this observation [118] and similar results were seen in the GUSTO IIa study for measurement of CK-MB mass. At present, a "second generation TnT immunoassay" is available [119] which may offer advantages.

Troponin I. The early articles on TnI focused on the diagnostic value of this marker in acute MI [77, 81, 84, 120, 121], and stressed the high cardiospecificity of TnI [81, 121, 122]. In the last couple of years the prognostic value of increased TnI levels in patients with acute coronary syndrome has triggered considerable interest [89, 92, 93, 123-125].

At present, three TnI immunoassays have been approved by the Food and Drug Administration in the United States ("Access", Sanofi; "Opus", Behring ; "Stratus", Dade). In addition, several other immunoassays are being developed. In view of this, a Committee was appointed in 1996 to standardize TnI measurements. The work of this Committee is essential and their conclusion urgently needed as at present each laboratory determines its own reference level and discriminatory value for each of the several

Table 2. Relationship between events and results of Troponin T measurements in patients with unstable angina or patients in whom an AMI was ruled out.

					Events (%)	
	No patients	Type of patients	Troponin T positive	Troponin T negative	Troponin T positive	Troponin T negative
Ravkilde 1992 [78]	127	Susp.AMI	44 (35%)[2]	83 (65%)[2]	6 (14%)	3 (4%)
Collinson 1993 [79]	40	Susp.AMI	6 (15%)[3]	34 (85%)[3]	4 (67%)	0 (0%)
Ravkilde 1995 [85]	124	Susp.AMI	25 (20%)[2]	99 (80%)[2]	6 (24%)	5 (5%)
Katus 1991 [72]	66	UAP	37 (56%)[1]	29 (44%)[1]	5 (14%)	1 (3%)
Wu 1995 [86]Ohman	131	UAP	27 (21%)[3]	104 (79%)[3]	8 (30%)	3 (3%)
1995 [90]	824	UAP	299 (36%)[3]	525 (64%)[3]	57 (19%)	47 (9%)
de Winter 1995 [87]	126	UAP	24 (19%)[2]	102 (81%)[2]	1 (4%)	2 (2%)
Lindahl 1996 [88]	581	UAP	206 (35%)[4]	375 (65%)[4]	29 (14%)	30 (8%)
Stubbs 1996 [91]	183	UAP	62 (34%)[2]	121 (66%)[2]	18 (29%)	21 (17%)
Lüscher 1997 [93]	309	UAP	61 (20%)[1]	248 (80%)[1]	7 (11%)	8 (3%)
Hamm 1992 [73]	84	UAP3	33 (39%)[2]	51 (61%)[2]	10 (30%)	1 (2%)
Seino 1993 [80]	22	UAP3	14 (64%)[2]	8 (36%)[2]	2 (14%)	0 (0%)
Burlina 1994 [82]	28	UAP3	16 (57%)[2]	12 (43%)[2]	5 (31%)	0 (0%)
Cumulative	2645		855 (32%)	1790 (68%)	158 (18%)	120 (7%)

Discrimination limit of TnT 0.50 µg/l[1], TnT 0.20 µg/l[2], TnT 0.10 µg/l[3], TnT 0.18 µg/l[4], see text for further details.
UAP = Unstable angina pectoris, UAP3 = Unstable angina pectoris, Braunwald class III,
Susp.AMI = Suspected acute myocardial infarction, Events= Cardiac death or non fatal MI

different TnI immunoassays. This clearly complicates the interpretation of results and makes comparison between studies difficult. In a retrospective study, Antman *et al* [89] found that TnI ≥ 0.4 µg/l ("Stratus", Dade) was associated to an increased mortality rate within 42 days (3.7 vs 1%), whereas Galvani *et al.* [92] reported a 30 day mortality rate of 9.1% vs 0% for TnI > 3.1 µg/l ("Stratus", Dade) (equals 0.6 µg/l in a newer version). Using a similar immunoassay ("Opus", Behring) Kano *et al* [124] identified an upper limit of normal of 0.2 µg/l, whereas Lüscher *et al* [93] used an upper reference level of 2.0 µg/l. Despite the limitations imposed by the different immunoassays TnI measurements appear to have prognostic value. However, more studies are needed to compare the prognostic value of TnI, TnT and CK-MB mass measurement.

Myosin light chains (MLC). Although there are numerous publications on the role of this marker in acute MI, only a few studies have focused on MLC in patients with unstable angina or patients with chest pain in whom an acute MI was ruled out [67-69, 83, 85]. Approximately 16 to 52% of these patients had increased MLC in different studies. Only two studies have evaluated the prognostic value of MLC, and both found an adverse outcome in patients with elevated MLC [68, 85]. This is consistent with studies using CK-

MB mass, TnT and TnI. However, a disadvantage of MLC is that it is not entirely specific for cardiac injury.

Biochemical markers versus the electrocardiogram. A significant correlation has been reported between transient ST-T wave changes and/or ST-T wave abnormalities on admission and the rise of CK-MB mass [70, 74, 85], TnT [72, 74, 85], and MLC [68, 85]. This is in agreement with the finding that ST segment and T wave changes on the admission ECG, transient ST-T shifts in serial ECGs or transient isolated negative T waves are associated with an unfavourable outcome in patients with unstable angina or suspected MI [9, 13, 94, 98, 99, 101, 105, 126]. This clearly shows that both ECG changes and elevations of cardiac enzymes/proteins are good markers of myocardial injury.

Several studies have investigated whether CK-MB mass, TnT, TnI and MLC have independent prognostic value [85, 88, 92, 118]. These studies demonstrated a similar role of these markers for the identification of patients with minor myocardial injury who had poor prognosis, comparable to that in patients with definite MI. In one study comparing CK-MB mass, TnT and MLC, all three markers provided similar independent prognostic information with regard to clinical findings and arrhythmias, but provided no additional prognostic information compared to ECG ST-T wave changes [85]. Other studies, however, have found that TnT adds independent prognostic information to the ECG ST-T findings [88, 90]. Elevated TnI levels seem also to possess independent prognostic value but a direct comparison with the ECG-findings has not been carried out [92]. Although the ECG is sensitive in diagnosing myocardial ischaemia, the biochemical markers appear to be even more sensitive. Furthermore, non-interpretable ECGs are common in patients with acute chest pains and therefore the use of sensitive biochemical markers has clear advantages.

Unstable angina and non-Q wave MI. The emerging concept of "microinfarction" deserves attention, particularly in view of the fact that some patients with unstable angina could be considered to have non-Q wave MI. Given the wide diagnostic time-window for TnT, TnI and MLC, abnormal levels may actually reflect, at least in some cases, the presence of recent MI.

Although an increasing number of studies have focused on the prognostic value of sensitive biochemical markers, only a few have presented survival/event-free curves (Figure 2) [74, 76, 78, 85, 91-93]. The concordant cardiac survival/event-free curves during the follow-up period for patients with definite MI according to conventional criteria and those with rise in CK-MB mass, TnT, TnI or MLC illustrate the prognostic similarities of these populations. Although the follow-up period in these studies ranged from 30 days to 4 years, it appears that the large majority of events take place within the first months. The majority of these events occurred in patients with unstable angina. The Braunwald classification is useful to identify mild and severe cases, and a recent study found that class IIIB i.e., rest angina within the latest 48 hours judged to be of cardiac origin, was associated with the poorest prognosis [52]. It is therefore important to bear in mind that several biochemical studies, in which all or the large majority of patients were in class IIIB, serological markers allowed further risk stratification (Table 1-3). At present, it appears that unstable angina patients with raised markers of myocardial injury considered as non-Q wave MI. Future prospective studies, several of which are ongoing, should determine whether these patients will also benefit from more aggressive medical treatment (e.g., thrombolytic, anticoagulant, antithrombotic agents) or invasive revascularization procedures, or both.

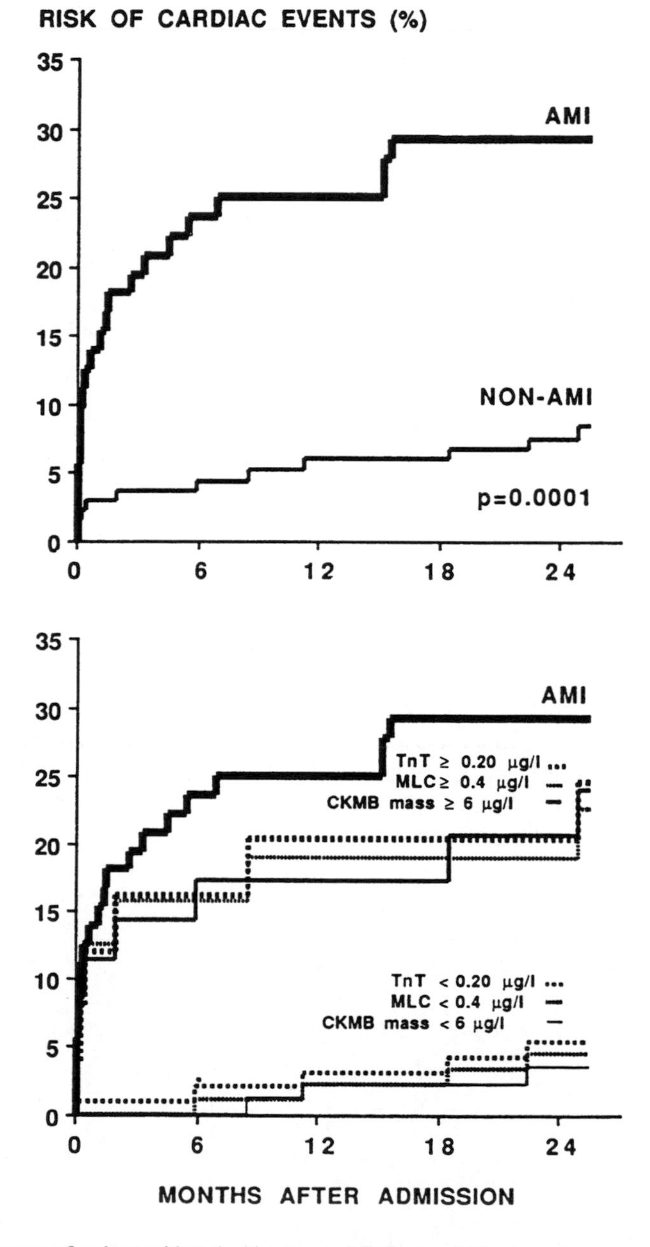

Figure 2. Clinical outcome of patients with and without acute MI (Top) and after two subgroups were assigned based on serum CK-MB mass, TnT or MLC levels (Bottom). End points: Cardiac death or nonfatal acute MI. Figure adapted from *[85]*.

Table 3. Relationship between events and results of Troponin I measurements in patients with unstable angina or in whom an AMI was ruled out.

	No patients	Type of patients	Troponin I positive	Troponin I negative	Events (%) Troponin I positive	Events (%) Troponin I negative
Galvani 1997 [92]	91	UAP3	22 (24%)[1]	69 (76%)[1]	6 (27%)	4 (6%)
Lüscher 1997 [93]	309	UAP	32 (10%)[2]	277 (90%)[2]	4 (13%)	11 (4%)
Kano 1997 [124]	75	UAP/Non-QMI	32 (40%)[3]	45 (60%)[3]	5 (17%)	0 (0%)

[1] Discrimination limit of TnI 3.1 μg/l (Stratus, Dade), see text for further details.
[2] Discrimination limit of TnI 2.0 μg/l (Opus, Behring), see text for further details.
[3] Discrimination limit of TnI 0.2 μg/l (Opus, Behring), see text for further details.
UAP = Unstable angina pectoris, UAP3 = Unstable angina pectoris, Braunwald class 3, MI = Myocardial infarction.
The studies by Antman et al.[89] and Ohman et al. [125] used the prognostic value based on admission blood sample.
This is in contrast to the above studies using the peak value in a sampling period.
Events = Cardiac death or non fatal MI

Conclusion

In summary, CK-MB mass, TnT, TnI and MLC have proven useful, not only for diagnosis of acute MI, but to identify different prognostic categories. Some high risk patients are not easily identified by current routine diagnostic tests. It has been shown that CK-MB mass, TnT, TnI and MLC are superior to CK and CK-MB activity for the detection of minor myocardial infarction. To date TnT and CK-MB mass seem to be the most useful markers for risk stratification of unstable angina patients. More studies on TnI, both prognostic and descriptive, are needed to elucidate the prognostic value of TnI in relation to TnT and CK-MB mass. Furthermore, the conclusions of the AACC standardization committee on TnI are eagerly awaited. The new sensitive biochemical markers could compliment current clinical practice guidelines regarding diagnosis and management of unstable angina as useful diagnostic tools and prognostic markers [6, 102]. An International task force is also warranted for the standardization of new criteria for the definition of myocardial damage and the diagnosis of acute coronary syndrome.

References

1. Conti CR, Brawley RK, Griffith LSC, *et al.* Unstable angina pectoris: Morbidity and mortality in 57 consecutive patients evaluated angiographically. Am J Cardiol 1973;32:745-50.
2. Maseri A. Pathogenetic classifications of unstable angina as a guideline to individual patient management and prognosis. Am J Med 1986;80(supll.4C):48-55.
3. Hugenholz PG. Unstable angina revisited once more. Eur Heart J 1986;7:1010-3.
4. Bashour TT, Myler RK, Andreae GE, Stertzer SH, Clark DA, Ryan CJM. Current concepts in unstable myocardial ischemia. Am Heart J 1988;115:850-61.

5. Braunwald E. Unstable angina. A classification. Circulation 1989;80:410-4.
6. Braunwald E, Jones RH, Mark DB, *et al.* Diagnosing and managing unstable angina. Circulation 1994;90:613-22.
7. Lindenfeld J, Morrison DA. Toward a stable clinical classification of unstable angina. J Am Coll Cardiol 1995;25:1293-4.
8. Harper RW, Kennedy G, DeSanctis RW, Hutter AM. The incidence and pattern of angina prior to acute myocardial infarction: A study of 577 cases. Am Heart J 1979;97:178-83.
9. Gazes PC, Mobley Jr EM, Faris Jr HM, Duncan RC, Humphries CB. Preinfarctional (unstable) angina - A prospective study - Ten year follow-up. Prognostic significance of electrocardiographic changes. Circulation 1973;48:331-7.
10. Heng M-K, Norris RM, Singh BN, Partridge JB. Prognosis in unstable angina. Br Heart J 1976;38:921-5.
11. Schroeder JS, Lamb IH, Hu M. Do patients in whom myocardial infarction has been ruled out have a better prognosis after hospitalization than those surviving infarction? N Engl J Med 1980;303:1-5.
12. Armstrong PW, Chiong MA, Parker JO. The spectrum of unstable angina: Prognostic role of serum creatine kinase determination. Am J Cardiol 1982;49:1849-52.
13. Mulcahy R, Al Awadhi AH, de Buitleor M, Tobin G, Johnson H, Contoy R. Natural history and prognosis in unstable angina. Am Heart J 1985;109:753-8.
14. Madsen JK, Thomsen BL, Sørensen JN, Kjeldgaard KM, Kromann-Andersen B. Risk factors and prognosis after discharge for patients admitted because of suspected acute myocardial infarction with and without confirmed diagnosis. Am J Cardiol 1987;59:1064-70.
15. Cairns JA, Singer J, Gent M, *et al.* One year mortality outcomes of all coronary and intensive care unit patients with acute myocardial infarction, unstable angina or other chest pain in Hamilton, Ontario, a city of 375.000 people. Can J Cardiol 1989;5:239-46.
16. Betriu A, Heras M, Cohen M, Fuster V. Unstable angina: Outcome according to clinical presentation. J Am Coll Cardiol 1992;19:1659-63.
17. Launbjerg J, Fruergaard P, Jacobsen HL, Madsen JK. Risk factors related to the 7-year prognosis for patients suspected of myocardial infarction with and without confirmed diagnosis. Cardiology 1992;80:294-301.
18. Liebson PR, Klein LW. The non-Q wave myocardial infarction revisited: 10 years later. Progress in Cardiovascular Diseases; 1997:XXXIX:399-444.
19. Allison RB, Rodriguez FL, Higgins Jr EA, *et al.* Clinicopathologic correlations in coronary atherosclerosis. Four hundred thirty patients studied with postmortem coronary angiography. Circulation 1963;27:170-84.
20. Falk E. Unstable angina with fatal outcome: dynamic coronary thrombosis leading to infarction and/or sudden death. Autopsy evidence of recurrent mural thrombosis with peripheral embolization culminating in total vascular occlusion. Circulation 1985;71:699-708.
21. Davies MJ, Thomas AC, Knapman PA, Hangarther JR. Intramyocardial platelet aggregation in patients with unstable angina suffering sudden ischemic cardiac death. Circulation 1986;73:418-27.
22. Fuster V, Badimon L, Badimon JJ, Chesebro JH. The pathogenesis of coronary artery disease and the acute coronary syndromes (First of two parts). N Engl J Med 1992;326:242-50.
23. Fuster V. Lewis A. Conner Memorial Lecture. Mechanisms leading to myocardial infarction: Insights from studies of vascular biology. Circulation 1994;90:2126-46.
24. Ambrose JA, Winters SL, Arora RR, *et al.* Angiographic evolution of coronary artery morphology in unstable angina. J Am Coll Cardiol 1986;7:472-8.
25. Gotoh K, Minamino T, Katoh O, *et al.* The role of intracoronary thrombus in unstable angina: Angiographic assessment and thrombolytic therapy during ongoing anginal attacks. Circulation 1988;77:526-34.
26. Constantinides P. Plaque fissures in human coronary thrombosis. J Atheroscler Res 1966;6:1-17.
27. Herrick JB. Clinical features of sudden obstruction of the coronary arteries. JAMA 1912;LIX:2015-20.
28. Levine SA, Brown CL. Coronary thrombosis: its various clinical features. Medicine 1929;VIII:245-418.
29. Davies MJ, Fulton WFM, Robertson WB. The relation of coronary thrombosis to ischaemic myocardial necrosis. J Path 1979;127:99-110.
30. Davies MJ, Thomas T. The pathological basis and microanatomy of occlusive thrombus formation in human coronary arteries. Phil Trans R Soc London B 1981;294:225-9.
31. Falk E. Plaque rupture with severe pre-existing stenosis precipitating coronary thrombosis. Characteristics of coronary atherosclerotic plaques underlying fatal occlusive thrombi. Br Heart J 1983;50:127-34.
32. Davies MJ, Thomas AC. Plaque fissuring - the cause of acute myocardial infarction, sudden ischaemic death, and crescendo angina. Br Heart J 1985;53:363-73.
33. Kragel AH, Gertz SD, Roberts WC. Morphologic comparison of frequency and types of acute lesions in

the major epicardial coronary arteries in unstable angina pectoris, sudden coronary death and acute myocardial infarction. J Am Coll Cardiol 1991;18:801-8.
34. Sherman CT, Litvack F, Grundfest W, *et al.* Coronary angioscopy in patients with unstable angina pectoris. N Engl J Med 1986;315:913-9.
35. Mizuno K, Satomura K, Miyamoto A, *et al.* Angioscopic evaluation of coronary-artery thrombi in acute coronary syndromes. N Engl J Med 1992;326:287-91.
36. de Feyter PJ, Ozaki Y, Baptista J, *et al.* Ischaemia-related lesion characteristics in patients with stable or unstable angina. A study with intracoronary angioscopy and ultrasound. Circulation 1995;92:1408-13.
37. Ambrose JA, Hjemdahl-Monsen CE, Borrico S, Gorlin R, Fuster V. Angiographic demonstration of a common link between unstable angina pectoris and non-Q wave acute myocardial infarction. Am J Cardiol 1988;61:244-7.
38. Hamm CW, Lorenz RL, Bleifeld W, Kupper W, Wober W, Weber PC. Biochemical evidence of platelet activiation in patients with persistent unstable angina. J Am Coll Cardiol 1987;10:998-1004.
39. Grande P, Grauholt A-M, Madsen JK. Unstable angina pectoris. Platelet behaviour and prognosis in progressive angina and intermediate coronary syndrome. Circulation 1990;81(Suppl I):I-16-9.
40. Merlini PA, Bauer KA, Oltrona L, *et al.* Persistent activation of coagulation mechanism in unstable angina and myocardial infarction. Circulation 1994;90:61-8.
41. Maseri A, L'Abbate A, Baroldi G, *et al.* Coronary vasospasm as a possible cause of myocardial infarction. A conclusion derived from the study of "preinfarction" angina. N Engl J Med 1978;299:1271-7.
42. Ambrose JA. Plaque disruption and the acute coronary syndromes of unstable angina and myocardial infarction: If the substrate is similar, why is the clinical presentation different? J Am Coll Cardiol 1992;19:1653-8.
43. Fuster V, Badimon L, Badimon JJ, Chesebro JH. The pathogenesis of coronary artery disease and the acute coronary syndromes (Second of two parts). N Engl J Med 1992;326:310-8.
44. Working Group on the establishment of ischaemic heart disease registers: Report of the fifth working group, WHO, Europ 8201 (5), Copenhagen, 1971.
45. Tunstall-Pedoe H, Kuulasmaa K, Amouyel P, Arveiler D, Rajakangas A-M, Pajak A on behalf of the WHO MONICA Project participants. Myocardial infarction and coronary deaths in the World Health Organization MONICA project. Registration procedures, event rates, and case-fatality rates in 38 populations from 21 countries in four continents. Circulation 1994;90:583-612.
46. Lee TH, Rouan GW, Weiberg MC, *et al.* Sensitivity of routine clinical criteria for diagnosing myocardial infarction within 24 hours of hospitalization. Ann Intern Med 1987;106:181-6.
47. Thaulow E, Erikssen J, Sandvik L, Erikssen G, Jorgensen L, Cohn PF. Initial clinical presentation of cardiac disease in asymptomatic men with silent myocardial ischemia and angiographically documented coronary artery disease (the Oslo ischaemia study). Am J Cardiol 1993;72:629-33.
48. Turi ZG, Rutherford JD, Roberts R, *et al.* Electrocardiographic, enzymatic and scintigraphic criteria of acute myocardial infarction as determined from study of 726 patients (A MILIS study). Am J Cardiol 1985;55:1463-8.
49. McQueen MJ, Holder D, El-Maraghi NRH. Assessment of the accuracy of serial electrocardiograms in the diagnosis of myocardial infarction. Am Heart J 1983;105:258-61.
50. Gorlin R. The puzzles of unstable angina pectoris. Mount Sinai J Med 1990;57:136-43.
51. Gersh BJ, Braunwald E, Rutherford JD. Chronic coronary artery disease. In Braunwald E, ed. Heart Disease. A Textbook of Cardiovascular Medicine. Fifth edition. Philadelphia: W.B. Saunders Company, 1997:1289-1365.
52. Van Miltenburg-van Zijl AJM, Simoons ML, Veerhoek RJ, Bossuyt PMM. Incidence and follow-up of Braunwald subgroups in unstable angina pectoris. J Am Coll Cardiol 1995;25:1286-92.
53. de Vreede JJM, Gorgels APM, Verstraaten GMP, Vermeer F, Dassen WRM, Wellens HJJ. Did prognosis after acute myocardial infarction change during the past 30 years? A meta-analysis. J Am Coll Cardiol 1991;18:698-706.
54. The GUSTO Investigators. An international randomized trial comparing four thrombolytic strategies for acute myocardial infarction. N Engl J Med 1993;329:673-82.
55. Fibrinolytic Therapy Trialists' Collaborative Group. Indications for fibrinolytic therapy in suspected acute myocardial infarction: collaborative overview of early mortality and major morbidity results from all randomised trials of more than 1000 patients. Lancet 1994;343:311-22.
56. Rawles J, on behalf of the GREAT Group. Halving of mortality at 1 year by domiciliary thrombolysis in the Crampian Region Early Anistreplase Trial (GREAT). J Am Coll Cardiol 1994;23:1-5.
57. Simoons ML, Vos J, Tijssen JGP, *et al.* Long-term benefit of early thrombolytic therapy in patients with acute myocardial infarction: 5 year follow-up of a trial conducted by the Interuniversity Cardiology

Institute of the Netherlands. J Am Coll Cardiol 1989;14:1609-15.
58. Pfeffer MA, Braunwald E, Moyé LA, *et al.* Effect of captopril on mortality and morbidity in patients with left ventricular dysfunction after myocardial infarction. Results of the survival and ventricular enlargement trial. N Engl J Med 1992;327:669-77.
59. Phibbs B. "Transmural" versus "subendocardial" myocardial infarction: An electrocardiographic myth. J Am Coll Cardiol 1983;1:561-4.
60. DeWood MA, Stifter WF, Simpson CS, *et al.* Coronary arteriographic findings soon after non-Q-wave myocardial infarction. N Engl J Med 1986;315:417-23.
61. O'Brien TX, Ross Jr J. Non-Q-wave myocardial infarction: Incidence, pathophysiology, and clinical course compared with Q-wave infarction. Clin Cardiol 1989;12:III-3-III-9.
62. Hutter Jr AM, DeSanctis RW, Flynn T, Yeatman LA. Nontransmural myocardial infarction: A comparison of hospital and late clinical course of patients with that of matched patients with transmural anterior and transmural inferior myocardial infarction. Am J Cardiol 1981;48:595-602.
63. Nicod P, Gilpin E, Dittrich H, *et al.* Short- and long-term clinical outcome after Q wave and non-Q wave myocardial infarction in a large patient population. Circulation 1989;79:528-36.
64. Huey BL, Gheorghiade M, Crampton RS, *et al.* Acute non-Q wave myocardial infarction asssociated with early ST segment elevation: Evidence for spontaneous coronary reperfusion and implications for thrombolytic trials. J Am Coll Cardiol 1987;9:18-25.
65. Kannel WB, Abbott RD. Incidence and prognosis of unrecognized myocardial infarction. An update on the Framingham study. N Engl J Med 1984;311:1144-7.
66. Ravkilde J. Creatine kinase isoenzyme MB mass, cardiac troponin T, and myosin light chain isotype 1 as serological markers of myocardial injury and their prognostic importance in acute coronary syndrome. Thesis. Da Med Bull 1997; in press.
67. Hoberg E, Katus HA, Diederich KW, Kübler W. Myoglobin, creatine kinase-B isoenzyme, and myosin light chain release in patients with unstable angina pectoris. Eur Heart J 1987;8:989-94.
68. Katus HA, Diederich KW, Hoberg E, Kübler W. Circulating cardiac myosin light chains in patients with angina at rest: Identification of a high risk subgroup. J Am Coll Cardiol 1988;11:487-93.
69. Hirayama A, Arita M, Takagaki Y, Tsuji A, Kodama K, Inoue M. Clinical assessment of specific enzyme immunoassay for the human cardiac myosin light chain II (MLC II) with use of monoclonal antibodies. Clin Biochem 1990;23:515-22.
70. Bøtker HE, Ravkilde J, Søgaard P, Jørgensen PJ, Hørder M, Thygesen K. Gradation of unstable angina based on a sensitive immunoassay for serum creatine kinase MB. Br Heart J 1991;65:72-6.
71. Gerhardt W, Katus H, Ravkilde J, *et al.* S-Troponin T in suspected ischemic myocardial injury compared with mass and catalytic concentrations of S-creatine kinase isoenzyme MB. Clin Chem 1991;37:1405-11.
72. Katus HA, Remppis A, Neumann FJ, *et al.* Diagnostic efficiency of troponin T measurements in acute myocardial infarction. Circulation 1991;83:902-12.
73. Hamm CW, Ravkilde J, Gerhardt W, *et al.* The prognostic value of serum troponin T in unstable angina. N Engl J Med 1992;327:146-150.
74. Ravkilde J, Hansen AB, Hørder M, Jørgensen PJ, Thygesen K. Risk stratification in suspected acute myocardial infarction based on a sensitive immunoassay for serum creatine kinase isoenzyme MB. A 2.5 year follow-up study in 156 consecutive patients. Cardiology 1992;80:143-51.
75. Markenvard J, Dellborg M, Jagenburg R, Swedberg K. The predictive value of CKMB mass concentration in unstable angina pectoris: preliminary report. J Intern Med 1992;231:433-6.
76. Pettersson T, Ohlsson O, Tryding N. Increased CKMB (mass concentration) in patients without traditional evidence of acute myocardial infarction. A risk indicator of coronary death. Eur Heart J 1992;13:1387-92.
77. Bodor GS, Porter S, Landt Y, Ladenson JH. Development of monoclonal antibodies for an assay of cardiac troponin-I and preliminary results in suspected cases of myocardial infarction. Clin Chem 1992;38:2203-14.
78. Ravkilde J, Hørder M, Gerhardt W, *et al.* Diagnostic performance and prognostic value of serum troponin T in suspected acute myocardial infarction. Scand J Clin Lab Invest 1993;53:677-85.
79. Collinson PO, Moseley D, Stubbs PJ, Carter GD. Troponin T for the differential diagnosis of ischaemic myocardial damage. Ann Clin Biochem 1993;30:11-6.
80. Seino Y, Tomita Y, Takano T, Hayakawa H. Early identification of cardiac events with serum troponin T in patients with unstable angina. Lancet 1993;342:1236-7.
81. Larue C, Calzolari C, Bertinchant J-P, Leclercq F, Grolleau R, Pau B. Cardiac-specific immunoenzymetric assay of troponin I in the early phase of acute myocardial infarction. Clin Chem 1993;39:972-9.
82. Burlina A, Zaninotto M, Secchiero S, Rubin D, Accorsi F. Troponin T as a marker of ischemic myocardial injury. Clin Biochem 1994;27:113-21.

83. Ravkilde J, Bøtker HE, Søgaard P, *et al.* Human ventricular myosin light chain isotype 1 as a marker of myocardial injury. Cardiology 1994;84:135-44.
84. Adams III JE, Schechtman KB, Landt Y, Ladenson JH, Jaffe AS. Comparable detection of acute myocardial infarction by creatine kinase MB isoenzyme and cardiac troponin I. Clin Chem 1994;40:1291-5.
85. Ravkilde J, Nissen H, Hørder M, Thygesen K. Independent prognostic value of serum creatine kinase isoenzyme MB mass, cardiac troponin T and myosin light chain levels in suspected acute myocardial infarction. Analysis of 28 months of follow-up in 196 patients. J Am Coll Cardiol 1995;25:574-81.
86. Wu AHB, Abbas SA, Green S, *et al.* Prognostic value of cardiac troponin T in unstable angina pectoris. Am J Cardiol 1995;76:970-2.
87. de Winter RJ, Koster RW, Schotveld JH, Sturk A, van Straalen JP, Sanders GT. Prognostic value of troponin T, myoglobin, and CK-MB mass in patients presenting with chest pain without acute myocardial infarction. Heart 1996;75:235-239.
88. Lindahl B, Venge P, Wallentin L for the FRISC Study Group. Relation between troponin T and the risk of subsequent cardiac events in unstable angina coronary artery disease. Circulation 1996;93:1651-7.
89. Antman EM, Tanasijevic MJ, Thompson B, *et al.* Cardiac-specific troponin I levels to predict the risk of mortality in patients with acute coronary syndromes. NEJM 1996;335:1342-9.
90. Ohman EM, Armstrong P, Christenson RH, *et al.* Cardiac troponin T levels for risk stratification in acute myocardial ischemia. NEJM 1996;335:1333-41.
91. Stubbs P, Collinson P, Moseley D, Greenwood T, Noble M. Prospective study of the role of cardiac troponin T in patients with unstable angina. BMJ 1996;313:262-4.
92. Galvani M, Ottani F, Ferrini D, *et al.* Prognostic influence of elevated values of cardiac troponin I in patients with unstable angina. Circulation 1997;95:2053-9.
93. Lüscher MS, Thygesen K, Ravkilde J, Heickendorff L, and the TRIM study group. The applicability of cardiac troponin T and I for early risk stratification in unstable coronary artery disease. Circulation 1997;96:2578-85.
94. Olson HG, Lyons KP, Aronow WS, Stinson PJ, Kuperus J, Waters HJ. The high-risk angina patient. Identification by clinical features, hospital course, electrocardiography and technetium-99m stannous pyrophosphate scintigraphy. Circulation 1981;64:674-84.
95. White RD, Grande P, Califf L, Palmeri ST, Califf RM, Wagner GS. Diagnostic and prognostic significance of minimally elevated creatine kinase-MB in suspected myocardial infarction. Am J Cardiol 1985;55:1478-84.
96. Gottlieb SO, Weisfeldt ML, Ouyang P, Mellits ED, Gerstenblith G. Silent ischaemia as a marker for early unfavourable outcomes in patients with unstable angina. N Engl J Med 1986;314:1214-9.
97. Hong RA, Licht JD, Wei JY, Heller GV, Blaustein AS, Pasternak RC. Elevated CK-MB with normal total creatine kinase in suspected myocardial infarction: Associated clinical findings and early prognosis. Am Heart J 1986;111:1041-7.
98. Granborg J, Grande P, Pedersen A. Diagnostic and prognostic implications of transient isolated negative T waves in suspected acute myocardial infarction. Am J Cardiol 1986; 57:203-7.
99. Swahn E, Areskog M, Berglund U, Walfridsson H, Wallentin L. Predictive importance of clinical findings and a predischarge exercise test in patients with suspected unstable coronary artery disease. Am J Cardiol 1987;59:208-14.
100. Yusuf S, Collins R, Lin L, Sterry H, Pearson M, Sleight P. Significance of elevated MB isoenzyme with normal creatine kinase in acute myocardial infarction. Am J Cardiol 1987;59:245-50.
101. Sclarovsky S, Rechavia E, Strasberg B, *et al.* Unstable angina: ST segment depression with positive versus negative T wave deflections - clinical course, ECG evolution, and angiographic correlation. Am Heart J 1988;116:933-41.
102. Ross Jr J, Gilpin EA, Madsen EB, *et al.* A decision scheme for coronary angiography after acute myocardial infarction. Circulation 1989;79:292-303.
103. Wilcox I, Freedman SB, McCredie RJ, Carter GS, Kelly DT, Harris PJ. Risk of adverse outcome in patients admitted to the coronary care unit with suspected unstable angina pectoris. Am J Cardiol 1989;64:845-8.
104. Clyne CA, Medeiros LJ, Marton KI. The prognostic significance of immunoradiometric CK-MB assay (IRMA) diagnosis of myocardial infarction in patients with low total CK and elevated MB isoenzymes. Am Heart J 1989;118:901-6.
105. Nyman I, Areskog M, Areskog N-H, Swahn E, Wallentin L, and the RISC Study Group. Very early risk stratification by electrocardiogram at rest in men with suspected unstable coronary heart disease. J Intern Med 1993;234:293-301.

106. Rizik DG, Healy S, Margulis A, *et al.* A new clinical classification for hospital prognosis of unstable angina pectoris. Am J Cardiol 1995;75:993-7.
107. Heller GV, Blaustein AS, Wei JY. Implications of increased myocardial isoenzyme level in the presence of normal serum creatine kinase activity. Am J Cardiol 1983;51:24-7.
108. Dorogy ME, Hooks S, Cameron RW, Davis RC. Clinical and angiographic correlates of normal creatine kinase with increased MB isoenzymes in possible acute myocardial infarction. Am Heart J 1995;130:211-7.
109. Hørder M, Gerhard W, Härkönen M, *et al.* Creatine kinase (EC 2.7.3.2) and creatine kinase B-subunit activity in serum in suspect myocardial infarction. The Committee on Enzymes of The Scandinavian Society for Clinical Chemistry and Clinical Physiology (SCE). The Nordic Clinical Chemistry Project (NORDKEM), Helsinki, Finland. 1981.
110. Murthy VV, Karmen A. Activity concentration and mass concentration (monoclonal antibody immunoenzymometric method) compared for creatine kinase MB isoenzyme in serum. Clin Chem 1986;32:1956-9.
111. Piran U, Kohn DW, Uretsky LS, *et al.* Immunochemiluminometric assay of creatine kinase MB with a monoclonal antibody to the MB isoenzyme. Clin Chem 1987;33:1517-20.
112. Fenton JJ, Brunstetter S, Gordon WmC, Rippe DF, Bell ML. Diagnostic efficacy of a new enzyme immunoassay for creatine kinase MB isoenzyme. Clin Chem 1984;30:1399-1401.
113. Katus HA, Remppis A, Looser S, Hallermeier K, Scheffold T, Kübler W. Enzyme linked immuno assay of cardiac troponin T for the detection of acute myocardial infarction in patients. J Mol Cell Cardiol 1989;21:1349-53.
114. Katus HA, Looser S, Hallermayer K, *et al.* Development and in vitro characterization of a new immunoassay of cardiac troponin T. Clin Chem 1992;38:386-93.
115. Katus HA, Remppis A, Scheffold T, Diederich KW, Kuebler W. Intracellular compartmentation of cardiac troponin T and its release kinetics in patients with reperfused and nonreperfused myocardial infarction. Am J Cardiol 1991;67:1360-7.
116. Mair J, Artner-Dworzak E, Lechleitner P, *et al.* Cardiac troponin T in diagnosis of acute myocardial infarction. Clin Chem 1991;37:845-52.
117. Ohman EM, Armstrong P, Califf RM, *et al.* Risk stratification in acute ischemic syndromes using serum troponin T. J Am Coll Cardiol 1995;25(suppl): 148A. Abstract.
118. Stubbs P, Collinson P, Moseley D, Greenwood T, Noble M. Prognostic significance of admission troponin T concentrations in patients with myocardial infarction. Circulation 1996;94:1291-7.
119. Müller-Bardorff M, Hallermayer K, Schroeder A, *et al.* Improved troponin T ELISA specific for cardiac troponin T isoform: assay development and analytical and clinical validation. Clin Chem 1997;43:458-66.
120. Cummins B, Auckland ML, Cummins P. Cardiac-specific troponin-I radioimmunoassay in the diagnosis of acute myocardial infarction. Am Heart J 1987;113:1333-44.
121. Cummins B, Cummins P. Cardiac specific troponin-I release in canine experimental myocardial infarction: Development of a sensitive enzyme-linked immunoassay. J Mol Cell Cardiol 1987;19:999-1010.
122. Adams III JE, Bodor GS, Dávila-Román VG, *et al.* Cardiac troponin I. A marker with high specificity for cardiac injury. Circulation 1993;88:101-6.
123. Wu AHB, Feng YJ, Contois JH, Azar R, Waters D. Prognostic value of cardiac troponin I in chest pain patients. Clin Chem 1996;42:651-2. Letter.
124. Kano S, Nishimura S, Tashiro Y, *et al.* Cardiac troponin I in diagnosis and prognosis of unstable coronary artery diasease. Clin chem 1997;43:S157. Abstract.
125. Ohman EM, Christenson RH, Peck S, *et al.* Comparison of troponin I and T for risk stratification in patients with acute coronary syndromes. Circulation 1996;94:I-323. Abstract.
126. Nordlander R, Nyquist O. Patients treated in a coronary care unit without acute myocardial infarction. Identification of high risk subgroup for subsequent myocardial infarction and/or cardiovascular death. Br Heart J 1979;41:647-53.

Chapter 11

TROPONIN T IN THE MANAGEMENT OF ISCHAEMIC HEART DISEASE PATIENTS: A CLINICAL CARDIOLOGIST'S VIEW

Peter Stubbs

Advances in the pharmacological and mechanical approaches to acute coronary syndromes have led to rapid changes in the management of patients admitted with acute coronary syndromes over recent years. These changes have been mirrored by the appearance of newer highly specific biochemical markers of myocardial damage and there is now an increasing number of markers available to clinicians for assessing the degree of myocardial damage that patients have sustained (Table 1). The "classic" role of conventional markers of myocardial damage is retrospective confirmation of myocardial infarction (MI). The standard ones used in the United Kingdom are creatine kinase, aspartate transaminase and hydroxybutyrate dehydrogenase. Newer markers such as CKMB mass and the troponins are now available and offer greater specificity than the traditional markers.

Table 1. Biochemical Markers of Myocardial Damage

Standard Biochemical Markers	Other markers
- Creatine Kinase - Asparate Transaminase - Hydroxybutyrate Dehydrogenase	- Myoglobin - CKMB concentration, activity, mass - CKMB isoforms measured by different methods - troponin T - troponin I - Myosin Light chains - Fatty acid binding protein

The role of new soluble markers

When new biochemical markers become available it is the absolute domain of the Clinical Chemist to evaluate them critically in terms of sensitivity, specificity, efficiency and analyser precision, in the rigid setting of quality control that laboratories practice, and to compare them with other markers. When the data are shown to Clinical

Cardiologists with statements such as 'useful management tool' and 'can be used for early diagnosis of Myocardial Infarction', a different set of questions need to be answered. The 'so what?' response is the most frequent and the most important hurdle that these newer biochemical markers have to overcome to convince physicians to change their current practice. This response is not unreasonable as most data presented fail to take account of the different roles for biochemical markers in differing patient groups presenting with chest pain. This is summarised in Table 2.

Table 2. Sensitivity and Specificity of Markers

The sensitivity, specificity and predictive value of biochemical markers of myocardial damage depends upon:
The population being studied
1. Known MI-ECG diagnostic or non diagnostic
2. Patients admitted to CCU with suspected acute coronary syndromes (selected on clinical and ECG criteria): MI, unsable angina, non-cardiac pain
3. Patients admitted to casualty with chest pain
The time point at which the marker is measured
1. On admission
2. At time point when maximum sensitivity has been reached
The cut-off used for the marker

If only patients with confirmed myocardial infarction are studied all biochemical markers will look good. Inclusion of patients who have received either thrombolysis or primary PTCA in sequential sampling studies will falsely increase the early sensitivity- particularly of the mainly cytoplasmic markers- due to the washout effect. If patients admitted to Coronary Care Units are studied, these will form a more heterogeneous group with final diagnoses of Myocardial Infarction (MI), Unstable Angina Pectoris (UAP) or Non Cardiac Chest Pain (NCCP) that have been selected on clinical grounds. Again, there will be a high incidence of heart damage. The specificity of the purely cytoplasmic markers however will begin to fall. Comparatively few data studying only patients admitted to casualty departments with chest pain have been presented, although large trials are due to report in the near future. Specificity becomes very important in this setting. The specificity of markers such as CK, and myoglobin in particular, plummet in this population. The time point at which the marker is measured is also important as all markers are time dependent into the circulation following myocardial damage. There will, therefore, be a variable sensitivity for the detection of myocardial damage on admission to hospital dependent upon the release kinetics of the markers studied. The cut-off values used to define positivity will also have an effect on both sensitivity and specificity of a biochemical marker. Other questions need to be addressed when examining the clinical usefulness of newer biochemical markers and these are summarised in Table 3.

Table 3. Clinical Usefulness of Biochemical Markers of Myocardial Damage

The clinical usefulness depends upon:
1. What additional information (ie. diagnosis and prognosis) does the measurement provide above and beyond clinical and ECG findings?
2. Does marker measurement have the ability to change patient management decisions?
3. How important are the new biochemical markers as a risk predictor compared to currently used variables?

ST segment elevation myocardial infarction

The role of biochemical markers in early diagnosis

The following is a case history examining the role of biochemical markers in the early diagnosis of patients with ST segment elevation myocardial infarction.

A 59 year old male attended our chest pain clinic with chest pain presumed to be of cardiac origin. The baseline ECG was normal and so he was exercise tested according to the Bruce protocol. He developed chest pain in stage III of the protocol with ST segment elevation inferiorly and with anterior ST segment depression on the ECG. These changes failed to respond to sublingual nitrates and so he was admitted to the CCU and thrombolysed. On admission to the CCU some ten minutes following the episode, venous blood was drawn for estimation of a number of biochemical marker concentrations. The results were as follows with their cut-off concentrations in brackets: myoglobin 57mcg/l (85), creatine kinase 103 mcg/l (120), CK MB mass 2.9 mcg/l (5), troponin I undetectable (0.1mcg/ml), troponin T undetectable (0.1mcg/ml). As expected they were all within normal limits. The injury current from the myocardium therefore is visible on the surface ECG within seconds and no biochemical marker can compete with the ECG for early diagnosis in this group.

In my view, in this patient group there is no clinical need for an early diagnostic biochemical marker. Laboratories should use the most cost-effective biochemical marker on a routine run to retrospectively confirm the Myocardial Infarction and, perhaps, give some indication of infarct size. In the reperfusion therapy era in which we now work it is probably best to avoid the purely cytoplasmic based markers because of the washout effect.

Other diagnostic uses of biochemical markers in patients with ST segment elevation myocardial infarction

Other authors in this book have addressed the role of biochemical markers as non-invasive tools to identify the effect of thrombolysis on reperfusion and the role of biochemical markers in assessing infarct size. In the clinical setting, markers of reperfusion, successful or otherwise, need to be able to provide a fast answer so that those who fail to reperfuse can receive an alternative strategy at a time point when it may still be possible to salvage myocardium, be cheap and easy to perform and have an acceptable sensitivity. Whether measuring rate of change of biochemical marker concentrations is superior to, and as convenient as, measuring ST segment change on serial ECG's remains to be established. Similarly, when assessing the clinical usefulness of biochemical markers in assessing infarct size, it must be realised that peak concentrations of these markers are not subtle tools. Rather, they tell the physician whether the patient has had a 'big' , 'medium' or 'small' myocardial infarction. They may therefore tell the physician what he already knows. If a patient is in cardiogenic shock or has developed Q waves anterolaterally on the ECG, it does not come as a surprise to find high peak concentrations of biochemical markers in the circulation!

The role of biochemical markers in early risk stratification in ST segment elevation myocardial infarction

Rather than use them for early ***diagnosis***, do biochemical markers have a role as clinical tools for early ***prognosis*** in this patient group at a time point at which we may be able to try different management strategies? Three studies have addressed this issue by examining the prognostic role of admission troponin T presence in patients admitted with ST segment elevation myocardial infarction (1-3). In routine clinical practice, the presence of troponin T at the 0.1mcg/l cut-off was associated with a 300% increase in 30 day cardiac death compared with its absence on admission (Table 4). This early excess in mortality agrees with both the GUSTO II troponin T study and the recently presented GUSTO III troponin T data (ESC Sweden 1997) and appears independent of the thrombolytic agent used. The admission troponin T concentration was superior to the admission creatine kinase activity and CKMB mass concentrations. The hazard continues to increase over time so that, by a median of three years follow up, there is nearly a four fold difference in survival (Figure 1).

Table 4. Cardiac death in STsegment elevation acute coronary syndrome according to admission TnT status

	GUSTO T	GUSTO III	STUBBS	STUBBS
FOLLOW UP	30 days (0.1)	30 days (0.2)	30 days (0.1)	3 years (0.1)
TnT NEG (>0.1mcg/l)	4.7%	6.1%	3.2%	6.4%
TnT POS (≥0.1mcg/L)	13%	15.7%	9.4%	23.4%

NEG: within normal limits
POS: revised levels

This adverse outcome appears to be due to a much lower achievement of TIMI III flow following thrombolysis in the admission troponin T positive group compared to the admission negative group (K.Ramanathan *et al*, AHA 1997). The presence of troponin T on admission, therefore, appears to more accurately define how long the patient has been infarcting for than the chest pain history and may also identify patients having large infarctions. Once it is detectable in the circulation, the optimal time window for a good result from thrombolysis has passed and these patients run a much more complicated clinical course.

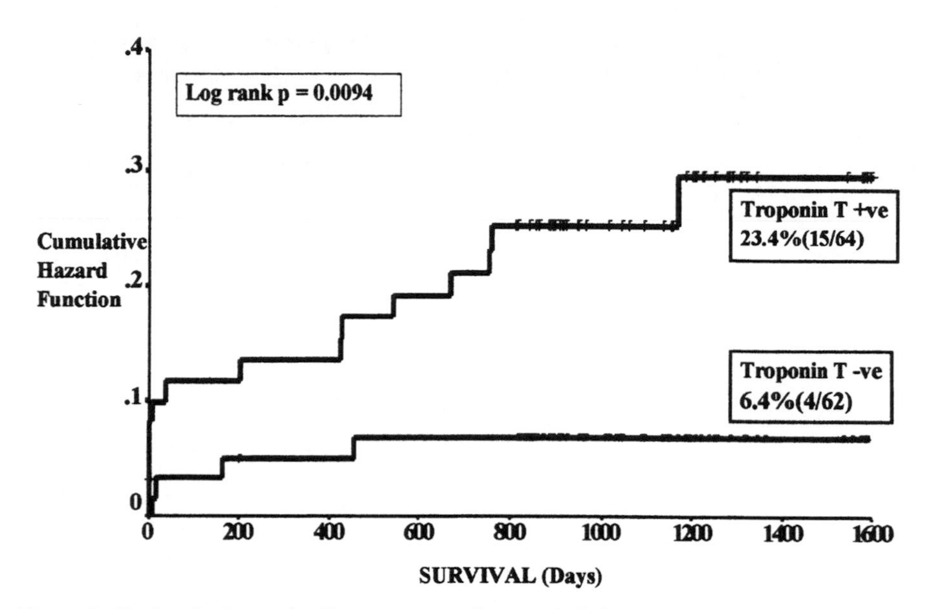

Figure 1. Cardiac death as the first event according to admission troponin T concentration in patients admitted with chest pain with ST segment elevation.

Whether direct PTCA or a combination of different antithrombotic agents can improve the current morbidity in this patient group awaits clinical study. troponin T can now be reliably and rapidly measured at the 0.1mcg/l with the current sensitive rapid assay, the ELECSYS analysers and the CARDIAC READER instruments and this should allow earlier risk stratification of this patient group. In patients with myocardial infarction and ST segment elevation then, troponin T survives the 'so what' question and the answers to the next clinical question 'what do I do differently for this patient group?' awaits clinical studies.

Non-ST segment elevation acute coronary syndromes

This patient group has non-diagnostic ECG's and is usually split retrospectively into a final diagnosis of either myocardial infarction or unstable angina using biochemical marker criteria. Specificity becomes important here to 'rule in' or 'rule out' myocardial damage and troponin T and troponin I have a significant role to play here in identifying minor degrees of myocardial damage in unstable angina patients missed by conventional markers. This role appears to be prognostically important.

There has been a move over recent times to try to risk stratify this patient group earlier. This has been driven, in the main, by drug trialists wishing to enter this patient group into drug trials earlier before the final diagnosis has been reached. These patients are now grouped together into what is called non-ST segment elevation acute coronary syndromes. Can the use of newer biochemical markers in this patient group provide any useful clinical information, again at a time point at which there is the potential to try different management strategies?

The prognostic significance of ADMISSION troponin concentrations in patients admitted with non ST segment acute coronary syndromes

Three studies have addressed the question of the prognostic significance of the admission concentration of biochemical markers in patients with non ST segment elevation acute coronary syndromes (2,4,5) (Table 5).

Table 5. Cardiac death in non-ST segment elevation acute coronary syndromes according to admission troponin status

	GUSTO T	TIMI III	STUBBS	STUBBS
FOLLOW UP	30 days	42 days	30 days	3 years
ASSAY CUT-OFF	(0.1)	(0.4)	(0.1)	(0.1)
TnT NEG	1.0%	----	1.9%	11.3%
TnI NEG	----	1.0%	----	----
TnT POS	9.0%	----	8.0%	25.0%
TnI POS	----	4.0%	----	----

The TIMI IIIb study used the Stratus Dade troponin I ELISA assay with a cut-off of 0.4mcg/l, the other two studies used the Boehringer Mannheim troponin T ELISA assay with a cut-off of 0.1mcg/l. The presence of either marker on admission was associated with an adverse outcome on short term follow up, with the admission troponin T concentration currently the better predictor when tested head to head (Christenen *et al* for the GUSTO II investigators, AHA 1996). The use of Hirudin in the GUSTO II study did not appear to offer any advantage over heparin with regards to outcome in the troponin T positive group. Revascularisation did appear to be beneficial in the TIMI IIIb trial (5) and in the study by Stubbs *et al* (6), for those who were troponin positive. These data are retrospective however.

The prognostic significance of PEAK troponin concentrations in patients admitted with non ST segment acute coronary syndromes

Two large randomised studies have examined this question (7,8). The FRISC study tested low molecular weight heparin and the TRIM study a thrombin inhibitor on outcome in this patient group. Essentially, both studies came to the same conclusion that if troponin T or troponin I is present it identifies high risk subgroups that do worse on short term follow-up than the group who were troponin negative. The TRIM and FRISC trials found mostly concordance with regards to outcome with troponin T and troponin I (Behring assay in TRIM, Sanofi I assay in FRISC), with troponin T being the better predictor of an adverse outcome (9,10). The thrombin inhibitor used in the TRIM trial however did not confer any advantage over heparin in the troponin positive group. The FRISC trial however did show a significant reduction in subsequent cardiac events in those patients who were positive for troponin and received low molecular weight heparin as opposed to placebo. This is the first prospective trial which may provide a useful alternative management strategy for this high risk subgroup (11). Whether early revascularisation is superior to low molecular weight heparin in this subgroup awaits the results of the ongoing FRISC II trial. No data are currently available on whether the newer glycoprotein IIb/IIIa anti-platelet agents may be more beneficial than standard therapy in troponin positive patients with non ST segment elevation acute coronary syndromes.

The role of Biochemical Markers in 'ruling out' myocardial damage

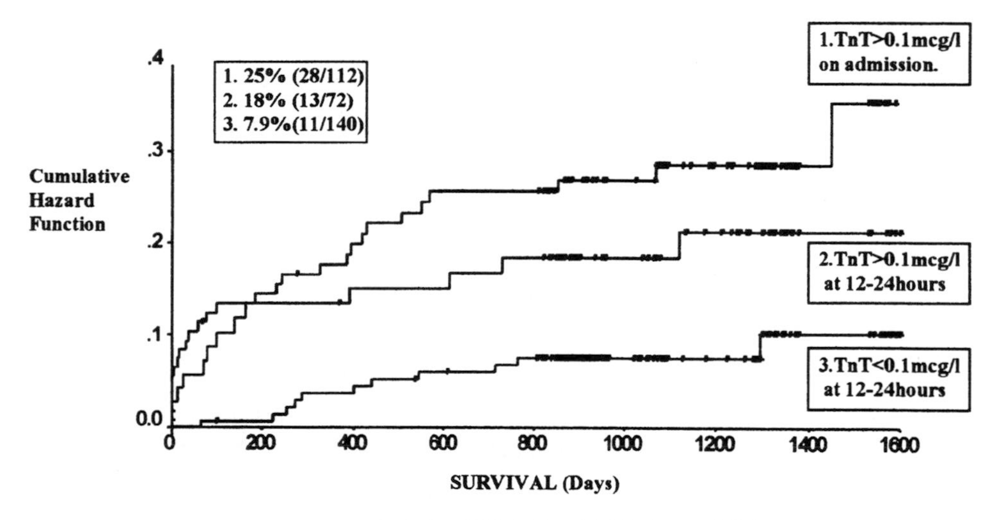

Figure 2. Cardiac death as first event according to timing of troponin T measurement and concentration in patients admitted with chest pain without ST segment evaluation

These newer markers also have a role to lay in 'ruling out' myocardial damage. All biochemical markers are time dependent into the circulation following myocardial injury, with considerable variations in the time to reach maximum efficiency. For troponin T and I, the absence of these markers in the circulation 12-24 hours following admission effectively rules out any myocardial damage. In all studies examining outcome in patients admitted with acute coronary syndromes this group is at low early risk for adverse cardiac events (2,5-8,12-14). As shown in Figure 2 however this risk is not negligible. The findings from published studies however do suggest that the troponin negative chest pain group can be stepped down from the coronary care unit into a lower dependency bed and a more conservative initial strategy practised.

Management algorithms for patients admitted with acute coronary syndromes using newer biochemical markers of myocardial damage

Based on the trial evidence that is currently available it is possible to construct integrated management algorithms using these new markers. Many questions remain unanswered and these have been included at the time points where they are relevant. The problem with any algorithm is that they are difficult to construct, easy to criticise and rapidly go out of date as more data become available. That said, I have tried to make my proposed algorithm clinically credible.

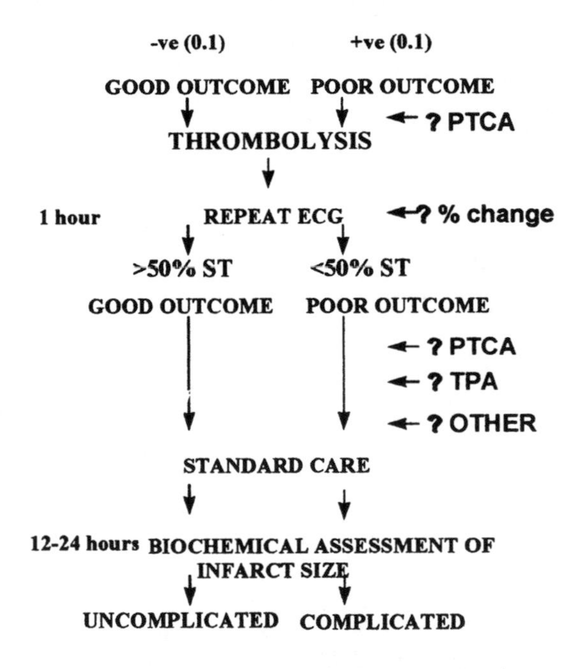

Figure 3. Management algorithm for patients admitted with ST segment elevation myocardial infarction

For patients admitted with ST segment myocardial infarction, the absence of Troponin T in the circulation on admission heralds a good overall outcome both in the short and long term. For those who are Troponin T positive other therapeutic strategies await study. As mentioned earlier, no data are available as to whether rate of change of biochemical markers is superior to a 50% reduction in the ST segment score at 60 minutes after the start of thrombolysis as a predictor of reperfusion. This is the second time point in those who fail to reperfuse that a different strategy may be tried and some of these possibilities are shown and await clinical study. After this time point the physician's routine management is practised.

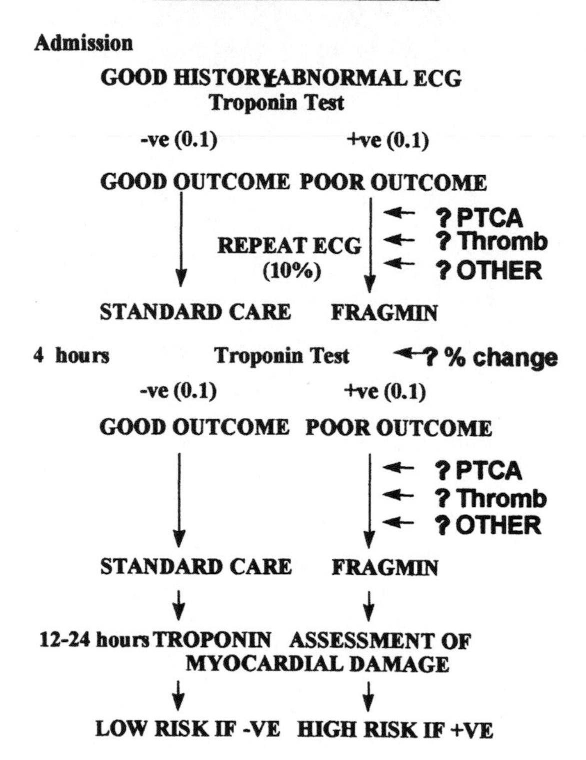

Figure 4. Management algorithm for patients admitted with Non ST segment elevation Acute Coronary Syndromes

In patients admitted with non-ST segment elevation acute coronary syndromes (ACS), in which Troponin T or I is detected, Fragmin should be administered according to the FRISC protocol. Whether the other strategies shown will turn out to be superior awaits study. If the ECG is non diagnostic on admission it tends to remain non-diagnostic, with only some 10% changing to ST elevation. For those patients who are troponin negative on admission but who are positive by 12 hours after admission, release curves show that some 85% will be positive by 4 hours from admission and this provides another timepoint at which myocardial salvage may still be possible if other strategies as shown are studied. If patients are positive for Troponin at this time point or at 12-24 hours from admission they should be switched from Heparin to Fragmin.

The measurement of Troponin T in particular is important at 12-24 hours from admission. At this timepoint if it is present it will identify all patients with myocardial infarction and also those patients with unstable angina who have sustained minor myocardial damage. The absence of the Troponins in the circulation at this time point effectively rules out cardiac damage and identifies a group at low risk of early cardiac events who can be stepped down to a lower dependency bed and treated and investigated in a more conservative manner.

Overall then, these new biochemical markers do give the physician something new to add into their current practice to risk stratify patients admitted with suspected acute coronary syndromes. It has taken 8 years of research so far to reach this point and a number of years are still required to clarify what final impact they will have on the management of patients with ischaemic heart disease.

References

1. Stubbs P, Collinson P, Moseley D *et al.* The prognostic significance of admission Troponin T concentrations in myocardial infarction. Circulation1996;94:1291-7.
2. Ohman E, Armstrong P, Christenson R *et al.* Cardiac Troponin T levels for risk stratification in acute myocardial ischaemia. N Engl J Med 1996;335:1333-41.
3. Ohman M. GUSTO III symposium. XIXth Congress of the European Society of Cardiology. Stockholm 1997.
4. Stubbs P. Troponin T and myocardial damage. MD thesis University of London 1995.
5. Antman E, Milenko M, Tanasijevic J *et al.* Cardiac-specific Troponin I levels to predict the risk of mortality in patients with acute coronary syndromes. N Engl J Med 1996;335:1342-9.
6. Stubbs P, Collinson P, Moseley D *et al.* Prospective study of the role of Troponin T in patients admitted with unstable angina. BMJ 1996, 313:262-264.
7. Lindahl B, Venge P, Wallentin L, for the FRISC Study Group. Relation between troponin T and the risk of subsequent cardiac events in unstable coronary disease. Circulation 1996;93:1651-7.
8. Luscher M, Ravkilde J, Thygesen K, Heickendorff L and the TRIM study group. Troponin T and Troponin I in detection of myocardial damage and subsequent cardiac events in 516 consecutive chest pain patients: A sub-study in the TRIM study. Eur Heart J 1996;17(Suppl):31
9. Lindahl B, Toss H, Venge P, Wallentin L for the FRISC investigators. Comparison of Troponin T and Troponin I for prediction of mortality in unstable coronary artery disease (Abs).Eur Heart J 1997;18:123.
10. Luscher M, Ravkilde J, Thygesen K *et al* on behalf of the TRIM Study group. Concordance between Troponin T and Troponin I values in 516 patients with unstable coronary disease. A TRIM sub-study (Abs).Eur Heart J 1997;18:638.
11. Lindahl B, Venge P, Wallentin L, for the FRISC Study Group. Troponin T identifies patients with unstable coronary artery disease who benefit from long-term antithrombotic protection. J Am Coll Cardiol 1997;29:43-8.
12. Hamm CW, Ravkilde J, Gerhardt W *et al.* The prognostic value of serum Troponin T in unstable angina. N Engl J Med 1992;327:146-50.
13. Collinson PO, Stubbs P. The prognostic value of serum Troponin T in unstable angina. N Engl J Med 1992;327:1760-1.
14. Ravkilde J, Horder M, Gerhardt W *et al.* Diagnostic performance and prognostic value of serum Troponin T in suspected acute myocardial infarction. Scand J Clin Lab Invest 1993;53:677-85.

Chapter 12

NEW SOLUBLE MARKERS FOR ASSESSMENT OF INFARCT SIZE

Naab M. Al-Saady and A. John Camm

It has been established both experimentally [1,2] and clinically [3,4] that the major determinant of acute and long-term prognosis following myocardial infarction is the extent of myocardial damage. It is useful, for risk stratification purposes, to estimate infarct size as early as possible, because infarct size correlates closely with mortality [5] and other prognostic indices such as cardiac failure [6], arrhythmia, and poor ventricular function. [7].

In the early 1970s there was a concerted effort to quantitate infarct size using a variety of techniques, such as electrocardiography [8], echocardiography, left ventriculography, radionuclide methods [9,10,5], or the kinetics of cardiac enzymes [11,5]. Clinically, the 12-lead ECG is convenient for estimating infarct size. However, because ST segment elevation and Q-wave amplitudes change spontaneously within 24 hours of the onset of acute myocardial infarction, and change rapidly after re-perfusion, [12,13,14] the estimation of infarct size from the ECG at the very early stage of myocardial infarction is unreliable. Bundle branch block may be present on the initial electrocardiograph and so disguise the site of infarction making the assessment of the ST segment unreliable.

Stunned myocardium has been reported in patients with successful re-perfusion of infarct related areas. Thus even in the presence of viable myocardium ventricular contractility may remain depressed for several days [15]. Therefore, the estimation of infarct size by echocardiography or left ventriculography at a very early stage is also of doubtful accuracy.

Myocardial ischaemia, if not reversed, leads to a chain of cellular events ultimately ending with disruption of cell membranes and cell death as discussed by Gary Baxter in chapter 1 of this book. When the cell membranes have lost their cellular integrity, cellular components diffuse into the interstitium and, from there, into the capillaries (70-90%) and lymphatics (10-30%) [16,17]. Whether the release of cytosolic markers always indicates cell death is controversial [18]. There are several reports indicating that release may also occur from reversibly damaged cells [19-21]. This topic is further discussed by W. Hermens in chapter 8 of this book. That persistent release of structurally bound markers such as the troponins, due to irreversible cell damage, is less controversial [22].

Enzymatic estimates of infarct size based on the assessment of their total release into the circulation [23] emerged as a technique to establish more accurately the amount of myocardial damage that occurs during myocardial infarction. CKMB was shown to correlate closely with clinical manifestations, impairment of ventricular function, [7] the incidence of ventricular arrhythmias, the incidence of heart failure [24], acute mortality rates and long-term mortality rate [3,25]. Enzymatic estimates of infarct size have been validated in animals [1,2]. Later the precision of enzymatic estimates was confirmed by post-mortem studies in humans showing a close correlation between histologic and enzymatic (CK-MB) estimates of infarct size [27].

Enzymatic estimates of infarct size were used experimentally and clinically to assess a variety of interventions designed to salvage ischaemic myocardium [28-32]. However, with the advent of thrombolytic therapy to induce re-perfusion, it was observed that there was a more rapid washout of CK, and that the ratio of CK released into the circulation to that depleted from the myocardium was altered Thus estimates of infarct size in the clinical setting may not be reliable [32]. In addition, the limitations of CKMB measurement, such as delayed response and short duration of elevation following a myocardial infarction (MI), as well as lack of specificity, stimulated the search for a more suitable diagnostic marker. In recent years efforts were directed towards the development sensitive, non-invasive, methods for sizing myocardial infarction.

Infarct size

The accurate determination of infarct size and the volume of myocardium at risk at the time of initial insult has important therapeutic and prognostic implications [33]. Myocardial ischaemia first appears in the sub-endocardial region. Initially, the ischaemic cells are reversibly injured and if these initial changes are not promptly reversed, they will result in the death of the myocytes. Thereafter, irreversible myocyte necrosis travels transmurally towards the epicardium in the form of a wave front [34]. Progress in the field of thrombolytic therapy has significantly affected management of patients with acute myocardial infarction (AMI) [35]. A timely and successful re-perfusion during an evolving infarction often reduces cell death and thus prevents immediate mortality, reduces the risk of larger infarction and preserves left ventricular function.

Quantitative histologic estimates of infarct size are regarded as the gold standard but obviously this technique has little clinical relevance. Attempts to quantitate infarct size attracted investigators from multiple disciplines and a variety of techniques have been proposed. The familiar triad, clinical history, ECG and serum enzyme assessment, is still of utmost significance in the diagnosis of myocardial infarction [36]. However, these criteria provide no precise quantitative information regarding the extent and location of the infarct. Efforts to relate infarct size to serum level changes of marker substances released in blood from necrosing myocytes have met with little success. In recent years, however, major advances have been made in the methods of enzymatic quantification based on immunological techniques to improve the detection and estimation of myocardial infarction. These methods are based on the production and availability of specific antibodies against intracellular, cardiac specific substances. The membranes of irreversibly injured and necrosing myocardial cells lose their property of semi-permeability and hence a number of intracellular cytoplasmic components become released into the circulation. The rise in serum concentration, as well as the activities of these substances, form the basis of the presently available biochemical tests for the diagnosis of MI.

Biochemical Markers

Myoglobin

As extensively described by J. Mair in chapter 5, myoglobin (Mb) is a cytosolic oxygen binding protein found in skeletal and heart muscle, constituting 2% of the total muscle proteins. It is cleared from the circulation by the kidneys [37]. Myoglobin has a low molecular weight and following muscle damage it is rapidly released into the blood stream. Myoglobin is often detectable 2 hours after the onset of coronary occlusion and peaks earlier than CK at 3-15 h [38]. Myoglobin is not specific for cardiac muscle because injury to skeletal muscle, and even strenuous exercise, can lead to the release of measurable amounts of myoglobin into the circulation [39]. Myoglobin levels are not different in normal human ventricles [40].

Several clinical studies have estimated infarct size from serial plasma Mb measurements. Stone *et al* [41] were the first to observe a significant correlation between peak Mb levels and the histological estimation of infarct size in an animal model, following persistent coronary artery occlusion. Maddison *et al* [42] collected blood samples at 4-hour intervals in 29 patients and showed a significant correlation between total Mb and total CKMB. Tommaso *et al* [43] demonstrated a good correlation between the area under the Mb concentration-time plot and total CK by blood sampling at 3-hour intervals in 8 patients. Groth *et al* [44] collected blood sample in 33 patients with AMI at 1-2 hour intervals after admission and reported a significant correlation between total Mb, calculated using a two-compartment model, and total CKMB. However, there were several problems with these clinical studies. Amongst these was the perfusion status of the infarct-related artery, which was not documented in these studies. This is important as they might have included patients with spontaneous reperfusion or treatment-related reperfusion. Several reports have suggested that reperfusion increases the release of CK from the infarcted area into the blood. A similar phenomenon may occur with Mb, therefore the infarct size determination by these enzymes may be an over estimation compared with the histological infarct size [45-47]. Another problem is the long intervals used in previous studies for blood sampling. However, recently in a study by Yamashita *et al* [48] most of these problems were addressed. They confirmed the perfusion status of the infarct-related artery by serial coronary angiography and accurately evaluated the peak levels and *K*d (exponential disappearance rate constant) of Mb from blood samples drawn at 15-minute intervals. In addition, they evaluated the severity of regional hypokinesis from follow-up ventriculograms and compared total Mb not only with total CK but also with the severity of regional hypokinesis. These authors found that infarct size could be estimated accurately 4 hours after reperfusion by calculating the total Mb.

The Troponins

The biochemistry and clinical significance of these markers have been addressed in other chapters of this book (see chapters 3, 4, and 11). Troponin is a thin-filament associated complex of the myocyte. There are three individual components in the complex, each of a single polypeptide chain. Troponin I, C, T form a complex that regulates the calcium-modulated interaction of actin and myosin in striated muscle. The different functions of the troponin subunits of T, C and I are associated with distinct amino acid sequences encoded by separate genes [49].

Troponin T: This is a contractile protein subunit of the myofibrillar complex that is released into the circulation when myocardial cell membranes are disrupted [50]. CTnT is compartmented in the contractile apparatus and has to dissociate from the troponin complex of thin filament to be released into the serum. This release process is slow and troponin-T is present in the serum for more than 120 h after myocardial infarction. Levels are detected in serum less than 4 h after the onset of infarction. The test is very sensitive and elevated levels can be detected in patients without diagnostic changes on the electrocardiogram or in CK levels. Serum cTnT release shows a biphasic response with peaks on the 1st. and 5th. days post coronary artery occlusion [50]. This response is thought to be caused by the partitioning of the protein into cytosolic and myofibril-bound fractions. The first peak, occurring within 24 hours of coronary artery occlusion, may be the result of the release of the minor cytosolic cTnT pool, although the plasma half-life is only 2 hours [50].

CTnT has been investigated extensively and found to be a sensitive marker of myocardial necrosis [50,52,53]. However, its specificity has not been fully defined and recent observations showed that increased plasma cTnT levels could be found in the absence of evidence of myocardial involvement in patients with polymyositis.[54].

Troponin-T release is not dependent on perfusion and serial measurements may provide an accurate assessment of infarct size [51].

Two recent studies have examined the correlation of serum cTnT to infarct size in humans using single-proton-emission computerised tomography imaging. Wagner *et al* [55] demonstrated correlation (r=0.73) using single-proton-emission computerised tomography / Tc-sestamibi estimates of infarct size and peak serum concentrations of cTnT and CKMB. However, they were unable to demonstrate a good correlation between the infarct size and the cTnT area under the curve (AUC) (r=0.54), possibly because they included the cytosolic cTnT peak in their area calculations. Omura *et al* [56] observed a similar correlation in humans using the second cTnT peak concentration in single-proton-emission computerised tomography/thallium estimates of infarct size (r=0.77). Serum cTnT was used for the estimation of infarct size in dogs [57]. Voss *et al* [57] found that infarct sizing in dog hearts initially did not correlate with serum cTnT or CKMB concentrations. However, when the data were separated by infarct location (e.g. right coronary artery; left circumflex coronary artery), the correlations improved dramatically, probably indicating different regional distribution of the marker.

There are clear differences in the distribution of both cTnT and CKMB in the human heart and the dog heart, being more uniform in the former.

Troponin I: (cTnI) is similar to cTnT, but it is not expressed in skeletal muscle and is not found in skeletal muscle during neonatal development or during adulthood, even after acute or chronic injury of skeletal muscle [58-60]. This makes cTnI unique among molecular markers for the detection of myocardial necrosis in these complex clinical situations. Its specificity for myocardium is very high [39, 61-63], yet its sensitivity for cardiac injury appears comparable to that of CKMB [59]. There is a 13-fold greater concentration of CTnI than CKMB in the heart [64].

CTnI, like cTnT, also exhibits a biphasic release profile with serum level elevations within 4-6 h reaching mean peak levels of 112 ng/ml (range from 20-550 ng/ml) at 18 hour and remaining above normal for up to 6-8 days post-infarction [59]. Moreover, cTnI levels in patients with chest pain were found to be unchanged or slightly elevated while they were always normal in patients with chest pain of non-cardiac origin. Recent publications showed that cTnI is not affected by hypothyroidism [65] or uraemia [66,67]. Thus, when cTnI is detectable in peripheral venous samples, the physician can confidently diagnose the presence of a cardiac insult.

Recently Mair *et al* [68] used cTnI to estimate myocardial infarct size. CTnI, creatine kinase, and creatine kinase MB activity were studied in 15 patients with first Q wave AMI. All patients received intravenous thrombolytic therapy. Scintigraphic estimates of myocardial infarction was carried out 5 weeks after MI. CTnI increased and peaked in parallel with CKMB activity, and the peak values correlated with each other ($r=0.76$, $p=0.002$). Significant correlations were also found between scintigraphic estimates of infarct size and area under cTnI curves ($r=0.53$, $p=0.042$).

Tumour necrosis factor - α (TNF-α)

As stated previously, estimation of infarct extent has been usually based on serum enzyme evaluation during the acute phase of infarction. However, no strong correlation has been identified which may have practical clinical use. The release of enzymes are likely to be influenced by the degree of reperfusion and the location of infarction [69-72].

Cytokines have pronounced effects on the cardiovascular system both locally and systemically, including promotion of inflammation, intravascular coagulation, cell adhesion, free radical generation, endothelial injury and possible progression of coronary atherosclerosis [73,74]. Following an MI several of these cytokines are released, particularly interleukins and TNF-α [75-78]. Although cytokines are present in many different cells, it has been suggested that their release following an ischaemic event is mainly caused by activation of monocytes [77]. TNF-α is a cytokine which enhances procoagulant activity in endothelial cells, activates neutrophils, stimulates fibroblast growth and suppresses lipoprotein lipase. TNF-α is a hormone-like polypeptide with a variety of activities involved in the defence against pathogenic micro-organisms and in the process of tissue repair [79]. In-vitro data reveal significantly higher release of TNF-α from macrophages in patients with stable or unstable angina compared to patients without evidence of coronary artery disease [75]. In an animal model it was shown that, after an ischaemic event, macrophages migrate into the ischaemic myocardium and release TNF-α locally [80]. The number of cells migrating into the myocardium increases until the 3rd day after vessel occlusion and remains constant over the next weeks. There is also clinical evidence of a significant increase of TNF-α in-patients with AMI [73,81]. These experimental and clinical data suggest that after an ischaemic event the macrophages that migrate into the myocardium are the source for the increased levels of TNF-α observed in patients with AMI. However, recent experimental data suggest that some of the TNF-α released during myocardial ischaemia is produced by the myocytes [82,83]. This is in contrast to clinical data illustrating that macrophages are the main source for the release of TNF-α into the coronary sinus of the heart allograft [84]. Therefore, the exact cellular source of TNF-α in patients with AMI remains controversial. In addition, TNF-α may also be produced by activated macrophages circulating in peripheral blood in ischaemia/reperfusion situations. However, there is evidence of increased surface expression of adhesion molecules, a sign of activation of monocytes and neutrophils, in samples taken from the aorta of patients with unstable coronary artery disease when compared with simultaneous control samples [85]. Therefore, this evidence suggests that TNF-α is produced either by cells invading the myocardium or by resident cells.

Hirschl *et al* [86] measured TNF-α in 50 patients following AMI. Their results demonstrated that the extent of changes in serum TNF-α concentration is significantly related to estimates of infarct size obtained scintigraphically. Increased TNF-α levels also correlated well with other parameters such as the occurrence of heart failure, the presence of rhythm disturbances, enzymatic estimated infarct size (alpha-hydroxybutyrate-dehydrogenase) and wall motion index obtained echocardiographically. Their data [86] revealed that Δ TNF-α may be reliable method of assessing infarct size. Furthermore, there was no influence of infarct location on the relation between Δ(TNF) and infarct size. This would offer a potential assessment advantage in the clinical setting but further studies are necessary to confirm these findings.

Fatty acid-binding protein

As discussed by J. Glatz in chapter 7, fatty acid-binding protein (FABP) is a relatively small protein that is abundantly present in the soluble cytoplasm of almost all tissue cells, including those from cardiac and skeletal muscles [87]. FABP is released following muscle injury from the damaged myocardial cells into blood plasma, in animal models of MI [88-90] and in patients with AMI [91-93]. FABP release curve closely resembles that of myoglobin. Both reach peak plasma concentrations within 4 h of AMI and return to normal within 24 h of AMI in patients treated with thrombolytics [94] and about 8 h up to 24-36 hours in untreated patients [94]. Despite this early release, only levels assessed at 72 h yield a comparable estimate of the extent of myocardial damage. Therefore, with frequent blood sampling and in the absence of renal failure, FABP could give a clinically useful estimate of myocardial infarct size.

Gylcogen Phosphorylase BB

Glycogen phosphorylase (GP) (see chapter 6) is an important enzyme in glycogenolysis and is normally bound to glycogen in the sarcoplasmic reticulum- glycogenolysis complex. Glycogen phosphorylase is known to possess 3 major isoenzymes: brain (BB), Liver (LL), & muscle (MM). They can be distinguished by functional, structural and immunological properties.

The first report that blood GP increases early after the onset of AMI, before CK increases, were by Krause *et al* [95] who used relatively insensitive enzymatic assays. Later Rabitzsch *et al* [96] measured GPBB isoenzyme in 27 patients with AMI. They found that GPBB increased within a few hours of the onset of MI, peaked significantly earlier ($P<0.034$) than CK and CKMB activity and returned to normal between 24 hour and 36 hour after the onset of MI. The peak values were observed at approximately 15 hours after the onset of MI [96]. The time course of GPBB in patients with AMI was markedly influenced by early reperfusion of the infarct-related coronary artery, in a similar fashion to other soluble markers, such as myoglobin and CKMB [97]. There is no report of the use of GPBB in the assessment of myocardial infarction size.

Conclusion

The evolution of myocardial infarction is a dynamic process and its rate of progression, as well as the ultimate extent of damage, can be modified favourably. It is generally accepted that ischaemic damage of cardiomyoctes increases the permeability of their cellular membranes and those of cellular organelles, and facilitates the release of cellular constituents into the circulation. The greater the cellular damage, the larger the molecules that can be released.

Cardiac enzymes have been used to estimate infarct size by measurement of cumulative release of the enzymes. Their estimates have been shown to correlate with histologic estimates of infarct size but this approach has limitations. The calculation is complicated and based on several assumptions, and may be affected by reperfusion therapy. The peak serum concentration of cardiac enzymes provides a crude estimate of infarct size but is even more vulnerable to interference by thrombolytic therapy.

The compounding effect of reperfusion is less marked with the new structurally bound markers such as cardiac isoforms of troponin T and I. The late peak of cTnT day 3-5 correlates well with infarct size with scintigraphy irrespective of reperfusion status.

For the early prediction of infarct size, no method has proved reliable in clinical evaluation. Attempts have been made to use the initial slope of the time curve for myoglobin and CK, as they are released faster than other constituents.

Using cardiac enzymes to estimate infarct size is also limited by variation in enzyme release independent of therapy. The kinetics of enzyme release may be affected by enzyme inactivation in living cells or by their retention in tissue which is necrosed and not reperfused.

These limitations in the use of cardiac enzymes in the assessment of myocardial infarct size have led to a search for alternatives. Novel candidates include structural proteins and cytokines, particularly TNF-α. Estimates of infarct size by this method have correlated closely with the results of scintigraphy. We do not yet posses enough data to predict the clinical usefulness of these markers. Early prediction of infarct size could have an important impact on the choice of therapy in acute MI. Current methods are imperfect and therefore the search for the ideal biochemical marker continues.

References

1. Kjekshus JK, Sobel BE. Depressed myocardial creatine phosphokinase activity following experimental myocardial infarction in rabbit. Circ Res *et al* 1970;27:403-414.
2. Shell WE, Kjekshus JK, Sobell BE. Quantitative assessment of the extent of myocardial infarction in conscious dog by means of analysis of serial changes in serum creatine phosphokinase activity. J Clin Invest 1970;50:2614-2625.
3. Sobell BE, Bresnahan GF, Shell WE, Yoder RD. Estimation of infarct size in man and its relation to prognosis. Circulation 1972;46:640-648.
4. Roberts R, Husain A, Ambos HD, Oliver GC, Cox JR Jr, Sobel BE. Relation between infarct size and ventricular arrhythmia. Br Heart J 1975;37:1169-1175.
5. Geltman EM, Ehsani AA, Campbell MK, Schechtman K, Roberts R, Sobel BE. The influence of the location and extent of myocardial infarction on long-term ventricular dysrhythmia and mortality. Circulation 1979;60:805-814.
6. Khan JC, Gueret P, Menier R, Giraudet P, Farhat MB, Bourdarias JP. Prognostic value of enzymatic (CPK) estimation of infarct size. J Mol Med 1977;2:223-231.
7. Rogers WJ, McDaniel HG, Smith LR, Mantle JA, Russel RO Jr, Rackley CE. Correlation of angiographic estimates of myocardial infarct size and accumulated release of creatine kinase MB isoenzymes in man. Circulation 1977;56:199-205.
8. Maroko PR, Libby P, Covell JW, Sobel BE, Ross J Jr, Braunwald E. Precordial S-T segment elevation maping: an atraumatic method for assessing alterations in the extent of myocardial ischemic injury. The effects of pharmacologic and hemodynamic interventions. Am J Cardiol 1972;29:223-230.
9. Geltman EM, , Roberts R, Sobel BE. Cardiac Positron tomography: Current status and future directions. herz 1980;5:107-119.
10. Sobel BE, Kjekshus KL, Roberts. Enzymatic estimation of infarct size. In: Hearse DJ, De Leiris (eds). Enzymes in Cardiology: Diagnosis and Research. Chichester, John Wiley and sons 1979;257-289.
11. Roberts R. Creatine kinase isoenzymes as a diagnostic and prognostic indices of myocardial infarction. In: Rattazzi MC, Scandalios JG, Whitt GS (ed): *Isozymes: Current Topics in Biological and Medical Research*. New Yourk, Alan R. Liss 1979;115-154.
12. Selwyn AP, Ogunro EA, Shillingford JP. Natural history and evaluation of ST segment changes and MB CK release in acute myocardial infarction. Br Heart J 1977;39:988-994.
13. Essen RV, Merx W, Effert S. Spontaneous course of ST-segment elevation in acute anterior myocardial infarction. Circulation 1979;59:105-112.
14. Hackworthy RA, Sorensen SG, Fitzpatrick PG, Barry WH, Menlove RL, Rothbard RL, Anderson JL, for the APSAC Investigators. Effect of reperfusion on electrocardiographic and enzymatic infarct size: results of a randomized multicenter study of intravenous anisolyated plasminogen streptokinase activator complex (APSAC) versus intracoronary streptokinase in acute myocardial infarction. Am Heart J 1988;116:903-914.
15. Braunwald E, Kloner RA. The Stunned myocardium: Prolonged postischemic ventricular dysfunction. Circulation 1982;66:1146-1149.
16. Spieckermann PG, Nordbeck H, Hearse DJ, ed. Enzymes in cardiology. Chichester: John Wiley & Sons, 1979;81-95.
17. van Kreel B, van der Veen FH, Willems GM, Hermens WT. Circulatory models in assessment of cardiac enzyme release in dogs. Am J Physiol 1993;254:H747-H754.
18. Adams JE, Bodor GS, Davila-Roman VG, Delmez JA, Apple FS, Ladenson, JH, Jaffe AS. Cardiac troponin I a marker with high specificity for cardiac injury. Circulation 1993;88:101-106.
19. Piper HM, Schwartz P, Spahr R, Hütter JF, Spieckermann PG. Early enzyme release from myocardial cells is not due to irreversible cell damage. J Mol Cell Cardiol 1984; 16: 385-8.
20. Heyndrickx GR, Amano J, Kenna T, Fallon JT, Patrick TA, Manders WT, Rogers GG, Rosendorff C, Vatner SF. Creatine kinase release not associated with myocardial necrosis after short periods of coronary artery occlusion in conscious baboons. J Am Coll Cardiol 1985;6:1299-1303.
21. Remppis A, Scheffold T, Greten J *et al.* Intracellular compartmentation of troponin T: Release kinetics after global ischemia and calcium paradox in the isolated perfused rat heart. J Mol Cell Cardiol 1995;27:793-803.
22. Mair J, Diensti F, Puschendorf B. Cardiac troponin T in the diagnosis of myocardial injury. Crit Rev Clin Lab Sci 1992;29:31-57.

23. Roberts R. Measurement of enzymes in cardiology. In Linden RJ (ed). Tchniques in the Life Sciences, Volume P3/I- Techniques in Cardiovascular Physiology: Part I. Ireland, Elsevier Scientific Publishers, 1983, pp P312/1-P312/24.
24. Bleifeld W, Mathy D, Hanrath P, Buss H, Effert S. Infarct size estimated from serial serum creatine phosphokinase in relation to left ventricular dynamics. Circulation 1977;55:303-311.
25. Gillespie IA, Sobel BE. A rationale for therapy of acute myocardial infarction: Limitation of infarct size. In: Stollerman GH (ed): *Advances in Internal medicine*. Chicago, Year Book Medical Publishers, 1977.
26. Kjekshus JK, Sobel BE. Depressed myocardial ceartine phosphokinase activity following experimental myocardial infarction in rabbit. Circ Res 1970;27:403-414.
27. Hackel DB, Reimar KA, Ideker RE, Mikat EM, Hartwell TD, Parker CB, Braunwald EB, Buja M, Gold HK, Jaffe AS, Muller JE, Raabe DS, Rude RE, Sobel BE, Stone PH, Robert R, MILIS Study Group. Comparison of enzymatic and anatomic estimates of myocardial infarction size in man. Circulation 1984;70:824-835.
28. Fukuyama T, Schechtman KB, Roberts R. The effects of intravenous nitroglycerin on hemodynamics, coronary blood flow and morphologically and enzymatically estimated infarct size in conscious dogs. Circulation 1980;62:1227-1238.
29. Mogelson S, Davidson J, Sobel BE, Roberts R. The effects of hyperbaric oxygen on infarct size in the conscious animal. Eur J Cardiol 1980;12:135-146.
30. Clark RE, Christlieb IY, Ferguson TB, Weldon CS, Marbarger JP, Sobell BE, *et al.* Laboratory and initial clinical studies of nifedipine, a calcium antagonist for improved myocardial preservation. Ann Surg 1981;193:719-732.
31. Jaffe AS, Geltman EM, Tiefenbrunn AJ, Ambos HD, Strauss HD, Sobel BE, *et al* Reduction of infarct size in patients with iferior infarction with intravenous nitroglycerin. Br Heart J 1983;49:452-460.
32. Roberts R, Croft C, Gold HK, Hartwell TD, Jaffe AS, Muller JE, *et al* Effect of propranolol on myocardial infarction size in a randomized, blinded, multicenter trial. N Engl J Med 1985;312:932-936.
33. Haider Kh H, Stimson WH. Cardiac troponin I: a biochemical marker of cardiac cell necrosis. Disease Markers 1993;11:205-215.
34. Alpert JS. The pathophysiology of acute myocardial infarction. Cardiology 1989;76:85-95.
35. Verstraete M. Advances in thrombolytic therapy. Cardiovascular Drugs & Therapy 1992;6:111-124.
36. Willerson JT. Clinical diagnosis of acute myocardial infarction. Hospital Practice 1989;24:65-77.
37. Klocke FJ, Cpley DP, Krawczyk JA, Reichlin M. Rapid renal clearance of immunoreactive canine plasma myoglobin. circulation1982;65:1522.
38. McComb JM, McMaster EA, MacKenzie G, Adgey AA. Myoglobin and creatine kinase in acute myocardial infarction. Br Heart J 1984;51:189-194.
39. Larue C, Calzolari C, Bertinchant JP, Leclercq F, Grolleau R, Pau B. Cardiac-specific immunoenzymometic assay of troponin I in the early phase of acute myocardial infarction. Clin Chem1993;39:972-979.
40. Lin L, Sylven C, Sotonyi P, Kaijser L, Jansson E. Myoglobin content and citrate synthase activity in different parts of the normal human heart. J Appl Physiol J Appl Physiol 1990;69:899-901.
41. Stone MJ, Waterman MR, Poliner LR, Templeton GH, Buja LM, Willerson JT. Myoglobinemia in an early and quantitative index of acute myocardial infarction. Angiology, 1978;29:386-392.
42. Maddison A, Craig A, Yusuf S, Lopez R, Sleight P. the role of serum myoglobin in the detection and assessment of myocardial infarction. Clin Chim Acta 1980;106:17-28.
43. Tommaso CL, Salzeider K, Arif M, Klutz W. Serial myoglobin vs CPK analysis as an indicator of uncomplicated myocardial infarction size and its use in assessing early infarct extension. Am Heart J 1980;99:149-154.
44. Groth T, Hakman M, Sylven C. Size estimation of acute myocardial infarction from serial serum myoglobin observations with due consideration of individual differences in basic kinetics. Scand J Clin Lab Invest 1984;44:65-78.
45. Jarmakani JM, Limbird L, Graham TC, Marks RA. Effect of reperfusion on myocardial infarct, and the accuracy of estimating infarct size from serum creatine phosphokinase in the dog. Cardiovasc Res 1976;10:245-253.
46. Vatner SF, Baig H, manders WT, Maroko PR. Effects of coronary artery reperfusion on myocardial infarct size calculated from creatine kinase. J Clin Invest 1978;61:1048-1056.
47. Roberts R. Enzymatic estimation of infarct size. Thrombolysis induced its demise: will it now rekindle its renaisance? Circulation 1990;81:707-710.

48. Yamashita T, Abe S, Arima S, Nomoto K, Miyata M, Maruyama I, Toda H, Okino H, Atsuchi Y, Tahara M *et al.* Myocardial infarction size can be estimated from serial plasma myoglobin measurements within 4 hours of reperfusion. Circulation 1993;87:1840-1849.
49. Bucher EA, Maisonpierre PC, Konieczny SF, Emerson CP. Expression of the troponin complex genes: transcriptional coactivation during myoblast differentiation and independent control in heart and skeletal muscles. Mol Cell Biol 1988;8:4134-4142.
50. Katus HA, Remppis A, Neumann FJ, Scheffold T, Diederich KW, Vinar G, *et al.* Diagnostic efficiency of toroponin T measurements in acute myocardial infarction. Circulation 1991;83:902-912.
51. Katus HA, Diederich KW, Schwartz F, Vellner M, Scheffold T, Kubler W. Influence of reperfusion on serum concentrations of cytosolic creatinine kinase and structural myosin light chains in acute myocardial infarction. Am J Cordial 1987;60:440.
52. Katus HA, Looser S, Hallermayer K, Remppis A Scheffold T, Borgya A, *et al* . Development and in vito characterization of a new immunoassay of cardiac troponin T. Clin Chem 1992;38:386-393.
53. Mair J, Artner-Dworzak E, Lechleitner P, Smidt J, Wagner I, Dienstl F, Puschendorf B. Cardiac troponin T immunoassay for diagnosis of acute myocardial infarction. Clin Chem 1991;37:845-852.
54. Kobayashi S, tanaka M, Tamura N, Hashimoto H, Hirose S. Lancet. Serum cardiac troponin T in polymyositis/dermatomyositis. 1992;340:726 (letter).
55. Wagner I, Mair J, Fridrich L, *et al.* Cardiac troponin T release, in acute myocardial infarction is associated with scintigraphic estimates of myocardial scar. Coron Artery Dis 1993;4:537-544.
56. Omura T, Teragaki M, Tani T, Yamagishi H, Yanagi S, Nishikimi T, *et al.* Estimation of infarction size using serum troponin T concentration in patitients with acute myocardial infarction. Jpn Circ J 1993;57:1062-1070.
57. Voss EM, Sharkey SW, Gernet AE, Murakama MM, Johnston RB, Hsieh CC, Apple FS. Human and canine cardiac troponin T and creatine kinase-MB distribution in normal and diseased myocardium. Infarct size using serum profiles. Arch Pathol Lab Med 1995;119:799-806.
58. Toyota N, Shimada Y. Differentiation of troponin in cardiac and skeletal muscles in chicken embryos as studied by immunofluorescence microscopy. J Cell Biol 1981; 91:497-504.
59. Cummins B, Auckland ML, Cummins P. Cardiac-specific troponin-I radioimmunoassay in the diagnosis of acute myocardial infarction. Am Heart J 1987;113:1333.
60. Martin AF, Orlowski J. Molecular cloning and developmental expression of the rat cardiac-specific isoform of troponin I. J Mol Cell Cardiol 1991;23:583-588.
61. Mair J, Wieser C, Seibt I, Arther-Dworzak E, Futwangler W, Waldenberger F, Balough D, Puschendorf B. Troponin T to diagbose myocardial infarction in bypass surgery. lancet 1991;337:434-435 (letter).
62. Mair J, Wagner I, Puschendorf B, Lechleitner P, Dienstl F, Calzolari C, Larue C. Cardiac troponin I to diagnose myocardial injury.Lancet 1993;341:838-839.
63. Adams JE, Bodor GS, Davila-Toman VG, Delmez JA, Apple FS, Ladenson JH, Jaffe AS. Cardiac troponin I. A marker with high specificity for cardiac injury. Circulation 1993;88:101-106.
64. Guest TM, Ramanathan AV, Tuteur PG, Schechtman KB, Ladenson JH, Jaffe AS. Myocardial injury in critically ill patients. A frequently unrecognized complication. JAMA 1995;273:1945-1949.
65. Cohen LF, Mohabeer AJ, Keffer JH, Jialal I. Troponin I in hypothyroidism. Clin Chem 1996;42:1494-1495.
66. Trinquier S, Flecheux O, Bullenger M, Castex F. Highly specific immunoassay for cardiac troponin I assessed in noninfarct patients with chronic renal failure or severe polytrauma. Clin Chem 1995;41:1675-1676 (letter).
67. Lofberg M, Tahtela R, Harkonen M, Somer H. Cardiac troponins in severe rahbdomyolysis. Clin Chem 1996;42:1120-1121
68. Mair J, Wagner I, Morass B, Fridrich L, Lechleitner P, Dienstl F, Calzolari C, Larue C, Puschendorf B. Cardiac troponin I release correlates with myocardial infarction size. Eur J Clin Chem Clin Biochem 1995;33:869-872.
69. Morrison J, Coromilas J, Munsey D, robbins M, Zema M, Chiaramida SU, Reiser P, Scherr L. Correlation of radionucleotide estimates of myocardial infarction size and release of ceartine kinase-MB in man. Circulation 1980;62:277-286.
70. Van der Laarse A, Hermens WT, Hollaar L, Jol m, Willems GM, Lemmers HE, Liem AH, Souverijn JH, Oudhof JH, de Hooge J *et al* Assessment of myocardial damage in patients with acute myocardial infarction by seriel measurement of serum alpha-hydroxybutyrate dehydrogenase levels.Am Heart J 1984;107:248-260.
71. Shell W, Mickle DK, Swan HJC. Effect of nonsurgical myocardial reperfusion on plasma creatine kinase kinetics in man. Am Heart J 1983;106:665-669.

72. Vatner SF, Baig H, Manders T, Maroko PR. Effects of coronary artery reperfusion on myocardial infarct size calculated from creatine kinase. J Clin Invest 1978;61:1048-56.
73. Maury CP. Monitoring the acute phase response: comparison of tumor necrosis factro (cachectin) and C-reactive protein responses in inflammatory and infectious disease. J Clin Pathol 1989; 42:1078-82
74. Vaddi K, Nicolini FA, Mehta P, Mehta JL. Increased secretion of tumor necrosis factor alpha and interferon-gamma by mononuclear leucocytes in patients with ischaemic heart disease. Circulation 1994;90:694-699.
75. Maury CPJ, Teppo A-M. Circulating tumour necrosis factor-α (cachectin) in myocardial infarction. J Intern Med 1989;225:333-336.
76. Struk A, Hack CE, Aarden LA, Brouwer M, Koster RRW, Sanders GTB. Interlukin-6 release and the acute-phase reaction in patients with acute myovardial infarction: a pilot study. J Lab Clin Med 1992;119:574-579.
77. Latini R, Bianchi M, Correale E, Dinarello CA, Fantuzzi G, Fresco C, Maggioni AP, Mengozzi M, Romano S, Sharpiro L, *et al.* Cytokines in acute myocardial infarction: selective increase in circulating tumor necrosis factor, its soluble receptor, and interlukin-1 receptor antagonist. J Cardiovasc Pharmacol 1994; 23:1-6.
78. Neuman F-J, Ott I, Gawaz M, Richardt G, Holzapfel H, Jochum M, Schowig A. Cardiac release of cytokines and inflammatory responses in acute myocardial infarction. Circulation 1995;92:748-755.
79. Arai K, Lee F, Miyajima A, Miyatake S, Arai N, Yokota T. Cytokines: cooridinators of immune and inflamatory response. Ann Rev Biochem 1990;59:783-836.
80. Arras M, Mohri M, sack S, Schwarz ER, Schaper J, Schaper W. Macrophages accumulate and release tumor necrosis factor-alpha in the ischaemic porcine myocardium. Circulation 1992;86(Suppl 1):1-33.
81. Tilg H, Mair J, Herold M. Acute phase response after myocardial infarction: correlation between serum levels of cytokines and C-reactive protein. Klin Wochenscher 1990;68:1083.
82. Yokoyama T, Vaca L, Rossen RD, Durante W, Hazarika P, Mann DL. Cellular basis for the negative inotropic effects of tumor nectrosis factor-α in adult mammalian heart. J Clin Invest 1993;92:2303-12.
83. Hershkovitz A, Choi S, Ansari A, Wesselingh S. Cytokine mRNA expression in postischemic/reperfused myocardium. Am J Pathol 1995;146:419-428.
84. Fyfe A, Daly P, Galligan L, Pirc L, Feindel C, Cardella C. Coronary sinus sampling of cytokines after heart transplantation: evidence of macrophage activation and interlukin-4 production within graft. J Am Coll Cardiol 1993;21;171-176.
85. Mazzone A, De Servi S, Ricevuti G, Mazzucchelli I, Fossati G, Pasotti D, Bramucci E, Angoli L, Marsico F, Specchia G, *et al* . Increased expression of neutrophil and monocyte adhesion molecules in unstable coronary artery disease. Circulation 1993;88:358-363.
86. Hirschl MM, Gwechenberger M, Binder T, Binder M, Graf S, Stefenelli T, Rauscha F, Laggner AN, Soh-chor H. Assessment of myocardial injury by serum tumour necrosis factor alpha measurements in acute myocardial infarction. Eur Heart J 1996;17:1852-1859.
87. Ockner RK, Manning JA, Poppenhausen RB, Ho WK. A binding protein for fatty acids in cytosol of intestinal mucosa, liver, myocardium, and other tissues. Science 1972;177:56-58.
88. Sohmiya K, Tanaka T, Tsuji R, *et al.* Plasma and urinary heart-type cytoplasmic fatty acid-binding protein in coronary occlusion and reperfusion induced myocardial injury model. J Mol Cell Cardiol 1993; 25: 1413-1426.
89. Knowlton AA, Apstein CS, Saouf R, Brecher P. Leakage of heart fatty acid binding protein with ischemia and reperfusion in the rat. J Mol Cell Cardiol 1989; 21: 577-583.
90. Volders PGA, Vork MM, Glatz JFC, Smits JFM. Fatty acid-binding proteinuria diagnosis myocardial infarction in the rat. Mol Cell Biochem 1993; 123: 185-190.
91. Tanaka T, Hirota Y, Sohmiya K, Nishimura S, Kawamura K. Serum and urinary human heart fatty acid-binding protein in acute myocardial infarction. Clin Biochem 1991; 24: 195-201.
92. Kleine AH, Glatz JFC, Van Nieuwenhoven FA, Van der Vusse GJ. Release of heart fatty acid-binding protein into plasma after acute myocardial infarction in man. Mol Cell Biochem 1992; 116: 155-162.
93. Tsuji R, Tanaka T, Sohmiya K, *et al.* Human heart-type cytoplasmic fatty acid-binding protein in serum and urine during hyperacute myocardial infarction. Int J Cardiol 1993; 41: 209-217.
94. Van Nieuwenhoven FA, Kleine AH, Wodzig KWH, *et al.* Discrimination between myocardial and skeletal muscle injury by assessment of the plasma ratio of myoglobin over fatty acid-binding protein. Circulation 1995; 92: 1848-1854.
95. Krause E-G, Will H, Böhm M, Wollenberger A. The assay of glycogen phosphorylase in human blood serum and its application to the diagnosis of myocardial infarction. Clin Chim Acta 1975;58:145-54.

96. Rabitzsch G, Mair J, Lechleitner P, Noll F, Hofmann U, Krause E-G, Dienstl F, Puschendorf B. Isoenzyme BB of glycogen phosphorylase b and myocardial infarction. Lancet 1993;341:1032-3.
97. Rabitzsch G, Mair J, Lechleitner P, Noll F, Hofmann U, Krause E-G, Dienstl F, Puschendorf B. Immunoenzymometric assay of human glycogen phosphorylase isoenzyme BB in diagnosis of ischemic myocardial injury. Clin Chem 1995;41:966-78.

Chapter 13

VALUE OF SOLUBLE MARKERS IN THE DIAGNOSIS OF REPERFUSION AFTER THROMBOLYSIS

Fred S. Apple

Clinical studies have demonstrated the benefits of thrombolytic therapy confirmed by reperfusion of infarct-related arteries [1,2,3]. Early and complete patency of infarct related arteries is an important therapeutic goal during the early hours after the onset of acute myocardial infarction (AMI). Data from numerous studies indicate that only a minority number of patients with AMI (<40%) are eligible for thrombolysis [4]. At present current indications for thrombolytic therapy include patients with chest pain consistent with AMI and at least 0.1 mm of ST segment elevation in at least 2 contiguous ECG leads in whom treatment can be initiated within 12 hours of pain onset, provided there are no contraindications to therapy [5]. Expanded indications for thrombolysis, however, may markedly increase the risk of haemorrhagic state [1,4]. While biochemical markers are sensitive and specific indications of AMI, [6,7] at present biochemical markers of myocardial injury do not serve as an indicator for thrombolytic therapy.

Intravenous thrombolysis is the most commonly used reperfusion method. Current regimens of intravenous thrombolysis fail to achieve complete Thrombolysis In Myocardial Infarction (TIMI flow grade 3). This is an important limitation since recent studies indicate that only early and complete patency (TIMI flow 3) provides full clinical benefit [4]. Therefore, the use of percutaneous transluminal coronary angioplasty (PTCA) for patients excluded from thrombolysis or rescue PTCA in patients with unsuccessful reperfusion following thrombolytic therapy can be used to provide reperfusion. ECG and clinical data have not been documented as successful indicators of infarct-related vessel patency. Angiography, while invasive, does provide early information on coronary anatomy and identifies patients with spontaneous reperfusion. Thus, the identification of an early biochemical marker of myocardial injury that could become an important part of a clinician's medical assessment of their AMI patient would both serve to identify successful reperfusion following thrombolytic therapy, as well as identify patients who fail to reperfuse and who would benefit from early invasive procedures. Advances in rapid technology and the advent of monoclonal antibodies specific to myocardial proteins (i.e. CK MB, myoglobin, cardiac troponin I, cardiac troponin T) currently allow the laboratory to assist clinicians in a prospective manner in the assessment of coronary reperfusion early (within 2 hours) after the administration of thrombolytic therapy.

It is accepted that the kinetics of myocardial protein appearance in the circulation following release from injured myocardium depends on infarct perfusion. [8] Several differences in serial protein patterns following AMI with and without reperfusion are evident as shown in Figure 1. Early reperfusion causes an earlier increase above the upper reference range and an earlier and greater protein peak after reperfusion. However, once the peak has occurred, there is no difference in the time of clearance of proteins between successful or unsuccessful reperfusion. Further, it is difficult to assess the amount of irreversible myocardial injury by biochemical infarct sizing because of the large variability in the amount of protein washout that appears after reperfusion [9].

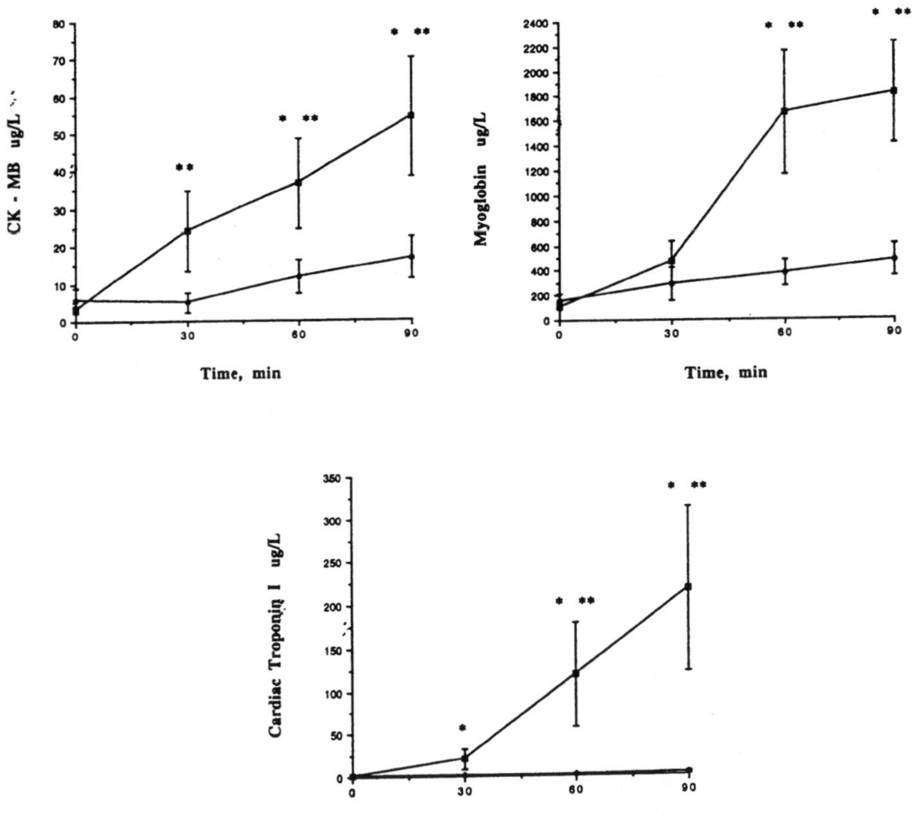

Figure 1. . Representative serial marker curves in acute myocardial infarction patients treated with thrombolytic therapy. Values represent the mean (SE bars) for 1 7 reperfused (□, TIMI flow 3) and 12 non-reperfused (◆TIMI flow 0, 1) patients for CK MB mass, myoglobin and cardiac troponin I.

The laboratory can best be utilised if a serum, plasma or whole blood test is available as an early marker to assess successful reperfusion after AMI. With early, frequent blood sampling, it may be possible to use a criterion, such as the rate of increase in marker levels, to determine the presence or absence of reperfusion following thrombolytic therapy as early as 90 minutes after initiation of therapy. Since there are inherent hospital time delays that have been documented using thrombolytic therapy for AMI [10], an organised medical subspecialty approach which incorporates the laboratory for treatment has been suggested to minimise delays. In addition, as part of the post-therapy organisation, the clinical laboratory also must become an integrated part of the overall plan to optimise the use and benefits that thrombolytic therapy may or may not provide.

Table 1. Representative guidelines used for detection of reperfusion using biochemical markers

Marker	Sensitivity	Specificity	Guideline	Ref. No.
Total CK	10%	82%	Time to peak <4hours	13
CK-MB	92%	100%	2.5 fold increase at 90 min.	16
	70%	70%	Appearance rate 0.4 μg/L/h	27
	88%	89%	Slope >17.5 μg/L/h	23
CK-MB isoforms	82%	78%	Peak ratio MB2/MB1 > 3.8 at 2hours	21
Myoglobin	94%	88%	Early initial slope at 90 min >150 μg/L/h	23
	77%	100%	90 min concentration/baseline concentration >5.0	25
	95%	100%	Increase at 90 min. vs 0 min >3	31
Cardiac Troponins				
I	82%	100%	90 min concentration/baseline concentration >6.0	25
T	80%	65%	Early initial slope at 90 min μg/L/h	23
	92%	100%	Increase at 60 mins following therapy ≥0.5μg/mL	23
	89%	83%	Relative increase at 90 mins following therapy >6.8	27

The clinical goal for successful reperfusion is to salvage myocardium in the early time period after acute coronary artery occlusion. Several strategies for the clinical laboratory utilising several accepted markers of myocardial injury have been studied. Unacceptable strategies because of the time delays which equate to loss of myocardium viability have included time to peak, clearance differences after peak and the estimation of area under the curve for peak markers. These strategies all involve greater than 10 hours after the initiation of therapy, and thus are not optimal. Several strategies do appear to provide a real workable time frame for alternate therapy. Numerous biochemical markers of myocardial injury have been studied and proposed for the non-invasive detection of coronary reperfusion [6,7,8]. Table 1 summarises sensitivity and specificity data from representative studies. The use of biochemical markers in assessing reperfusion status and their diagnostic performance usually has been determined in retrospective analysis of AMI patients undergoing angiography (60 to

120 minutes) following thrombolytic therapy. The rapid increase of total CK, CK-MB, myoglobin, cardiac troponin T (cTnT), and cTnI after successful thrombolytic therapy induced reperfusion have been reported [11,12]. The early appearing kinetics of these markers in successfully reperfused AMI patients have been shown to be very similar. In nine TIMI grade 3 flow patients monitored every 30 minutes over 120 minutes following initiation of therapy, similar time course appearance have been noted, as shown in Figure 2 [11]. The following sections will summarise studies categorised by each biochemical marker.

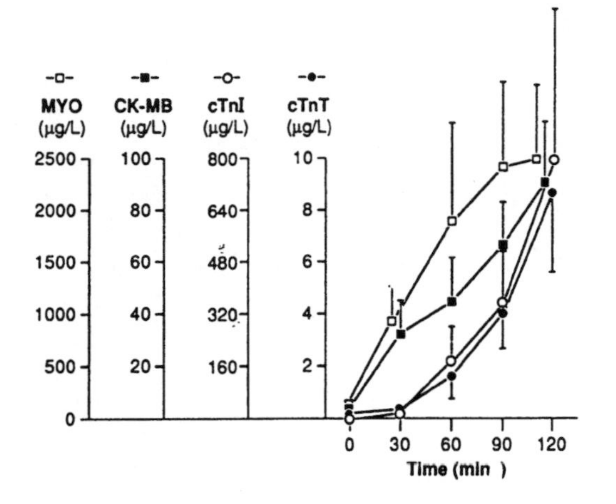

Figure 2. Time course (mean and SE) of serum cTnI, cTnT, CK-MB and MYO concentrations after initiation of thrombolytic therapy on 9 patients with TIMI Grade 3 reperfusion flow.

Creatine kinase: Release of total CK does not provide early detection of reperfusion. Clinical studies have demonstrated clinical specificities at 80% with poor clinical sensitivities at 10 to 40% using time to peak of total CK at <4 hours or >16 hours [13]. Early detection of reperfusion, however, has been documented for total CK-MB [14,15,16,17,18] and CK-MB isoforms [19,20,21,22]. Several studies have now demonstrated that (a) following thrombolytic therapy in AMI patients, greater than two fold increases in CK-MB occur within 90 minutes of reperfusion; (b) the rate of increase of CK-MB within the first 4 hours separate reperfused versus non-reperfused patients; and (c) after myocardial reperfusion in AMI patients, washout of CK-MB parallels the washout of the tissue CK-MB2 isoform. The rate of appearance of CK-MB is preferable to the practice of assessing reperfusion by an earlier time to peak of serum total CK or CK-MB levels. Further, changes in CK-MB values in association with reperfusion have also been described by the slope of CK-MB or the ratio of CK-MB at baseline and approximately 30 to 90 minutes after thrombolytic therapy. This approach also looks very promising. These findings support the use of a sensitive, specific, and rapid CK-MB mass assay that can be available on an emergency basis [8].

CK-MB isoform analysis should provide a more specific marker for myocardial reperfusion compared to CK-MM isoforms [19-22] in reperfused AMI patients, CK-MB isoform activity in serum peaked earlier than in non-reperfused patients and the rate of release for each isoform in the serum of reperfused patients was from three to four fold greater than in non-reperfused patients. CK-MB isoforms were eliminated from blood at the same rate for both groups of patients. The MB2/MB1 ratio demonstrated the earliest peak (from 0.75 to 2.25 hours after initiation of therapy) compared to AMI patients who did not reperfuse (>6h). The MB2/MB1 ratio also provided the best discrimination within one hour after treatment. Thus, the rate at which MB2 isoform is released from the myocardium into the circulation appears to be the most useful index of coronary reperfusion.

Studies have also demonstrated that the rate of CK-MB release in patients with successful reperfusion had a higher predictive value versus unsuccessful reperfusion, whereas the isoform ratio had a weaker association with reperfusion status [20]. However, the technology for a rapid, 24 hour a day CK-MB isoform testing service in a clinical laboratory still requires dedicated personal and still requires a need for improvement in technology.

Troponin

Measurement of cTnT and cTnI in serum has been proposed for evaluating the success of AMI patients treated with thrombolytic therapy such as streptokinase and tissue plasminogen activator. As shown in Figures 1 and 3, AMI patients with successful reperfusion had significantly different release patterns of cTnT and cTnI versus time as compared to patients with unsuccessful reperfusion. Patients with successful reperfusion have peak concentrations of cardiac troponins within 24 hours after onset of AMI. In patients with unsuccessful reperfusion release of troponin is more gradual and prolonged.

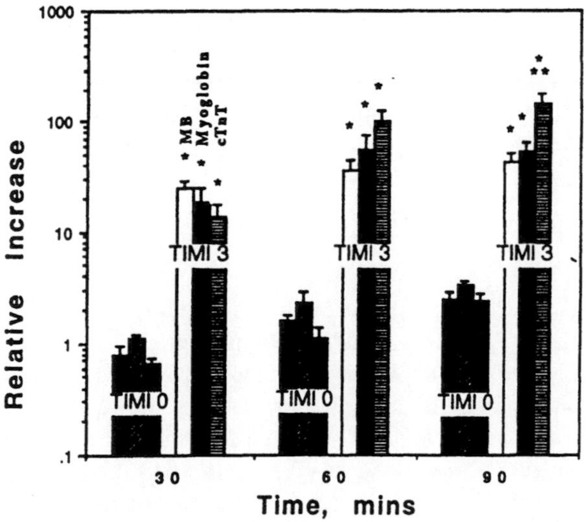

Figure 3. Bar graph shows the plots of relative increase (log scale) vs. time for CK MB (☐), myoglobin (☐) and cTnT (☐), 30, 60, and 90 min following rt.-PA therapy in nine acute myocardial infarction patients. Results are means with SE bars; * = P <0.05 for thrombolysis in myocardial infarction TIMI 3 (n=5) vs. TIMI 0,1 (n =4); * * = P < 0.05 for cTnT vs. CK-MB and myoglobin within TIMI 3 group.

Cardiac troponin investigators have also suggested that the rate of cardiac marker release may be useful in determining reperfusion status. In two studies [24,26] cardiac troponin T appeared to have advantages within 90 minutes after initiation of thrombolytic therapy, because of its earlier relative increase, improved sensitivity, and cardiac specificity compared to CK-MB and myoglobin. In a similar study, concentrations of cardiac troponin I, CK-MB and myoglobin were measured in early, serial serum samples from 25 consecutive patients who were given front loaded rt-PA during acute myocardial infarction as part of a trial requiring angiography [25]. The relative increases for cTnI, CK-MB, and myoglobin were greater at 30, 60, 90 minutes in patients with TIMI 3 (reperfusion) flow versus patients with TIMI 2 (partial reperfusion) to TIMI 0,1 flow (no reperfusion). In patients with TIMI 3 flow, the relative increase and clinical sensitivity (82 to 86%) of cTnI for detecting reperfusion was greater at 30, 60, and 90 minutes compared to CK-MB (sensitivity 59 to 77%) and myoglobin (sensitivity 63 to 70%). Therefore, measurement of cTnI was the better non-invasive predictor of early, coronary artery reperfusion in this study. In contrast to these findings, a study of 63 patients in which serum cardiac troponin T, myoglobin, and CK-MB were measured showed that myoglobin was the best predictor while cardiac troponin T offered no advantages for non-invasive prediction of coronary reperfusion compared to CK-MB and myoglobin when early slopes were determined.

Elevations of cTnI and cTnT are highly specific for myocardial injury. In individuals without myocardial disease, very low to undetectable concentrations of cTnI and cTnT are present in the serum. This contrasts to the low but measurable concentrations of CK-MB and myoglobin detected in serum from skeletal muscle turnover in non-cardiac related diseases and normal individuals. [6,7]. The initial rapid release of cardiac troponin subunits I and T following successful reperfusion is most likely derived from the soluble, cytosolic myocardial fraction (6% cTnT; 3% cTnI). The only study that has directly compared cTnI and cTnT release kinetics following reperfusion showed a parallel rise with over 120 minutes after thrombolytic therapy (Figure 2) [11]. Therefore, release of cTnI or cTnT from myocardium into the blood following acute myocardial infarction and following the washout that accompanies successful reperfusion generates an excellent signal compared to no detectable baseline levels prior to myocardial damage.

Myoglobin

The measurement of the rapid release of myoglobin into the circulation has also been described for monitoring the success of thrombolytic therapy in patients with AMI. [27-33]. In one of the early studies with myoglobin, serial measurement of serum myoglobin in patients undergoing thrombolytic therapy showed that myoglobin peaks significantly earlier after reperfusion (111 minutes) compared to patients who did not reperfuse (360 minutes). A rapid rise in myoglobin after successful reperfusion was evident by over a 4.6-fold rise over the first 2 hours after reperfusion with a predictive accuracy of 88%. Other studies that have compared patterns of myoglobin to CK-MM isoforms, total CK activity or cTnI and cTnT release have confirmed that myoglobin can be used for very early (within 90 minutes of thrombolytic therapy) non-invasive detection of reperfusion with high specificity (>80%). It has also been shown that a myoglobin to total CK activity ratio of greater than 5.0 was indicative of reperfusion, with a sensitivity, specificity and accuracy of 75%, 96%, and 92% respectively. Further, it has been concluded that reperfusion status might be satisfactorily predicted by a single sample obtained at the time of admission to assess spontaneous reperfusion very early (90 minutes) after thrombolytic therapy.

Two recent larger studies measured myoglobin and CK-MB (n = 96) [33] and myoglobin, CK MB and cTnT (n = 97) [27], respectively, 90 minutes following therapy. All AMI patients were treated with thrombolytic agents ≤6 hours after the onset of symptom. In the first study, a single myoglobin measurement at 90 minutes after the start of thrombolytic therapy, combined with clinical variables (including time from chest pain to therapy) improved the prediction of reperfusion (TIMI 3) after AMI compared to TIMI 0, 1 flow patients (no reperfusion). Median baseline myoglobin levels in the reperfused group (n=71) and the non-reperfused groups were similar (112 μg/L; 81μg/L, respectively). However, samples collected at angiography were different; 915μg/L; 273 μg/L, respectively. The combined model gave a C-index value of 0.82, indicating a potential strategy for discriminating open from closed groups. In the second study in 97 patients with AMI an effective, early determination of reperfusion after thrombolysis was possible in patients treated 3 hours after onset of chest pain when the relative increase of either myoglobin or cTnT was determined at 90 minutes. The relative increase was defined as: 90 minute concentration - 0 min concentration /0 min concentration. Sensitivity for detection of reperfusion ranged from 79% to 89% and specificity for non-reperfusion ranged from 83% to 84%. Overall, the diagnostic performance of myoglobin was less susceptible to small changes in diagnostic threshold cut-offs.

In conclusion, increased concentrations of any of the five most studied markers of myocardial injury (CK MB mass, CK MB isoforms, myoglobin, cTnI and cTnT) 90 minutes after initiation of thrombolytic therapy, appears to provide accurate assessment of reperfusion. Both cTnI and cTnT represent cardiac specific proteins with the largest tissue to circulation gradient, thus eliminating the potential of false positive increases from non-cardiac sources would be possible for the other markers. Rapid, commercial assays are now available for serum and plasma quantitation within 20 minutes and developmental work on whole blood bedside quantitative testing is underway. These platforms should thus allow for 24 hour a day testing. Thus, the laboratory can play an integral role in assisting clinicians in the differential decisions for the need of alternative therapy after initiation of thrombolytics. Larger, prospective studies involving a comparison of both cardiac troponins with myoglobin are still desired.

References

1. Simes RJ, Topol EJ, Holmes DR, White HD, Rutsch WR, Vahanian A, *et al.* Link between the angiographic substudy and morality outcomes in a larger randomised trial of myocardial reperfusion. Circulation 1 995;91:1 923-8.
2. Kennedy JW. Optimal management of acute myocardial infarction requires early and complete reperfusion. Circulation 1 995; 91:1 905-7.
3. Gersh BJ, Anderson JL. Thrombolysis and myocardial salvage. Circulation 1993;88:296-306.
4. Juliard JM, Himbert D, Goimard JL, Aubry P, Karrillon GJ, Boccara A, *et al.* Can we provide reperfusion therapy to all in selected patients admitted with acute myocardial infarction? J Am Coil Cardiol 1 997;30:1 57-64.

5. WHO Task Force. Reperfusion in acute myocardial infarction. Circulation 1994;90;2091-2102.
6. Adams Ill JE, Abendschein DR, Jaffe AS. Biochemical markers of myocardial injury. Circulation 1 993;88:750-63.
7. Apple FS. Acute myocardial infarction and coronary reperfusion: serum cardiac markers for the 1 990s. Am J Clin Path 1 992;97:21 7-6.
8. Apple FS, Preese LM. Creatine kinase MB: detection of AMI and monitoring reperfusion. J Clin Immunoassay 1994;17:24-9.
9. Blanke H, Vonhardenberg D, Cohen M, Kaiser H, Karsch KR, Holt J, *et al.* Patterns of creatine kinase release during acute myocardial infarction after non-surgical reperfusion. Comparison with conventional treatment and correlation with infarct size. J Am Coil Card 1 984;675-80.
10. Sharkey SW, Brunette DD, Ruiz E, Hession WT, Wysham DG, Hodges M. An analysis of time delays preceding thrombolytics for acute myocardial infarction. JAMA 1989;262:3171-4.
11. Apple FS, Sharkey SW, Henry TD. Early serum cardiac troponin I and T concentrations after successful thrombolysis for acute myocardial infarction. Clin Chem 1995;41:1197-8.
12. Lewis B, Ganz W, Laramee P, Cercek B, Hod H, Shah PK, Lew AS. Usefulness of a rapid initial increase in plasma creatine kinase activity as a marker of reperfusion during thrombolytic therapy for acute myocardial infarction. Am J Cardiol 1 988;62:20-4.
13. Gore JM, Roberts R, Ball SP, Montero A, Goldberg RJ, Dalen JE. Peak creatine kinase as a measure of effectiveness of thrombolytic therapy in AMI. Am J Cardiol 1987; 59:1234-8.
14. Ong L, Coromilas J, Zimmerman JM, Green S, Padmanabhan V, Reiser P, Bigger JT, Morrion J. A physiologically based model of creatine kinase MB release in reperfusion of acute myocardial infarction. Am J Cardiol 1 989;64:1 1-5.
15. Sharkey SW, Apple FS, Eisperger KJ, Tilbury RT, Miller S, Fjeidos K, Asinger RW. Early peak of creatine kinase MB in acute myocardial infarction with a nondiagnostic electrocardiogram. Am Heart J 1 988; 116:1 207-1 1.
16. Garabedian HD, Gold HK, Yasuda T, Johns JA, Finkelstein DM, Gavin RJ, *et al.* Detection of coronary artery reperfusion with creatine kinase - MB determination during thrombolytic therapy: Correlation with acute angiography. J Am Coil Cardiol 1988;1 1:729-34.
17. Christenson RH, Clemmensen P, Ohman EM, Toffaletti J, Silverman LM, Grande P, *et al.* Relative increase in CK MB isoenzyme during reperfusion after MI is method dependent. Clin Chem 1990;36:1444-9.
18. Ohman EM, Christenson RH, Califf RM, George BS, Samaha JK, Kereiakes DJ, *et al.* Non-invasive detection of reperfusion after thrombolysis based on serum creatine kinase MB changes and clinical variables. Am Heart J 1 993; 1 26:81 9-26.
19. Christenson RH, Ohman EM, Topol EJ, O' Hanesian MA, Sigmon KN, Duh SH, *et al.* Creatine kinase MM and MB isoforms in patients receiving thrombolytic therapy and acute angiography. Clin Chem 1 995;41:844-52.
20. Christenson RH, Ohman EM, Clemmenson P, Grande P, Toffaletti J, Silverman LM, Volimer RT, Wagner GS. Characteristics of creatine kinase MB and MB isoforms in serum after reperfusion in acute myocardial infarction. Clin Chem 1989;35:21 79-88.
21. Puleo PR, Perryman MB. Non-invasive detection of reperfusion in acute myocardial infarction based on plasma activity of creatine kinase MB subforms. J Am Coil Card 1991;17:1047-52.
22. Abe S, Nomoto K, Arima S, Migata M, Yasashita T, Maruyama 1, *et al.* Detection of reperfusion 30 and 60 minutes after coronary recanalisation by a rapid new assay of creatine kinase isoforms in acute myocardial infarction. Am Heart J 1993; 1 25:649-56.
23. Zabel M, Hohniosen SH, Koster W, Prinz M, Kasper W, Just H. Analysis of creatine kinase, CK MB, myoglobin and troponin T time-activity curves for early assessment of coronary artery reperfusion after intravenous thrombolysis. Circulation 1 993;87:1 5642-50.
24. Apple FS, Voss E, Lund L, Preese L, Berger CR, Henry TD. Cardiac troponin, CK MB, and myoglobin for the early detection of acute myocardial infarction and monitoring of reperfusion following thrombolytic therapy. Clin Chim Acta 1 995; 237:59-66.
25. Apple FS, Henry TD, Berger CR, Landt YA. Early monitoring of serum cardiac troponin I for assessment of coronary reperfusion following thrombolytic therapy. Am J Clin Path 1 996; 1 05:6-1 0.
26. Abe S, Arima S, Yamashita T, Miyata M, Okino H, Toda H, *et al.* Early assessment of reperfusion therapy using cardiac troponin T. J Am Coil Cardiol 1994;23:1382-9.
27. Laperche T, Steg G, Dehoux M, Benessiano J, Grollier G, Alliot E, *et al.* A study of biochemical markers of reperfusion early after thrombolysis for acute myocardial infarction. Circulation 1995;92:2079-86.

28. Katus HA, Diederich KW, Scheffold T, Ueller M, Schwarz W. Non-invasive assessment of infarct reperfusion: The predictive power of the urine to peak value of myoglobin, CK MB and CK in serum. Eur Heart J 1988; 9:619-24.
29. Laperche T, Steg G, Benessiano J, Dehoux M, Juliard JM, Gourgon R. Patterns of myoglobin and MM creatine kinase isoforms release early after intravenous thrombolysis or direct percutaneous transluminal coronary angioplasty for acute myocardial infarction. Am J Cardiol 1 992;70:1 1 29-34.
30. Ellis AK, Little T, Masud AZ, Kiocke FJ. Patterns of myoglobin release after reperfusion of injured myocardium. Circulation 1985;72:639-47.
31. Abe J, Yamaguchi T, Isshiki T, Naka J, Ishizaka N, *et al.* Myocardial reperfusion can be predicted by myoglobin-creatine kinase ratio of a single blood sample obtained at the time of admission. Am Heart J 1 993; 1 26:279-85.
32. Lshii J, Nomura M, Ando T, Hasegaura H, Kimura M, Kurokama **H,** *et al.* Early detection of successful coronary reperfusion based on serum myoglobin concentration. Comparison with serum creatine kinase MB. Am heart J 1 9 94; 1 2 8:641 -8.
33. Christenson RH, Ohman EM, Sigmon KN, O'Haneain, Atonsozana sGL, Newby KL, *et al.* Combining myoglobin and coronary reperfusion after thrombolytic therapy. J Am Coil Card 1 995; 25:148A (Abstract).

Chapter 14

THE ASSESSMENT OF MYOCARDIAL DAMAGE IN HEART TRANSPLANTATION

Jon R. Anderson

The central aim of cardiac surgery is to preserve myocardial function during the peri-operative period whilst operating on the heart. Different strategies have evolved to protect the heart during an obligatory ischaemic insult which have resulted in safer surgery and acceptable myocardial function in the post-operative period [1]. It is likely that all cardiac operations cause some degree of cardiac injury but until the development of the newer markers of myocardial damage such as troponin-T, troponin-I and CK-MB isoforms, the assessment of injury was difficult to define [2]. With the use of these novel markers the assessment of pre, peri and post-operative injury has been more precise. These new markers of myocardial injury also allow comparisons of different strategies of myocardial protection [3,4]. In the field of cardiac transplantation there is the potential for myocardial damage in the donor organ prior to transplantation (pre-operative), during the "obligatory" ischaemic time between procurement and reperfusion in the recipient (peri-operative) and during the rejection process in the post-operative period. The role of newer markers of myocardial injury has been explored in each of these settings in patients undergoing transplantation at St. George's Hospital and the results of these studies form the basis of this chapter.

Donor myocardial injury

The cardiac function of brain dead organ donors is an important prognostic factor in the clinical outcome of cardiac transplantation [5]. Numerous experimental and clinical studies suggest that brain death may induce segmental or global myocardial dysfunction [6]. Animal studies of brain death have shown the presence of ECG changes (including acute ST segment depression) and also that induction of brain death causes profound haemodynamic changes [7]. Primary organ dysfunction following cardiac transplantation is responsible for the majority of early postoperative deaths and may be due to poor organ preservation at the time of procurement, prolonged ischaemic time or pre-existing myocardial damage in the donor [8]. Poor organ preservation and prolonged ischaemic times are avoidable but myocardial injury in the donor heart is not easy to assess. Most donor hearts are deemed to have good function by haemodynamic and ECG criteria but subtle degrees of myocardial injury may be missed [9,10]. Biochemical evidence of myocardial injury in cardiac donors has received little attention until recently, probably due to the lack of reliable and sensitive markers. Brain death is often associated with multiple trauma and/or cardiopulmonary resuscitation and therefore there is often a rise in the CK-MB isoenzyme even in the absence of ischaemic myocardial injury [11]. Myosin light chain analysis in paediatric donors demonstrated that elevated levels in the donor were associated with a poor outcome [12] The advent of more specific markers of myocardial injury may help in the detection of pre procurement myocardial damage.

Table 1. Troponin T and Troponin I levels in donors undergoing cardiectomy for transplantation. (expressed as median values and range)

All donors (n=68)		
Tn-T	0.31(0-7.5ng/ml)	
Tn-I	0.1(0-2.9ng/ml)	
Aetiology of brain death	Trauma (n=24)	Intracranial haemorrhage (n=38)
Tn-T	0.1 (0-7.5ng/ml)	0.1(0-1.5ng/ml)
Tn-I	0.02(0-2.9ng/ml	0.09(0-1.9ng/ml)
Post-operative function	No inotropes (n=43)	Inotropes required (n=25)
Tn-T	0.07 (0-7.5ng/ml)	0.4 (0.07-4.8ng/ml)
Tn-I	0(0-0.07ng/ml)	0.4(0-2.9ng/ml)
	Survival (n=60)	Early graft failure (n=8)
Tn-T	0.1(0-7.5ng/ml)	0.9(0.07-4.8ng/ml)
	0.02(0-2ng/ml)	0.4(0-2.9ng/ml)

Cardiac troponin T (TnT) and troponin I (TnI) in cardiac donors fulfilling standard haemodynamic and ECG criteria were measured prior to organ procurement and the levels were correlated with the clinical course of the recipient (Table 1). Overall the median Tn-T was 0.31 ng/ml (range 0-7.5 ng/ml) and median Tn-I was 0.1 ng/ml (range 0-2.9 ng/ml). There was no difference in the troponin levels of donors suffering intracranial haemorrhage compared to those suffering traumatic head injury. Analysis of CK-MB in these groups demonstrated a higher level in those suffering traumatic head injury. Of the 68 donors studied, 43 had good primary function in the first 48 hours post-operatively. All of the remainder (n= 25) required some degree of cardiac support within the first 48 hours post-operatively in the form of intravenous adrenaline, intra-aortic balloon pump counterpulsation or ventricular assist and 8 recipients died of primary organ failure. Both Tn-T and Tn-I levels were higher in this latter group compared to those with good primary function ($p < 0.01$) When the patients with primary organ failure resulting in death (n=8) were considered separately, the difference was still significant ($p < 0.01$) compared to the survivors (n=60). Analysis of CK-MB in these patient sub groups, however, revealed no differences and there was no difference in ischaemic time that could explain the poor function. The median peak Tn-T within the first 48 hours post operatively was higher in the group requiring adrenergic support but the difference failed to reach significance (4.25 ng/ml v 3.22 ng/ml, $p = 0.13$). There was a certain amount of overlap in troponin values between those with good function and those requiring inotropic support. However, using a cut off point of 0.2 ng/ml for donor TnT, the sensitivity for diagnosing poor primary function in the recipient was 82.6% with a specificity of 82.9%. A cut off point of 0.1 ng/ml for donor Tn-I to diagnose poor primary function gave a sensitivity of 69.2% with a specificity of 86.3%. Donor biopsy at the time of implantation showed no evidence of myocardial damage and recipients who died within 48 hours showed no evidence of rejection at post mortem although in some cases ischaemic necrosis was present. Recipients with an uncomplicated course who did not require adrenergic support had significantly lower TnT levels (0.36 ng/l v 0.07 ng/l) compared to those patients with early allograft failure (on mechanical and inotropic support) and those requiring prolonged adrenergic support ($p < 0.01$). Patients on adrenergic support had a higher peak, TnT concentrations than those not requiring adrenergic support. Similar findings were seen with cTn-I measurements. Cardiac troponins (T and I) form part of the tropomyosin complex in the myocyte and are released into the circulation upon myocyte injury [13,14]. Both markers have a long half life (about 6 hours) and have a diagnostic window of up to 6 days reflecting previous damage despite normal CK-MB isoenzyme levels [15]. All of the donors in our series had normal haemodynamics and ECG but there was a surprising degree of myocardial injury as assessed by Tn-T and Tn-I measurements. We have previously reported that donor Tn-T levels predict the need for inotropes in the early post-operative period. In this larger series, an elevated Tn-T also predicted primary graft failure [16]. A raised Tn-I in paediatric donors has shown to be a significant predictor of early graft dysfunction and we have confirmed this finding also in adult practice [17]. As more hospitals begin to use the cardiac troponins for the assessment of myocardial injury, it may be possible to perform the assay as part of the donor workup. This would allow the retrieving team to base their judgement on both haemodynamic criteria and reliable biochemical markers of myocyte injury. Both assays take under 2 hours to perform in the laboratory but a rapid bedside assay has recently become available. The observation that a raised level of cardiac troponin seems to be a strong predictor of morbidity and mortality in the recipient has a number of implications. A raised troponin level may influence the decision to procure the heart for transplantation, especially if the donor is in the so-called 'marginal group'. Moreover, if the heart is used in the knowledge that the troponins are raised, inotropic or mechanical support may be implemented at an early stage before primary dysfunction becomes irreversible. The results mentioned previously also have pathophysiological implications as they suggest that brain death induced myocardial dysfunction is associated with some degree of myocardial cell damage. This observation is compatible with the hypothesis that the massive sympathetic activity secondary to brain death may be responsible for myocardial dysfunction. A study has confirmed these findings by prospectively measuring TnT in brain dead patients and assessing `cardiac function using transoesophageal echocardiography [18]. Elevated TnT levels were found to be associated with myocardial dysfunction and therefore abnormal TnT is considered to be a useful predictor of poor primary function post-transplantation. The percentages of patients with elevated CK-MB or CK-MB/CK were not significantly different when donors with good, moderate or poor myocardial function were compared. In conclusion, it is likely that biochemical methods of detecting myocardial damage in the donor will become part of the routine assessment prior to organ procurement for cardiac transplantation.

Perioperative cardiac damage during heart transplantation

Theoretically, perioperative ischaemic damage in the donor heart begins with the application of the aortic cross clamp but should end when the aortic cross clamp is released in the recipient. Unfortunately, this is an oversimplification brain death, pre-existing donor myocardial injury, cardioplegia, suboptimal heart preservation, prolonged ischaemic time, cardiopulmonary bypass and reperfusion injury all contribute to impaired ventricular performance [7]. However, under ideal conditions, the most important variable should be the ischaemic time prior to reperfusion. A time of 4 to 6 hours following donor cardiectomy is considered safe according to standard laboratory and clinical experience [19]. A marked decrease in left ventricular function, high energy phosphate content, histological integrity and increase in metabolic waste products, myocardial water content and coronary vascular resistance have been found after prolonged ischaemia [20]. Different strategies of myocardial preservation have been developed but clinical studies have failed to show significant differences between myocardial preservation and storage techniques [21]. If ischaemia persists beyond 6 hours, myocardial performance may be compromised. In the setting of heart transplantation objective physiological data from such tests as echocardiography and pulmonary flotation catheters are operator dependent and may be unreliable due to the anatomical distortion of the heart associated with the biatrial anastamotic technique [22]. Although advances in myocardial preservation have greatly improved the outcome following cardiac surgery, peri-operative myocardial damage remains an important complication of transplantation and requires objective assessment. The early detection and quantification of ischaemic myocardial damage is important both for the initiation of specific therapy and to improve methods of myocardial preservation. Measurement of CK-MB does not correlate well with post operative myocardial performance. The development of newer markers of myocardial injury such as CK-MB isoforms, Tn-T and Tn-I has triggered renewed interest in the question of how best to assess myocardial injury [23]. The relationship between time activity curves of serum CK-MB mass activity, MB2/MB1 isoform activity, troponin T and duration of ischaemic preservation was explored in a number of patients undergoing orthotopic heart transplantation. We tested the hypothesis that analyses of time activity curves of CK-MB and troponin T release could be useful markers of myocardial cell damage after heart transplantation.

CK - MB Isoforms

As previously mentioned, the isoforms of CK-MB are sensitive markers of myocardial injury. These are detectable even when the isoenzymes of CK-MB are within the normal range [24] Experimental studies suggest that several cardioplegic techniques and reperfusion techniques may decrease the ischaemic injury caused by heart preservation but this is difficult to measure in the absence of a reliable marker of myocardial injury [25]. The clinical value of creatine kinase MB2 isoform separation as a highly sensitive measure of myocardial damage has been investigated in patients undergoing orthotopic cardiac transplantation. Changes were monitored in the activity of CK and its isoenzyme CK-MB. The ratio of CK-MB isoforms and the area under the time activity curve (AUC) were used as an index of peri-operative myocardial damage. The peak MB2/MB1 ratio, at the time of aortic cross clamp release, occurred sooner than the peak activities of CK-MB2, CK-MB or total CK. The MB2/MB1 ratio also returned to normal values more rapidly than the other 3 variables. At the time of peak MB2/MB1 ratio, the CK-MB activity was due almost entirely to the MB2 isoform. The results for such samples are given as 100% MB2 (Table 2).

Table 2 Median peak and return to baseline values and corresponding times for total activities and ratios following cardiac transplantation.

	CK		CK-MB		CK-MB2		MB/MB1	
Peak value (range) I.U./L	1103	(595-5780)	107	(40-224)	85	(40-165)	100% MB2	
Peak time (h)	8.5	(2-24)	4.2	(1-24)	2.0	(1-12)	0	(0-1)
Return to baseline time HRS	120	(48-120)	120	(48-120)	72	(24-120)	24	(12-120)

CK-MB2 activity seemed suitable as an objective measure of myocardial damage for studies comparing different methods of myocardial preservation. The higher sensitivity of the CK-MB2 isoform measurements compared with total CK and CK-MB activity were confirmed by the results of this study. The return of MB2/MB1 ratio to normal value within 12 hours from reperfusion is useful in the differential diagnosis of enzyme release due to intra-operative myocardial damage as opposed to enzyme leaks due to other causes. The very early peak and rapid return to baseline values of CK-MB2 release, however, has the potential to overcome some of the problems experienced with other markers which have a longer half life (Tn-T for example). The AUC, which is proportional to the amount of enzyme released, could be used as an end point in studies comparing different forms of myocardial protection. However, to be fully accurate, the AUC would have to be corrected for the myocardial weight either by weighing the donor heart or estimating its mass by echocardiography.

Troponin release following cardiac transplantation

The pattern of TnT release during the first 24 hours following transplantation fell into 2 groups. In group 1, the mean (±SD) TnT concentration reached a maximum of 4.15 ± 2.63 ng/ml at a mean of 2.6 ± 1.5 hours following cross clamp release. In group 2, the mean TnT concentration was still rising : 7.47 ± 6.55 ng/ml at 24 hours post cross clamp release. In group 1, the mean TnT concentration reached a second peak of 3.22+/- 1.8 at a mean of 9 ± -6.3 days post transplantation, but did not reach the diagnostic cut off value of 0.2 ng/ml in between the 2 peaks. In group 2, the peak TnT concentration was reached between 5 and 15 days in the remaining 2 patients. Twenty nine patients were followed for a mean of 87 ±- 32 days post transplantation. The mean TnT concentration reached a maximum of 3.5 ±- 1.62 ng/ml at 6.1 ±- 3.7 days and returned to baseline values of 0.2 ±- 0.23 ng/ml at 47 ±- 18 days post transplantation. Troponin I analysis showed a median peak of 0.99 (0.45 - 9.9) ng/ml at 3 ±- 2 days and returned to a baseline value of 0.06ng/ml (< 0.003-0.01) at 14 ±- 5 days post transplantation.

The area under the TnT time concentration curve was calculated for up to 40 days post-transplantation, the period during which the maximum number of samples was collected in all patients. The patients were then divided into 2 groups. Those with no episodes of rejection (n=12) and those with one or more episodes of severe rejection (Grade 3a or more, n=17). The median (range) AUC for each group was 52.8 (13.6-120) ng/ml.days and 38 (17.1-139) ng/ml.days, respectively.

The AUC was not related to the total ischaemic time before transplantation. However, patients requiring adrenergic support (n=11) and those with renal dysfunction (serum creatinine > 200 umol/L) (n=6) in the first 6 weeks following transplantation, 3 of whom also required adrenergic support, had significantly higher AUC (TnT) than those who did not require adrenergic support or those who had no renal dysfunction (p=0.0001). Zimmermann *et al* [26] characterized the behaviour of cardiac troponin-T during the first 3 months following cardiac transplantation. In their study they found that the peak TnT occurred at 7.1 (+/- 4.2) days following transplantation and that the levels of circulating TnT did not return to below twice the detection level of their assay for a mean of 50 days post-operation. The reason for this is uncertain but it may be the result of an initial ischaemic insult followed by an immune mediated phenomenon of gradually decreasing intensity as immunosuppressive therapy is maximized and immune tolerance occurs. Their results were similar to those obtained in our studies. The release of TnT post heart transplantation lasts longer than the release of TnT following coronary artery surgery, valve replacement or acute myocardial infarction (AMI). Following AMI, TnT appears in the circulation 3.5 hours after the onset of chest pain and the concentrations remain elevated in the circulation for up to 6 days [27]. Small amounts of TnT are detectable in the circulation after cardiac surgery and this is related to the ischaemic time [28]. The behaviour of TnT in the circulation following cardiac transplantation is different. The changes seen are consistent with an immune mediated phenomenon rather than persistent immunological damage. This immune activation may be humoral, cellular or a combination of both. Scintigraphic studies using indium-111 labelled antimyosin found evidence of prolonged sarcolemmal damage after transplantation [29]. These studies showed considerable myocardial myosin uptake soon after transplantation with a gradual reduction over the ensuing months. The reason for the discrepancy between myosin and TnT which are both markers of ischaemic damage is unclear. The amount of circulating TnT release was not correlated with the iscahemic time or the number of acute rejection episodes. These findings may be explained by leakage of TnT from damaged but viable myocytes as well as prolonged disintegration of irreversibly damaged myocytes in the first weeks following cardiac transplantation. Carrier *et al* [30] found that TnT peaked at an early stage. The reason for this discrepancy is not clear but they did not carry out prolonged sampling following transplantation and may not have appreciated the longer term behaviour of this marker. Troponin I has an earlier peak that TnT and is simpler to measure than CK-MB isoform. The stability of TnI means that it is likely to be the marker of choice in comparing myocardial preservation strategies in cardiac transplantation [4].

Myocardial damage in the post-operative period

Rejection following transplantation is a major cause of morbidity and mortality. There are a number of components to rejection and it is likely that more than one process occurs at once. A full discussion of the recognised forms of rejection is beyond the scope of this chapter but the most common form is acute cellular rejection. This is an immune mediated phenomenon primarily involving the lymphocyte cell system [31,32]. The histological features seen in this type of rejection have been standardized by the International Society for Heart and Lung Transplantation [33] (Table 3). The process is initiated by the presence of proteins (in this case the donor heart) which are recognized as "foreign" by the immune system. A sequence of events follows, that is complex and not fully understood, the end result of which is the infiltration of the myocardium by lymphocytes and ultimately myocyte destruction. The allograft initially displays good function even whilst the rejection is occurring. However, if this is left untreated graft dysfunction will occur. Prompt treatment with augmentation of immunosuppression may reverse the graft dysfunction and lead to resolution of the histological findings [34]. Ideally, surveillance biopsies would detect rejection at an early stage before clinical signs occur so that treatment may be given before the occurrence of graft dysfunction. It is known from post mortem studies that endomyocardial biopsy underestimates the degree and severity of allograft rejection [35]. The biopsy grading is entirely arbitrary (the current gold standard for the diagnosis of rejection) from Grade 0 to 4. The timing of routine biopsy sampling is also arbitrary and the grading system does not take clinical features or humoral markers of rejection into account. In spite of these problems, endomyocardial biopsy is still the most reliable and

Table 3. The grading system proposed by the International Society for Heart and Lung Transplantation [33].

Biopsy Grade	Histological findings
0	No evidence of rejection (normal biopsy)
1a	Focal (perivascular or interstitial) infiltrate without necrosis
1b	Diffuse but sparse infiltrate without necrosis
2	One focus only with agressive infiltration and/or focal myocyte damage
3a	Multifocal agressive infiltrates and/or myocyte damage.
3b	Diffuse inflammatory process with necrosis
4	Diffuse aggressive polymorphous ± infiltrate ± oedema, ± haemorrhage, ± vasculitis, with necrosis.

universally agreed method of rejection surveillance, diagnosis and monitoring of response to treatment. Nearly 10% of mild rejection episodes are associated with severe allograft dysfunction and require aggressive treatment [36]. It has been suggested that myocyte necrosis is not a prerequisite for the impairment of graft function [37]. Interleukin-1, IL-2 and TNF, all found in increased levels during acute rejection, cause allograft dysfunction even in the absence of myocyte necrosis. Acute rejection is a cause of death in over 30 % of patients although the risk of death from rejection declines with time [38]. Much research effort, time and money has been channelled into the search for a marker of rejection that could replace biopsy but none has proved as reliable as biopsy [39]. Although patients at high risk of acute rejection can be identified, rejection can occur in any patient at any time and therefore, a noninvasive marker for rejection is required. By developing such a marker, the potential risks of endomyocardial biopsies, expense and inconvenience would be reduced or eliminated. However, the complex mechanisms that contribute to the process of acute rejection mean that there is unlikely to ever be one single test that will act as a reliable marker for rejection. The end result of acute cellular rejection however is myocardial necrosis and therefore a sensitive marker of myocardial damage may be a useful marker for rejection. Ideally, such a marker would begin to rise before myocyte necrosis occurs, which will allow for treatment to start at an early stage. The traditional marker of myocardial injury, CK-MB isoenzymes, proved disappointing in this regard [40]. Other markers of myocardial injury have also been tried but they appear to lack sensitivity, probably as a reflection of the complexities of the process of rejection [41,42]. Myocyte necrosis is often cited as a criterion indicating the need for additional immunosuppression and for the classification of acute cellular rejection (Grade 2 or 3a). The light microscopic changes of myocytes during rejection may look irreversible but under electron microscopy these changes are consistent with reversible injury [43]. The features of irreversible injury, lysis or loss of sarcolemmal membranes, mitochondrial swelling and intramitochondrial densities were identified by using experimental models of ischaemic injury. The recovery of left ventricular function following successful treatment of rejection adds support to the argument that the presence of some features of necrosis may not necessarily imply cell destruction . Several questions regarding the use of more sensitive markers of myocardial injury as non invasive markers of acute rejection, should be answered.

Among these are the following

1. Can these markers of myocardial injury diagnose acute cellular rejection?
2. Can these markers predict the onset of acute rejection ?
3. Can these markers dispense with or reduce the need for endomyocardial biopsy?

Creatine kinase MB isoform release in acute cardiac allograft rejection

In a study from our institution the histological changes seen with acute cellular rejection on myocardial biopsies were correlated with levels of the isoforms of the cardiospecific CK-MB iso-enzyme MB1 and MB2 [44]. Moderate to severe cellular rejection is known to be associated with myocyte necrosis which should, theoretically, result in enzyme leakage from the myocardial cells. Isoforms of CK-MB are able to detect myocardial damage at levels below that of current "standard" enzyme assays (CK-MB and LDH) and could, potentially, detect cellular necrosis associated with the more severe grades of rejection [45]. In our study patients were followed up to 12 months after transplantation. There were 284 myocardial biopsies available for interpretation (Table 4), the majority of which were associated with rejection grades 0 and 1 with only a small proportion of biopsies demonstrating clinically significant rejection (Grade 2 and 3, 18 and 24, respectively). Fifty percent (n=25) of the patients had no significant rejection as assessed by endomyocardial biopsies. Of the remaining patients; 16 had a single episode of rejection, 6 had two rejection episodes and the remainder had three rejection episodes. The mean MB2/MB1 ratio was 0.88 (± 0.42) in plasma samples taken from normal controls (n=30) and only one control subject had an MB2/MB1 ratio greater than 1.5. An MB2/MB1 ratio of 1.5 was, therefore, chosen as a cutoff point for the detection of acute rejection. The ratio MB2/MB1 in plasma samples taken at the time of biopsy from patients with biopsy grades 2 or 3 was not significantly different to the ratio in the samples from patients with rejection grades 0 or 1 (1.65 v 1.33, p = ns). The sensitivity for diagnosing a moderately severe rejection was 52% with a specificity of 52%. However, in patients with significant acute rejection (grades 2 and 3) in whom consecutive samples were collected, the MB2 /MB1 ratio was significantly increased prior to histological changes seen on biopsy (13/16 episodes, $p<0.001$) by a mean of 14 days. The sensitivity for predicting rejection (Grade 2 or 3) prior to biopsy was 60% with a specificity of 68%. The sensitivity for prediction was increased to

67% if only grade 3 episodes were analysed. Plasma CK and CK-MB remained within the normal range and the levels showed no correlation with the biopsy findings.

Table 4. Markers of myocardial injury and ISHLT biopsy grade.

	ISHLT grade			
Marker	0	1	2	3a
CK-MB (I.U./L)	3.9± 0.4	3± 0.6	3.6± 0.8	2.9± 0.8
MB2/MB1 (ratio)	1.9± 0.1	1.9± 0.6	1.8± 0.2	1.6± 0.3
Troponin T(ng/ml)	1.1± 1	0.9± 0.2	1.3± 0.4	0.7± 1
Troponin I(ng/ml)	0.04	0.03	0.02	0.01

Cardiac Troponin as a noninvasive marker of rejection

The role of both cardiac TnT and TnI as non invasive markers of acute cellular rejection was also investigated in our study. Thirty one consecutive patients, 26 men, 5 women, median age 53 years, range 28-60 years were studied following orthotopic cardiac transplantation. Twenty nine patients were followed for a mean of 87 ± 32 days post transplantation. The mean TnT concentration reached a maximum of 3.5 ± 1.62 ng/ml at 6.1 ± 3.7 days following transplantation and returned to baseline values of 0.2 ± 0.23 ng/ml at 47 ± 18 days post transplantation. From the time that the TnT concentration returned to baseline values to the end of the study period, 6 patients developed 12 episodes of severe rejection. TnT concentration was > 0.2 ng/ml (0.9, 2.6, 0.4, 0.4 ng/ml) in 4 episodes of rejection. In the remaining episodes, TnT concentration was below 0.2 ng/ml. (Table 4). Again there was no correlation with biopsy grade and the level of TnT. Unlike CK-MB isoforms, the TnT levels did not appear to have a predictive role. In a sub group of these patients (n=17), serum samples were obtained before biopsy, during in hospital stay and at outpatient visits and were analysed for TnI. Patients were followed for a mean of 61 ± 16 days post-transplantation. Eight patients each had an episode of rejection associated with myocytolysis (Grade 2 and 3). At the time of biopsy, there was no significant difference in TnI concentration between patients with grade 2 and 3 rejection and those with grade 0 and 1 rejection (0.015 ± 0.025 ug/L vs 0.036 ± 0.073 ug/L, respectively; p = ns). In addition TnI did not appear to have predictive value in the diagnosis of acute rejection.

The changes seen on endomyocardial biopsies probably represent a late stage in the rejection process. Treatment of severe histological rejection before the occurrence of symptoms appears to reduce the risk of allograft loss [45]. Biochemical markers of rejection depend on the detection of molecules that are released in response to cellular damage from their intracellular location to the serum. The MB isoenzyme of CK has proved of little value and the recently described cardiac troponin-T remains elevated for 3 months following cardiac transplantion effectively negating its role as a marker of acute rejection in a period when rejection is most likely to occur [26]. Subforms of the CK-MB isoenzyme (MB2 and MB1) have the potential to act as non-invasive markers of rejection. These subforms are mainly used for the detection of acute myocardial infarction, reperfusion injury and peri-operative myocardial damage during cardiac surgery [46, 47]. Its potential as a marker of acute rejection in cardiac transplant patients is based on the premise that the MB2/MB1 ratio is raised when cytolysis occurs as part of the rejection process (as in ISHLT rejection grade 2 and 3). The assay, however, has failed to diagnose acute rejection with the required sensitivity and specificity to replace biopsy. It is of interest that the MB2/MB1 ratio was raised in a number of patients prior to the appearance of histological changes in the biopsy. It remains to be seen if this test could be used in addition to other diagnostic tests (biochemical, immunological or biophysical) to substitute endomyocardial biopsies in these patients.

The assay for CK-MB isoforms is difficult to perform and is unlikely to gain widespread acceptance. CK-MB isoforms have a short half life in the serum and therefore their presence only gives an indication of cell damage at that point in time. Structural proteins on the other hand, are comparatively easy to measure and the molecules are very stable in the serum Serum myoglobin has been investigated as a noninvasive marker of rejection but the absolute level or directional change was of no benefit in identifying rejection [41]. Whilst both TnT and TnI are reliable markers of ischaemic myocardial injury, neither seem to be particularly useful in the detection or prediction of acute cellular rejection. There is a difference between the release kinetics of TnT and TnI following cardiac transplantation. Why this should be so is not clear. TnT exists in intracellular compartments as well as structurally bound to the tropomyosin complex. Following acute ischaemic injury, there is a rapid release and peak of TnT followed by a slower more sustained rise and peak [47,48]. This is an entirely different pattern to that seen following transplantation. It is known that disruption of the myocyte cell membrane is one of the primary events leading to myocytolysis in acute cellular rejection. This occurs before structural changes become apparent on light microscopy. The prolonged release of TnT may be a marker of immune mediated damage and its gradual return to normal values may suggest the onset of host tolerance to the transplanted organ. Animal studies have shown that TnT may be a reliable noninvasive marker of rejection. However we have shown that this does not extrapolate to the human situation. In conclusion, it seems that even extremely sensitive markers of myocardial injury have a limited role to play in the diagnosis of acute rejection probaly due to the complex nature of the rejection process.

Summary

Novel markers of myocardial injury have an application in the field of cardiac transplantation. Troponin-T and Troponin-I can identify sub clinical myocardial injury in donor hearts and may predict recipient cardiac function. Isoforms of CK-MB and Troponin-I both seem to be reliable markers of injury in the peri-operative phase and will undoubtedly be used in studies of myocardial preservation. Disappointingly, no marker is sufficiently reliable at present to dipense with biopsy in detecting acute cellular rejection and the search for such a reliable marker(s) is ongoing.

References

1. Jain U. Myocardial infarction during coronary artery bypass surgery. J Cardiothorac Vasc Anaesth 1992;6:612-6.
2. Birdi I., Angelini G. D., Bryan A. J. Biochemical markers of myocardial injury during cardiac operations. Ann.Thorac Surg 1997;63:897-84.
3. Anderson J. R, Hosssein-Nia M., Kallis P., Pye M., Holt D.W., Murday A. and Treasure T. Comparison of two strategies for myocardial management during coronary artery surgery. Ann Thorac Surg 1994;58:768-73.
4. Etievent J-P., Choron S., Toubin, Taberlet C., Alwan K., Clement F., Cordier A., Schipman N., Kantelip J-P. Use of Cardiac Troponin I as a marker of peri-operative ischaemia. Ann Thorac Surg 1995;95:1192-4.
5. Darracott- Cankovic S., Stovin P.G.I., Headlong D., Wallwork J., Wells F., English T.A.H. Effect of donor heart damage on survival after transplantation. Eur J Cardiothor Surg 1989;3:525-532.
6. Bittner H.B., Kendall S.W.H., Campbell K.A., Montine T.J., Van Trigt P. A valid experimental brain death organ donor model. J Heart Lung Transplant 1995;14:308-17.
7. Bittner H.B., Kendall S.W.H., Chen E.P., Davis R.D., Van Trigt P. Myocardial performance after graft preservation and subsequent cardiac transplantation from brain dead donors. Ann Thorac Surg 1995;60:47-54.
8. Keck B.M., Bennett L.E, Fiol B.S., Daily O.P., Novick R.J., Hosenpund J. D., Worldwide thoracic organ transplantation: a report from the UNOS/ISHLT International Registry for Thoracic Organ Transplantation. Clin Transplant 1995:35-48.
9. Baldwin J.C, Anderson J.L., Boucek M.M., Bristow M.R., Jennings B., Ritsch M.E., Silverman N.A.. Task Force 2: Donor Guidelines. J. Am. Coll. Cardiol. 1992 22:15-20.
10. Potter C.D.O., Wheeldon D.R., Wallwork J. Functional assessment and management of heart donors: a rationale for characterization and a guide to therapy. J Heart Lung Transplant 1995;14:59-65.
11. Donnelly R., Hillis W.S., Cardiac Toponin T. Lancet 1993:341/8842:410-411.
12. de Begona J.A., Gundry S.R., Razzouk A.J., Boucek M.M., Kawauchi M., Bailey L.L.. Myosin Light Chain efflux after heart transplantation in infants and it correlation with ischaemic time. J Thorac Cardiovasc Surg 1993;106:458-462.
13. Katus H.A., Looser S., Hallermayer *et al.* Development and in vitro characterization of a new immunoassay of cardiac troponin T. Clin Chem 1992;38:386-70.
14. Adams J.E., Bodor G.S., Davilla V.G., *et al.* Cardiac troponin-I: a marker with a high specificity for cardiac injury. Circulation 1993;88:101-6.
15. Bhayana V., Henderson A.R. Biochemical markers of myocardial damage. Clin Biochem 1995;28:1-4
16. Anderson J.R., Hossein-Nia M., Brown P., Holt D.W., Murday A. Donor cardiac troponin-T predicts subsequent inotrope requirements following cardiac transplantation. Transplantation 1994;58:1056-7.
17. Grant J.W., Canter C.E., Spray T.L., Landt Y., Saffitz J., Ladenson J., Jaffe A.S. Elevated donor cardiac troponin I. A marker of acute graft failure in infant heart recipients. Circulation 1994;90:2618-2621.
18. Grant J.W., Canter C.E., Spray T.L., Landt Y., Saffitz J., Ladenson J., Jaffe A.S. Elevated donor cardiac troponin I. A marker of acute graft failure in infant heart recipients. Circulation 1994;90:2618-2621.
19. Carrier M., Solymoss B.C., Cartier R., Lerclerc Y., Pelletier L.C. Cardiac troponin T and creatine kinase MB isoenzyme as biochemical markers of ischaemia after heart preservation and transplantation. J Heart Lung Transplant 1994;13:696-700.
20. Burt J.M., Larson D.F., Copeland J.G. Recovery of heart function following 24 hours preservation and ectopic transplantation. J Heart Transplant 1986;5:298-303.

21. Stein D.G., Drinkwater D.C., Laks H., *et al.* Cardiac preservation in patients undergoing cardiac transplantation: a clinical trial comparing University of Winsconsin Solution and Stanford Solution. J Thorac Cardiovasc Surg 1991;102:657-65.
22. Seivers H.H., Leyh R., Jankhe A, Petry A., Kraatz E.G., Hermann G., Simon R., Bernhard A.. Bicaval versus atrial anastomoses in cardiac transplantation. J Thorac Cardiovasc Surg 1994;108:780-4.
23. Hamm C.W., Katus H.A. New biochemical markers for myocardial cell injury. Curr Opin Cardioil 1995;10:335-9.
24. Puleo P.R., Guadagno P.A., Roberts R., *et al.* Sensitive rapid assay of subforms of creatine kinase MB in plasma. Clin Chem 1989;35:1452-6.
25. Carrier M., Tourigny A., Thoribe N., *et al.* effects of cold and warm blood cardioplegia assessed by myocardial pH and release of metabolic marker. Ann Thorac Surg 1994;58:764-9.
26. Zimmerman R., Baki S., Dengler T.J., *et al.* Troponin T release after heart transplantataion. Br Heart J 1993;69:395-7.
27. Ravkilde J, Horder M., Gerhardt W., *et al.* Diagnostic performance and prognostic value of serum trroponin T in suspected acute myocardial infarction. Scand J Clin Lab Invest. 1993;53:667-71.
28. Katus H., Schoeppenthau M., Tanzeem A., *et al.* Non-invasive assessment of perioperative myocardial cell damage by circulating cardiac troponin T. Br Heart J 1991;65:259-62.
29. Ballester M., Obrador D., Carrio I., *et al.* Indium 111-monoclonal antimyosin antibody studies after the first year after heart transplantation. Circulation 1990;82:2100-8.
30. Carrier M., Solymoss B.C., Cartier R., Lerclerc Y., Pelletier L.C. Cardiac troponin T and creatine kinase MB isoenzyme as biochemical markers of ischaemia after heart preservation and transplantation. J Heart Lung Transplant 1994;13:696-700.
31. Duquesnoy P., Demetris A.J. Immunopathology of cardiac transplant rejection. Current Opinion in Cardiology 1995;10:193-206.
32. Ansari A.A., Sundstrom J.B., Kanter K., Mayne A., Villinger F., Gravinas M.B., Herskowitz A.. Cellular and molecular mechanisms of human cardiac myocyte injury after transplantation. J Heart Lung Transplant 1995;14:102-12.
33. M.E Billingham, N.R Cary, M.E Hammond. A working formulation for the standardization of nomenclature in the diagnosis of heart and lung rejection. Heart rejection study group. J Heart Transplant 1990;9:587-93.
34. Olivari M.T, Jessen M.E, Baldwin BJ, Horn VPH, Yancy CW, Ring WS, Rosenblatt RL. Triple drug immunospression with steroid discontinuation by six moths after heart transplantation. J Heart Lung Transplant 1995;14:127-35.
35. Nakhleh RE, Jones J, Goswitz JJ, Anderson EE, Titus J. Correlation of endomyocardial biopsy findings with autopsy findings in human cardiac allografts. J Heart Lung Transplant 1992 11(3 part1) 479-85.
36. Yeoh T-K, Frist WH, Eastburn TE, Atkinson J. The clinical significance of mild rejection of the cardiac allograft. Circulation 1992;86[suppl II]: II 267- II-27121.
37. D, Torres F, Concha M, Valles F. Repetitive non-treated episodes of Grade 1B or 2 acute rejection impair long term cardiac graft function. J Heart Lung Transplant 1995;14:452-460.
38. Kirklin JK, Naftel DC, Bourge RC, White-Williams C, Caulfield JB, Tarkka MR, Holman WL, Zorn GI. Rejection after cardiac transplantation. A time related risk factor analysis. Circulation 1992;86(Suppl II): II-236-II-241.
39. Hosenpud JD. Noninvasive diagnosis of acute allograft rejection. Another of many searches for the holy grail. Circulation 1992;85:368-71.
40. Ladowski JS, Sullivan M, Schatzlein MH, Peterson AC, Underhill DJ, Scheeringa RH. Cardiac isoenzymes following transplantation. Chest 1992;102:1520-21
41. Gash AK, Kayne FK, Morley D, Fitzpatrick JM, Alpern JB, Brozena SC. Serum myoglobin does not predict cardiac allograft rejection. J Heart Lung Transplant 1994;13:451-4.
42. Ballester M, Obrador D, Carrio I, Moya C, Auge JM, Bordes R, Marti V, Bosch I, Berna-Roqueta, Estorch M, Pons-Llado G, Camara JM, Aris A, Caralps JM. Early postoperative reduction of monoclonal antimyosin antibody uptake is associated with absent rejection-related complications after heart transplantation. Circulation 1992;85:61-68

43. Ratliff NB, Myles JL, McMahon JT *et al.* Myocyte injury in acute cardiac transplant rejection and in lymphocytic myocarditis is similar and reversible. Transplant Proc 1987;19:2568-72.
44. Hook S, Caple JF, McMahon JT, Myles JL, Ratliff NB Comparison of myocardial cell injury in acute cellular rejection versus acute vascular rejection in cyclosporin-treated heart transplants. J Heart Transplant 1995;14:351-8.
45. Anderson JR, Hossein-Nia M, Brown P, Corbishley C, Murday A, Holt DW. Creatine kinase MB isoforms- a potential predictor of acute cardiac allograft rejection. J Heart Lung Transplant 1995;14:666-70.
46. Anguita M, Lopez-Rubio F, Arizon JM, Latre JM, Casares Lopez-Granados A, Mesa D, Gimenez D, Torres F, Concha M, Valles F. Repetitive non-treated episodes of Grade 1B or 2 acute rejection impair long term cardiac graft function. J Heart Lung Transplant 1995;14:452-460.
47. Puleo PR, Perryman B. Noninvasive detection of reperfusion in acute myocardial infarction based on plasma activity of creatine kinase MB subforms. J Am Coll Cardiol 1991;17:1047-52.
48. 16. Katus HA, Remppis J, Scheffold D, Diederich KW, Keubler W. Intracellular compartmentation of cardiac troponinT and its release kinetics in patients with reperfused and non reperfused myocardium. Am J Cardiol 1991;67:1360-67.

Chapter 15

ECONOMIC ASPECTS OF NEW BIOCHEMICAL MARKERS FOR THE DETECTION OF MYOCARDIAL DAMAGE

Role of biochemical markers in the management of patients with chest pain

Paul O. Collinson

Biochemical confirmation of myocardial infarction is integral to the diagnostic process in the patient presenting with suspected acute coronary syndromes (ACS). The electrocardiogram (ECG) is the initial investigation, but diagnostic sensitivity is only 55-75% for acute myocardial infarction (AMI), rising to 80% when the serial ECG is used [1-4]. However, ST segment elevation on the ECG is 99.7% sensitive and specific for a final diagnosis of AMI [4]. Since this group is the only one which will benefit from thrombolytic therapy [5], the initial ECG is not a diagnostic but a management tool. This will however be the minority of patients presenting with chest pain. The typical casemix in patients presenting to a District General Hospital Emergency Department (ED) who undergo investigations to confirm or exclude acute coronary syndromes ACS is illustrated in Figure 1a. Biochemical testing is required for the confirmation of nonQ wave AMI in 20% of those with a final diagnosis of myocardial infarction, and for the exclusion of cardiac damage in the remainder, whose final diagnosis will be of either unstable anginal pain (UAP), stable angina or non-ischaemic chest pain (NICP).

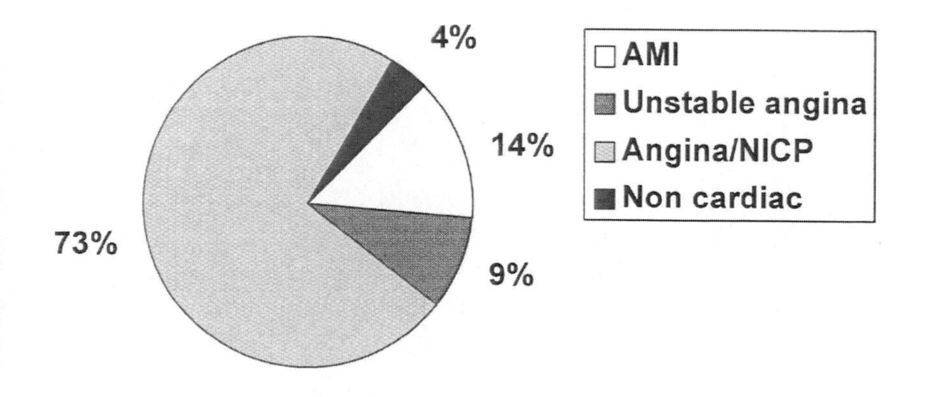

Figure 1a. Casemix of patients presenting to the Emergency Department with Chest Pain.

It has been shown that as many as 11.8% of patients with AMI are sent home from the Accident and Emergency Department [6], although a figure of around 6% is more usual. The casemix of patients admitted to a District General Hospital who undergo investigations to confirm or exclude ACS is illustrated in Figure 1b, with the casemix of those admitted to CCU illustrated in Figure 1c. It can be seen that clinical selection improves the proportion of patients with AMI but still 14.0% of patients occupying CCU beds have a final diagnosis of NICP. In the USA the proportion of people without IHD occupying CCU beds is higher.

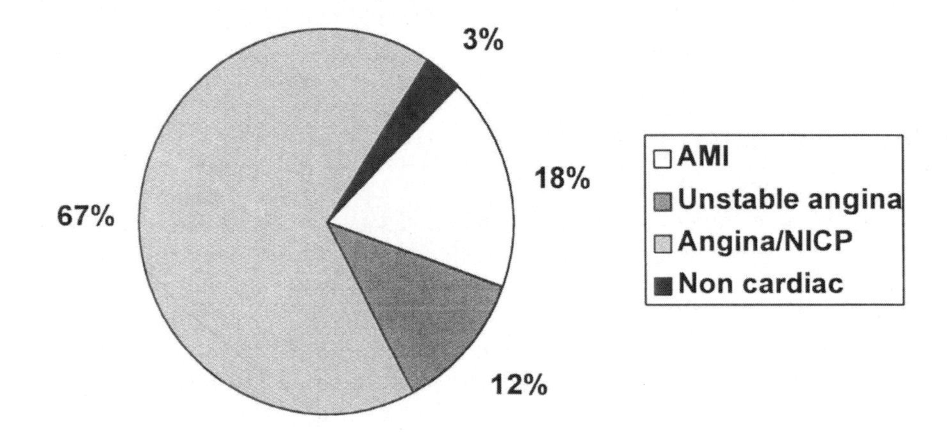

Figure 1b. Casemix of patients admitted to hospital for investigation of suspected acute coronary syndromes.

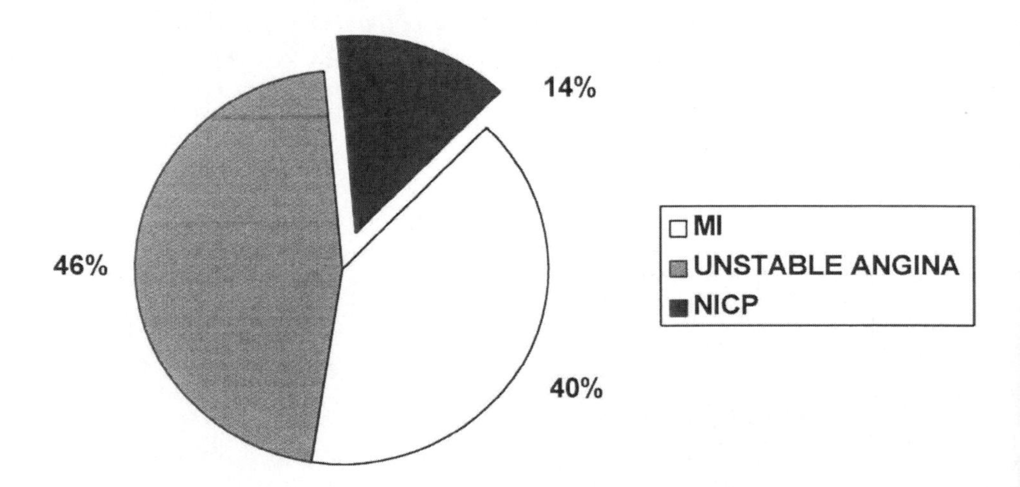

Figure 1c. Casemix of patient admitted to CCU by final diagnosis.

Recognition of the problem of missed diagnosis and inappropriate admission has led to the development of a number of approaches to improve clinical decision making. Diagnostic algorithms, including computer based systems, have been developed to improve the initial clinical selection [7,8]. Rapid diagnostic protocols based on early "rule-out" strategies for ACS can be used for early categorisation of patients into high and low risk groups by rapid identification of patients without AMI. These can be based on conventional diagnostic approaches [9-11], or using rapid sampling protocols utilising change in value, rather than exceeding a single reference or diagnostic threshold [12-15]. Application of such protocols can reduce length of stay on the CCU [16].

The problem is not simply one of rapidly ruling in or ruling out AMI. The currently available markers are able to do this within 8-12 hours of admission, and the technology exists, either by the use of point of care testing (POCT) or rapid delivery sample transit systems plus electronic data links to produce biochemical test results rapidly. The problem is that of the management of the patient with unstable angina. As Eagle has said: *The challenge is not always simply to rule in or rule out myocardial infarction, but rather to distinguish patients with acutely unstable coronary lesions from those with either stable coronary disease or none.* Eagle NEJM 1991; 325:1250

In the absence of a reliable biochemical protocol to identify myocardial damage 4-6% of patients with minor but significant cardiac damage, who are at high short term risk of death or major infarction, are sent home from the emergency room. Patients with a final diagnosis of unstable angina (UAP) or non-Q wave AMI have a worse short term (3-12 month) prognosis than patients with a classical Q wave AMI .The 12 month re-admission rate in this patient group with recurrent unstable angina, myocardial infarction or cardiac death is 20%.

The standard diagnostic test in Europe, North America and many centres in the UK for the confirmation of myocardial damage suspected on clinical grounds is measurement of creatine kinase (CK) and of its MB isoenzyme (CK-MB). This represents the standard against which all other tests should be judged [17]. Measurement of the cardiac structural protein cardiac troponin T (cTnT) has been evaluated in a number of studies covering cardiac and non-cardiac populations [18-22]. There have also been developments in POCT allowing rapid, accurate measurement of cTnT at the bedside [23-25]. Recent publications [26-30] have established that cardiac troponin T measurement is significantly more sensitive and specific than CK-MB and has an unique and established prognostic value in CK negative unstable angina. The combination of initial ECG plus cardiac troponin measurement allows patients to be efficiently categorised into risk groups.

Economic aspects of current diagnostic strategies

Missed diagnosis

The question of the missed diagnosis and its impact on the health care system will depend on the tendency to litigation. In the UK, the tort system and the use of judges to set the size of awards for negligence has meant that the number of cases in which an inappropriate diagnosis results in litigation is uncommon and awards are relatively modest. However, this climate is changing, with increasing sizes of awards. Funds currently assigned for medical litigation are estimated at £20 million nationally (for the equivalent of a national managed care system) and increasing at the rate of 20% per annum. In the USA, the litigation cost of missed diagnosis is high, because this patient group is at high risk of subsequent cardiac death. Ischaemic heart disease is an area in which it has been estimated that at least 20% of malpractice dollars are awarded to plaintiffs [31].

Inappropriate admission

The cost of inappropriate admission is high. Patients without IHD occupying CCU beds represent a waste of an expensive resource. From the data in Fig 1c, assuming a two day stay at a typical cost of £480 per day in the UK or $1000 in the USA, for a hospital admitting 3000 chest pain patients per year to the CCU this represents a cost of £39,7048 or $82,7184. In the USA it has been estimated that 5-10 billion dollars are spent annually investigating patients for chest pain, of whom only 30%-40% are ultimately diagnosed as having acute coronary syndromes. This means that 3-6 billion dollars annually are allocated to patients with NICP [31]. The problem of inappropriate admission has been highlighted as an area for study [32].

Diagnostic accuracy

Accurate diagnostic and prognostic categorisation in patients admitted with ACS is also required. The cost-benefit of thrombolytic therapy in patients with ST segment positive AMI has been defined. The economic aspects of management of patients with unstable angina have been less well studied. Morbidity and mortality in this patient group is high. Studies have been performed to compare revascularisation strategies in patients with both symptomatic and asymptomatic coronary artery disease [33,34]. The benefit of accurate prognostic categorisation with appropriate revascularisation can, therefore, be used to target resources.

Accurate diagnostic classification is not required solely to define the short term management of the patient, but also to plan longer term therapy. The value of secondary prevention strategies in patients with ischaemic heart disease has been unequivocally proved for both survival and need for intervention. There is, therefore, an economic cost of missed morbidity not only for the patient who is misdiagnosed at first presentation, but also for those admitted to hospital and misdiagnosed. The annual cost of inappropriate therapy in a patient incorrectly diagnosed as nonQ AMI on the basis of current biochemical tests will be at least £420 ($750) per year in pharmaceutical costs alone for a combination of aspirin plus a statin. As the majority of patients will fall into this category, accurate diagnosis becomes essential. As can be seen from figure 1b, potentially 3% of all patients may be misclassified on the basis of diagnosis using conventional markers, at an annual cost of £56,756 (for 4,500 cases per annum) and a 10 year treatment cost of £567,567

The advent of newer and more costly options for management of both AMI and unstable angina, including primary angioplasty, low molecular weight heparins and glycoprotein IIb/IIIa antagonists merely serves to underscore the need to improve current diagnostic approaches. These approaches must fit into an efficient decision making strategy to optimise therapy, which is both clinically and cost efficient.

The use of cardiac specific proteins, although efficient in identifying patients at high risk of cardiac death and in need of intervention, identifies a further population, missed by current strategies. This, in itself will have economic consequences by identifying additional patients at need of investigation and treatment.

Types of economic analysis

Cost minimisation analysis

In cost minimisation analysis the consequences of two interventions being compared are identical. Therefore, the analysis reduces to a comparison of costs alone.

Cost effectiveness analysis

There are few instances in which two health care interventions produce identical effects; the normal situation is one in which a difference in costs needs to be compared with a difference in consequences. Where this difference can be expressed in terms of a change in one main parameter, the difference in costs is related to the main difference in effects, measured in natural units. This may include cost per life year saved, disability days avoided or cases detected [35-37]. Comparison of the effects of intensive versus non intensive treatment of diabetes shows equivalent survival, but marked differences in terms of incidence of retinopathy and amputation.

Cost utility analysis

Interventions may differ in more than one way, with differences obtained at the expense of other effects. This is usually seen when there are differences in length and quality of life. In such instances, changes in health status are valued relative to one another to produce an overall index of health gain. The most widely used is the quality adjusted life year (QALY). Differences between efficacy and side effects may occur, producing differences in direct and indirect benefits. These are usually deducted from differences in direct and indirect costs and the results expressed as a ratio. Hence, interventions are compared in terms of the incremental cost per life year or QALY

Cost Benefit analysis

In cost benefit analysis, attempts are made to value all costs and consequences in the same units, usually financial. The results of the evaluation are then expressed solely as net economic benefit and net cost. It is therefore possible to assess whether total value of extra benefits of one intervention exceeds the extra costs. The problem with this approach is that not all qualities (such as length and quality of life) can be expressed in financial terms.

When assessing testing (and technology in general) a total model of care is required to identify the costs and cost consequences of alterations in the diagnostic process.

Economic analysis of new biochemical markers

It is apparent from the above that there are two different factors to be reconciled when considering the economic analysis of the use of the new cardiac markers. These are the actual cost of the tests itself and the impact on diagnostic and prognostic categorisation. To these must then be added the current diagnostic strategies in use and the clinical situations to which these are applied. The use of newer test may enable modification of current strategies, which may have an impact on costs, both positively and negatively.

The cost of an individual troponin measurement is greater than that of either CK, CK-MB activity, aspartate transaminase (AST) or lactate dehydrogenase (LDH), but comparable to CK-MB mass. Illustrative costs are shown in Table 1. Cost minimisation analysis can be applied when the utilisation of troponin will replace a range of other tests. This is critically dependant on the testing strategy used. In the situation in which the laboratory utilises multiple measurements of CK, CK-MB mass and lactate dehydrogenase (LDH) and LD isoenzymes, simple substitution of one or two troponin measurement for CK, CK-MB and LDH will provide an immediate cost saving. In the situation in which the cardiac enzyme strategy for the hospital utilises serial measurement of a combination of CK, CK-MB activity, aspartate transaminase (AST) and LDH or hydroxybutyrate dehydrogenase (HBD) as cardiac markers, transferring to troponin estimation will be more expensive.

Table 1. Performance Characteristics, Time Frame and Costs of available tests.

Test	Sensitivity	Specificity	Time frame	Cost/ Patient
Creatine Kinase (CK)	90	80	daily for 3 days	3.30
Asparate Transaminase (AST)	90	65	daily for 3 days	3.30
Lactate Dehydrogenase (LDH)	90	80	daily for 3 days	3.30
CK plus AST	90	90	daily for 3 days	6.60
CK,ASI and LDH	95	90	daily for 3 days	9.90
Creatine Kinase (CK) slope (by laboratory testing)	100	80	4 hours	2.20
Creatine Kinase (CK) slope (by point of care testing)	100	80	4 hours	4.80
Rapid serial CK plus CK-MB activity	95	90	8 hours	13.08
Rapid serial CK plus CK-MB mass	100	95	8 hours	22.08
CK plus CK-MB activity	95	90	daily for 2-3 days	13.08
CK plus CK-MB mass	100	95	daily for 2-3 days	22.08
Troponin T	100	100	once 12-24 hr Post admission	7.5
Troponin T (POCT)	100	100	once 12-24 hr Post admission	10.0
Troponin T	100	100	admission and 12-24 hr Post admission	15

However, this assumes that the diagnostic information provided is equal. This is clearly not the case, with troponin measurement identifying a further subset of patients at high risk. The outcomes are therefore not equivalent. It would, therefore, seem that the appropriate method would be a cost benefit analysis. This ignores two other factors, the clinical situation in which the markers will be used and the current clinical practice which the markers will alter. The clinical utility of the new markers will depend on the role to which they are to be put. This in turn will depend on the prior probability of IHD in the patient population.

The major contribution to hospital costs comes not from the cost of biochemical testing but from the length of hospital stay (and interventions). At this institution, in formulating a strategy for utilisation of troponin, we have also considered the diagnostic time windows of the markers. A single measurement of cardiac troponin T, 12-24 hours from admission is adequate to provide a definitive diagnosis. However, this will not allow the timing of the event. Hence, troponin measurement needs to be performed serially or combined with a second marker which has a shorter time window, or an earlier rise. In selecting the combination (one short halflife marker plus one long halflife marker) the choice will depend on available technology. We have combined CK with troponin.

In formulating the diagnostic algorithms we have developed and costed, we have considered the process of care. We have compared current strategies, the best strategy using available markers and the best strategy for rapid diagnosis using the available technology and the additional prognostic data available from troponin measurement. Data are based upon information obtained following a 3 month audit (data in Figure 1b above), projected to 4500 cases per annum. The end results of the protocol analysis can be costed, and can then be compared by cost minimisation analysis to demonstrate the effect of use of new markers and technologies.

1. Patients in the Emergency Department

A rapid investigation protocol in the Emergency Department that will accurately fast track admission for patients at high risk of cardiac disease combined with fast track out-patient consultation and investigation has been suggested. We have implemented a policy based on previous studies using rate of change of CK [15]. This strategy although very sensitive, has resulted in the admission of patients solely on the basis of an elevated CK. Because of the low prevalence of ischaemic heart disease among patients presenting with atypical chest pain and the low specificity of CK, over 80% of such patients are found to have a non-cardiac cause of elevation of CK and to have been an unnecessary admission. As cTnT is always elevated when a raised CK is of cardiac origin, failure to detect cTnT in the presence of an elevated CK allows exclusion of cardiac damage with 100% sensitivity and specificity. Hence measurement of cTnT in patients who present solely with an elevated CK allows exclusion of cardiac damage. These patients can be safely sent home.

The routine measurement of cTnT to screen for AMI on all patients in the ED not scheduled for admission on clinical or ECG grounds was compared with serial CK measurement at this institution and was found not to be clinically effective. None of the patients studied had a myocardial infarction which had not been detected initially by serial CK measurement. This would be expected from the known kinetics of cardiac marker release, and is consistent with studies showing poor early diagnostic performance of cTnT. Follow up of all patients sent home similarly revealed a detection rate of missed AMI of less than 1%., Hence, cTnT should only be measured in the ED to confirm or exclude AMI in patients with an elevated CK.

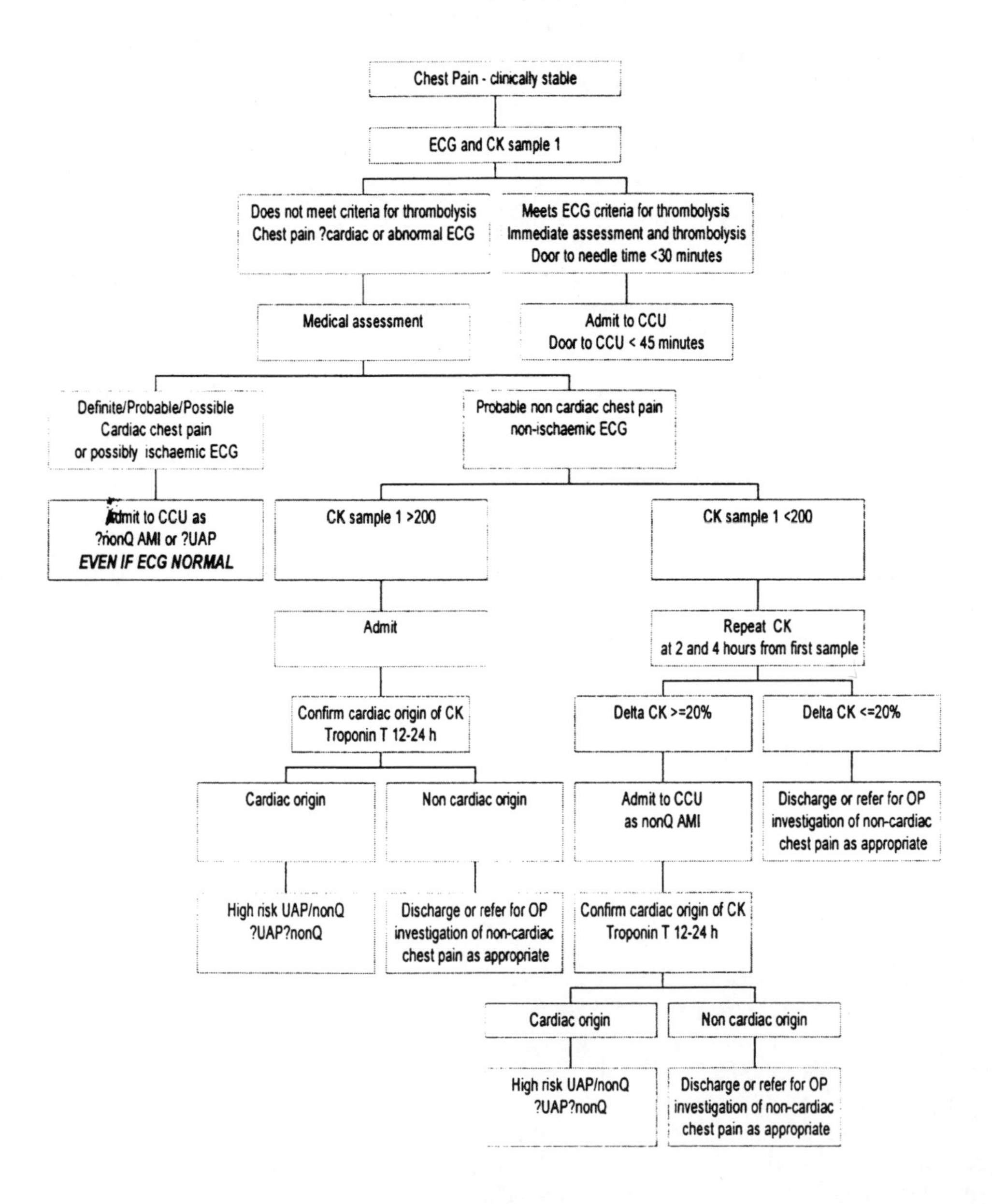

Figure 2. Mayday University Hospital - Chest Pain Protocol

The basic and enhanced protocol for the management of patients presenting to the ED with chest pain is shown in Figures 2, 3. The essential feature of this protocol is that all chest pain patients whose immediate admission is not indicated on clinical grounds, and who do not have an established non-cardiac cause for their chest pain, should undergo a point of care test for CK and, when indicated, cTnT. Patients whose sole criteria for admission are atypical chest pain and a CK of > 200 have a Troponin T analysis performed prior to admission, to exclude the 80% of these patients in whom the raised CK is non cardiac in origin. Predicted test performance is shown in Table 2 with cost analysis of three possible alternatives in Table 3. Comparing the two protocols by cost minimisation analysis, assuming each admission for ?AMI will occupy a CCU bed for 1.5 days @ £405 per day, medical bed for 2.5 days @ £163. Cost per admission is therefore £1015. Total cost per 100 ED CK only chest pain protocol patients is £480 + £19,285 = £19,765, for CK plus CK-MB £6547, for CK plus cTnT, £720.

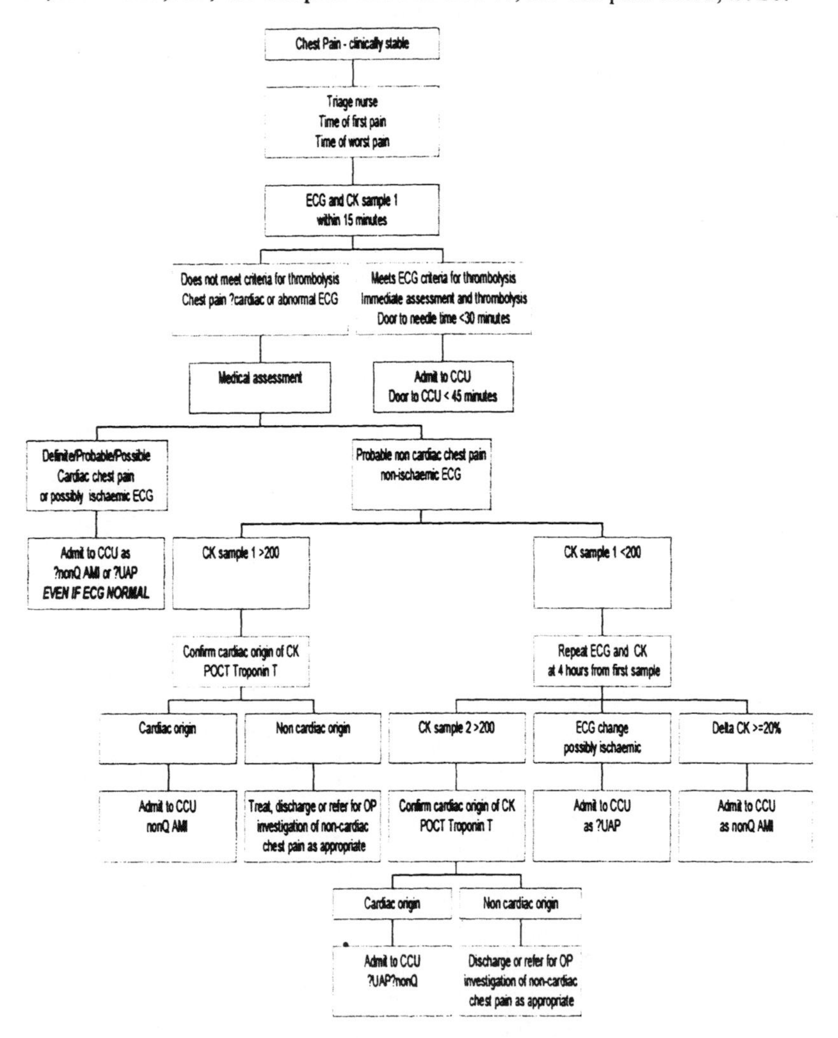

Figure 3. Mayday University Hospital - Chest Pain Protocol with POCT

Table 2. Test performance in Accident and Emergency Department patients with atypical chest pain and without ECG changes, estimated for prevalence of AMI in these patients of 5%.

	AMI Correctly Diagnosed	**AMI Patients Missed**	**Incorrect Diagnosis of MI**	**AMI Correctly Excluded**
CK 0 and 4 hours	5	0	19	76
CK plus CK-MB 0 and 4 hours	5	0	5	90
CK 0 and 4 h plus cTnT	5	0	0	95

Table 3. Diagnostic options and cost analysis per 100 ED patients with elevated CK

Strategy	Test cost	Inappropriate admissions	Cost
CK 0 and 4 hours	480	19	19,285.
CK plus CK-MB 0 and 4 hours	1472	5	5075
CK 0 and 4 h plus cTnT	720	0	0

Cost minimisation analysis

Applying these figures to the diagnostic protocols yields the following:

Total number of cases seen per annum	990
Number expected to have CK> 200 (11.8%)	120
Number expected to be admitted by CK only	120

Measuring CK plus CK-MB

Number with AMI (CK true positives) will be	22/120
Number without AMI (CK false positives) will be	47/120
Inappropriate admissions avoided	51/ year

Measuring Troponin T with CK> 200

Number with AMI (CK true positives) will be	22/120
Number without AMI (CK false positives) will be	0/120
Inappropriate admissions avoided	98/ year

Total costs are then for CK, £103,934; CK plus CK-MB, £52,920 and for CK plus cTnT £5,664. Thus, for an additional cost of £1,200 per annum for POCT measurement of Troponin T in the ED 147 CCU and 245 Medical bed days per year can be saved, equivalent to a cost saving of £99,470 on bed days and a total cost saving of £98,270.

2. Strategies for the investigation of in-patients with suspected acute ischaemic heart disease (myocardial infarction or unstable angina)

A 3 month audit was carried out of patients admitted to this institution with a diagnosis of definite or possible cardiac chest pain. During this time both daily CK x 3 and Troponin T were routinely available. This yielded data on the total numbers and also on the proportion of patients with Q-wave MI, non-Q wave MI, unstable angina and non-ischaemic chest pain, which was consistent with those registered in numerous studies in the literature. The proportion of patients in each category has been applied to the total number of in-patients in Mayday University Hospital for whom "cardiac enzymes" were requested over a 12 month period. The results are set out in the Table 4. Patients admitted with suspected AMI without characteristic ST changes and for identification of high risk unstable angina patients comprise 74.4% of all requests for "cardiac enzymes". The protocol models are daily CK for 3 days, CK + CK-MB mass daily for 3 days, CK x 2 (0 and 4 hours) by POCT in CCU plus cTnT at 12-24 hours and CK + CK-MB mass over 12 hours (3 samples). Costs are estimated in Table 5. The protocol assumes 100% specificity for CK-MB mass in the nonQ AMI group with characteristic CK changes, 95% specificity in patients with a high CK without nonQ AMI and detection of 65% of nonQ AMI in patients without a diagnostic CK value. The prevalence of non-Q wave MI in this group of patients is 607 out of 3,350 patients (18.1%), 2,743 of whom do not have AMI.

Table 4. Total number of patients in Mayday University Hospital in whom CK and cTnT were requested over 12 months divided into sub-groups by extrapolation of audit data.

		Number	Percent
Total number of patients with CK requests 1996		4,500	100
Total Admitted with definite or ? AMI		3,912	86.9
Typical AMI with ST-segment elevation		562	12.5
? Cardiac pain - Rule out AMI		3350	74.4
(i)	Non-Q wave MI (CK and cTnT positive)	141	3.1
(ii)	High risk unstable angina (CK negative, cTnT positive)	466	10.4
(iii)	Incorrect diagnosis of non-Q AMI (CK positive, cTnT negative)	135	3
(iv)	NICP/low risk unstable angina	2608	58.0
Miscellaneous **in-patients** (e.g. fall with raised CK, ? post-operative AMI etc) already admitted for non cardiac reasons		588	13.1

Using the current strategy, all 3350 patients will take 3 days for cardiac enzymes (CK ± CK -MB mass). Patients with non-Q wave AMI, will have an average stay of 5 days (1.5 days in CCU). Patients without non-Q wave MI will wait 3 days for enzyme results to confirm lack of myocardial damage and will stay a fourth day for their pre-discharge stress test, giving an average duration of stay of 4 days (1.5 days in CCU). The accelerated strategy comprises POCT CK x2 (0 and 4h) in all patients admitted to the CCU plus cTnT x1 at 12-24h post admission, or 3 CK plus CK-MB over 12 hours. Patients can be referred for immediate stress testing on clinical judgement if their CK's are negative. Patients will be identified as non-Q wave AMI, high risk unstable angina or NICP/low risk unstable angina within 24 hours. NICP/low risk unstable angina patients and can have their stress test and be discharged within 36-48 hours of admission (average 2 days). Thus, average CCU stay reduces to 1.0 day for patients with NICP/low risk unstable angina, and medical occupancy will average only 1.0 day. This approach has been demonstrated previously to be feasible [16]. The cost analysis of the protocol is shown in Table 6.

Table 5. Costs analysis in patients admitted to rule out AMI

Strategy	nonQ AMI Correctly Diagnosed	nonQAMI Patient Missed	Incorrect Diagnosis of AMI	AMI Correctly Excluded	Annual Cost £
CK daily x 3	141/607	466/607	135/2743	2608/2743	11,055.00
POCT Ck x2 0 +4 hrs. Lab cTnT x 1 12-24 hrs	607/607	0/607	0/2743	2743/2743	41,205.00
CK + CK-MB Mass x 3	444/607	163/607	7/2743	2736/2743	73968

Costs Minimisation analysis

The rapid diagnosis option uses 1372 (CK plus cTnT) or 1450 (CK plus CK-MB) fewer CCU bed days and 3784 (CK plus cTnT) or 4175 (CK plus CK-MB) fewer medical bed days. Potential savings are £1,142,113.51 (CK plus cTnT) or £1,296,148.99 (CK plus CK-MB)

Cost benefit analysis

With the CK plus cTnT strategy 466 high risk patients are identified who would be missed by the current strategy: the cost of identifying these high risk patients is given by:

$$\frac{\text{Cost of Recommended Strategy - Cost of Current Strategy}}{\text{466 high risk patients}}$$

$$= \frac{41{,}205 - 11{,}055}{466} = \text{£64.70 per high risk patient}$$

This identifies 466 high risk patients with an expected 12 month mortality of 20% (93 deaths), which could probably be reduced to 5% with early revascularisation.
With the CK plus cTnT strategy 135 patients without AMI are identified who would be misdiagnosed as nonQ AMI, potential treatment cost (statin plus aspirin) £56,700 per year.

Table 6. Cost analysis of protocol in pounds sterling

	number	CCU bed cost	Medical bed cost	Total	Saving
		£	£	£	£
CK x 3					
Non Q	276	167473	157273	3247456	
No AMI	3074	1867652	1252787	3120439	
Total patients	3350			3445185	
Difference in bed days			0		
Lab cost	11055				
Lab cost difference	0				
Total cost				3456240	0
CK + CK-MB					
Non Q	450	273580	256918	530498	
No AMI	2900	1761545	1181612	2943157	
Total patients	3350			3473655	
Difference in bed days			175		
Total lab cost	73968				
Lab cost difference	62913				
Total cost				3547623	-91383
POCT CK x 2+ cTnT					
Non Q	607	36861	346155	714759	
No AMI	2743	1111014	447149	1558162	
Total patients	3350			2272922	
Difference in bed days					
Total lab cost	41205				
Lab cost difference	30150				
Total cost				2314127	1142114
Rapid rule out with CK plus CK-MB					
Non Q	450	273580	256918	530498	
No AMI	2900	1174363	472645	1647008	
Total patients	3350			2177506	
Difference in bed days					
Total lab cost	73968				
Lab cost difference	62913				
Total cost				2251474	1296149

Conclusion

The new biochemical markers have a major role to play for cost effective patient management of patients with chest pain. When applied as part of an integrated decision making strategy they can be used to identify efficiently both high risk and low risk patients in a cost efficient manner.

Acknowledgements

I would like to thank the assistance of my colleagues in developing, evaluating and utilising these models. In particular Dr Peter Stubbs (Cardiology Department), Dr Rudolph Canepa-Anson (Cardiology Department) and Mr Kambiz Hashemi (Emergency Department)

References

1. McQueen M, Holder D. El-Maraghi N. Assessment of the accuracy of serial electrocardiography in the diagnosis of acute myocardial infarction. Am Heart J. 1983; 105: 258 - 61
2. Zarling EJ, Sexton H, Milnor P. Failure to diagnose acute myocardial infarction. JAMA 1983; 250: 1177-81
3. Brush JE, Brand DA, Acampora A, Chalmer B, Wackers FJ. Use of the admission electrocardiogram to predict in-hospital complications of acute myocardial infarction. NEJM 1985; 312: 1137-41
4. Yusuf S, Pearson M, Sterry H, Parish S, Ramsdale D, Rossi P, Sleight P. The entry ECG in the early diagnosis and prognostic stratification of patients with suspected acute myocardial infarction. European Heart Journal. 1984; 5: 690-6
5. Fibrinolytic Therapy Trialists Collaborative Group. Indications for fibrinolytic therapy in suspected acute myocardial infarction: collaborative overview of early mortality and major morbidity results from all randomised trials of more than 1000 patients. Lancet 1994; 343:311-22
6. Emerson PA, Russel NJ, Wyatt J, Crichton N, Pantin CFA, Morgan AD, Fleming PR. An audit of doctors management of patients with chest pain in the accident and emergency department. Quarterly Journal of Medicine. 1989; 70: 213-20
7. Pozen MW, D'Agostino RB, Selker HP, Sytkowski PA, Hood WB. A predictive instrument to improve coronary care unit admission practices in acute ischaemic heart disease. a prospective multi-centre clinical trial. NEJM 1984; 310:1273-8
8. Goldman L, Cook EF, Brand DA, et al. A computer protocol to predict myocardial infarction in emergency department patients with chest pain. NEJM 1988; 318: 797-803
9. Lee TH, Rouan GW, Weisberg MC et al. Sensitivity of routine clinical criteria for diagnosing myocardial infarction within 24 hours of hospitalisation. Ann Int Med. 1987; 106: 181-6
10. Lee TH, Juarez G, Cook EF, et al. Ruling out acute myocardial infarction - a prospective multi-centre validation of a 12-hour strategy for patients at low risk. NEJM 1991; 324: 1239-46
11. Hedges JR, Kobernick MS. Detection of myocardial Ischaemia/Infarction in the emergency department patient with chest discomfort. Emergency Medicine Clinics of North America. 1988; 6: 317-40.
12. Collinson PO, Rosalki SB, Flather M, Wolman R, Evans T. Early diagnosis of myocardial infarction by timed sequential enzyme measurements. Ann Clin Biochem 1988; 25: 376-82
13. Dufour RD, LaGreanade A, Guerra J. Rapid serial enzyme measurements in evaluation of patients with suspected myocardial infarction. Am J. Cardiol. 1989; 63: 652-5
14. Collinson PO, Rosalki SB, Kuwana T, Garratt HM, Ramhamadamy EM, Baird IM, Greenwood TW. Early diagnosis of acute myocardial infarction by CK-MB mass measurements Ann Clin Biochem 1992; 29: 43-7
15. Collinson PO, Stubbs PJ. 4 hour "rule-in" diagnosis of acute myocardial infarction for early risk stratification by creatine kinase increment (CK change). Ann Clin Biochem 1996; 33: 308-13
16. Collinson PO, Ramhamadamy EM, Stubbs PJ, Rosalki SB, Garrat HM, Mosely D, Evans DH, Fink RS, Baird IM, Greenwood TW. Rapid diagnosis of patients with acute chest pain reduces patient stay in the coronary care unit. Ann. Clin. Biochem. 1993; 30: 17-22.
17. Ryan TJ, Anderson JL, Antman EM, Braniff BA, Brooks NH, Califf RM et al. ACC/AHA Guidelines for the management of patients with acute myocardial infarction. Circulation 1996; 94:2341-50

18. Katus HA, Remppis A, Neumann FJ et al. Diagnostic efficiency of Troponin T measurements in acute myocardial infarction. Circulation 1991; 83:902-12.
19. Gerhardt W, Katus HA, Ravkilde J, et al. S-troponin T in suspected ischaemic myocardial injury compared with mass and catalytic concentrations of s-creatine kinase isoenzyme MB. Clin Chem 1991;37:1405-11.
20. Collinson PO, Moseley D, Stubbs P, Carter D. Troponin T for the differential diagnosis of ischaemic myocardial damage. Ann Clin Biochem 1993;30:11-6.
21. Wu AHB, Valdes R, Apple FS, Gornet T, Stone MA, Mayfield-Stokes, et al. Cardiac troponin T immunoassay for diagnosis of acute myocardial infarction. Clin Chem 1994; 40:900-7
22. Collinson PO, Chandler HA, Stubbs PJ, Moseley DS, Lewis D, Simmons MD. Cardiac troponin T and CK-MB concentration in the differential diagnosis of elevated creatine kinase following arduous physical training. Ann Clin Biochem 1995; 32: 450-3.
23. Mach F, Lovis C, Chevrolet J-C, Urban P, Unger P-F, Bouillie M et al. Rapid bedside whole blood cardiospecific immunoassay for the diagnosis of acute myocardial infarction. Amer J. Cardiol 1995; 75:842-5
24. Collinson PO, Thomas S, Siu L, Vasudeva P, Stubbs PJ, Canepa-Anson R. Rapid troponin T measurement in whole blood for detection of myocardial damage. Ann Clin Biochem 1995; 32: 454-8
25. Collinson PO, Gerhardt W, Katus HA, Muller-Bardorff M, Braun S, Schricke U et al. Multicentre evaluation of an immunological rapid test for the detection of troponin T in whole blood samples. Eur J. Clin Chem Clin Biochem. 1996; 34:591-98
26. Hamm CW, Ravkilde J, Gerhardt W, et al. The prognostic value of serum Troponin T in unstable angina. New Engl J Med 1992;**327**:146-50.
27. Ravkilde J, Horder M, Gerhardt W, Ljungdahl L, Pettersson Tryding N, Moller BH, Hamfelt A, Graven T, Asberg A et al. Diagnostic performance and prognostic value of serum troponin T in suspected acute myocardial infarction. Scandinavian Journal of Clinical & Laboratory Investigation. 1993; 53:677-85.
28. Wu AHB, Lane PL. Meta-analysis in clinical chemistry: Validation of cardiac troponin T as a marker for ischaemic heart disease. Clin Chem 1995; 41(B): 1228-33
29. Stubbs PJ, Collinson PO, Moseley D, Greenwood TW, Noble MIM. Prospective study of the role of cardiac troponin T in patients admitted with unstable angina. BMJ 1996; 313: 262-4
30. Lindahl B, Venge P, Wallentin L. Relation between troponin T and the risk of subsequent cardiac events in unstable coronary artery disease. Circulation .1996; 93:1651-7
31. Werne CS. Cost effective triaging, diagnosis and treatment of patients with chest pain in the emergency medicine department. Critical Interval 1997; 3.
32. Lee TH, Goldman L. The coronary care unit turns: Historical trends and future directions. Ann Int. Med 1988; 108: 887-94
33. The Bypass Angioplasty Revascularisation Investigation (BARI) Investigators. Comparison of coronary bypass surgery with angioplasty in patients with multivessel disease. NEJM 1996; 335:217-25
34. Hlatky MA, Rogers RJ, Johnstone I, Boothroyd MS, Brooks MM, Pitt B et al. Medical care costs and quality of life after randomisation to coronary angioplasty or coronary bypass surgey, NEJM 1997; 336:92-9
35. Sculman KA, Lynn LA, Glick HN, Eisenberg JM. Cost-effectiveness of low dose zidovudine therapy for asymptomatic patients with human immunodiffieciency virus (HIV) infection. Annals of Internal Medicine 1991; 114: 798-801.
36. Davies LM, Drummond MF. Assessment of costs and benefits of drug therapy for treatment resistant schizophrenia in the United Kingdom. Brit. J. Psych. 1993; 162:38-42
37. Hull RD, Hirsch J, Sackett DL, Stoddart GL. Cost effectiveness of clinical diagnosis, venography and non-invasive testing in patients with symptomatic deep-vein thrombosis. NEJM 1981; 304: 1561-7

Chapter 16

ROLE OF SERUM BIOCHEMICAL MARKERS IN CLINICAL TRIALS

Alan H. B. Wu

The field of cardiovascular medicine is changing rapidly with the development of new medical and surgical approaches to treat high risk and diseased patients. Placebo-controlled clinical trials are necessary to determine the efficacy of experimental therapies. Trials in coronary heart disease are usually identified by mnemonics abbreviations. The endpoints of these trials include presence or absence of ischaemia, rates of reinfarction, death, coronary artery patency, measurements of left ventricular function, infarct size, and degree of stenosis. Because many clinical trials involve thrombolysis with inhibition of thrombus and platelet function, the incidence of bleeding is also, frequently, a measured outcome. An assessment of the "quality of life" is very important in prospective outcomes analyses, but is rarely considered in these trials. Beyond these endpoints, there may be others that address specific questions or needs, on a subfraction of the overall population studied. In the past, cardiac markers, such as creatine kinase and CK-MB isoenzyme, have played a minor adjunctive role in these studies. With the development of new markers for ischaemia, minor myocardial injury, infarct sizing, and congestive heart failure, the use of novel cardiac markers such as glycogen phosphorylase BB, C-reactive protein, cardiac troponins T and I, and myosin light and heavy chains is likely to be more prominent. This chapter will review the objectives of various clinical trials, particularly in those areas in which cardiac markers have played an important role in the evaluation of data. Studies in which markers might have been useful, had they been known and available at the time, are also discussed.

Design of clinical trials in cardiology

Clinical trials in cardiology are essentially the same as those conducted for other medical specialities. A comprehensive textbook on clinical trials has been compiled by Spilker [1]. As summarized in Table 1, clinical trials can be subdivided into four major chronological phases. Most cardiology studies involving the efficacy of drugs are randomized, prospective, double blind, and placebo-controlled. In this manner, biases are minimized as neither the patient nor the attending physician know which subjects receive the experimental medication in question. In interventional studies, trials must be conducted as open-label since these procedures cannot be made hidden to patients. There have been some clinical trials whereby the advantages of the experimental procedure or drug is evident upon initial review, before the intended number of patients have been enrolled. When this occurs, the clinical trial may be stopped, as it may be unethical to deny the placebo group patients the access to the experimental procedure.

Table 1. Description of Phases in Clinical Trials[a]

Phase	Description
I	Initial safety trials conducted on healthy volunteers or ill patients. Basic pharmacokinetic data are obtained during this phase.
IIa	Pilot clinical trials in selected patient populations with the disease in question.
IIb	Controlled trials to evaluate and demonstrate clinical efficacy.
IIIa	Trials conducted prior to and for the purpose of submission to the Food and Drug Administration (FDA) as a New Drug Application (NDA).
IIIb	Supplementary clinical trials conducted after NDA submission but prior to FDA approval.
IV	Studies conducted after FDA approval to provide additional data on efficacy or safety. Trials can be conducted on new dosage formulations, different patient populations, or new clinical indications.

[a]Abridged from Spilker B. Guide to Clinical Trials. New York: Raven press, 1991:xxii-xxiii.

The sample size in cardiology studies differ greatly depending on the magnitude of the effect expected, variability of parameters analyzed, and the desired level of statistical significance to be achieved. For studies comparing the mortality associated with the use of different intravenous thrombolytic agents, thousands of enrollments have been necessary in the past. In order to enlist this large number of subjects, many of these trials are conducted at many sites, and are often multinational. These multicentre trials must be carefully planned and monitored so that data from different sites can be effectively pooled. Table 2 lists some of the large major clinical trials that have been conducted over the last few years, and the original purpose of these trials [2-12].

Requirements for use and measurement of cardiac markers in clinical trials

When cardiac markers are used to determine specific outcomes in clinical trials, there must be strict adherence to sample collection times and measurement techniques. The interpretation of results must also be made with knowledge of the limitations of the assays used. Perhaps the most essential point in conducting a valid trial is the timing of blood samples and the selection of the proper laboratory tests for the intended endpoint. Biochemical markers differ from one another in the patterns of release from ischaemic or necrotic tissue, and their appearance and clearance from the peripheral circulation. The window of optimal testing for many of the markers in use, or being proposed, is described in Table 3. Tests of the pathophysiology of acute coronary syndromes include C-reactive protein and amyloid protein A, soluble fibrin monomers and thrombus precursor protein, P-selectin, and glycogen phosphorylase BB isoenzyme [13]. These markers are useful to determine the presence of inflammation, thrombosis, platelet activation, and ischaemia, respectively. Other early markers of myocardial necrosis include myoglobin, and its ratio to carbonic anhydrase III, creatine kinase MB isoforms, and free fatty acid binding protein. Definitive biochemical tests for AMI diagnosis include creatine kinase-MB isoenzyme, and cardiac troponins T and I. For these tests, it is important that blood is not collected too soon or too late after injury, as this may result in a falsely negative result.

Table 2. Recent Clinical Trials in Cardiology

Name (abbreviation)	Purpose	No. Pts.	Ref.
Gruppo Italiano per lo Studio della Streptochinasi nell'Infarcto Miocardico (GISSI)	streptokinase and mortality	11,806	2
Thrombolysis in Myocardial Infarction (TIMI), Phase I	IV tPA vs. streptokinase	290	3
Thrombolysis in Myocardial Infarction, Phase II	invasive vs. conservative treatment after tPA	3262	4
Thrombolysis and Angioplasty in Myocardial Infarction (TAMI)	elective vs. immediate PTCA after tPA	386	5
Global Utilization of Streptokinase and Tissue Plasminogen Activator for Occluded Coronary Arteries (GUSTO)	four thrombolytic strategies for AMI	26,003	6
Second International Study of Infarct Survival (ISIS-2)	streptokinase and aspirin for AMI	17,187	7
Thrombolysis in Myocardial Infarction, Phase IIIB	tPA in unstable angina	1473	8
Beta-blocker Heart Trial (BHAT)	oral ß-blockers after AMI	3837	9
Captopril-Digoxin Multicentre Research Group	ACE inhibitors vs. digoxin	300	10
COhort of Rescue Angioplasty in Myocardial Infarction (CORAMI)	rescue PTCA after failed thrombolytics	87	11
Evaluation of 7E3 for the Prevention of Ischaemic Complications (EPIC)	ReoPro in PTCA	2099	12

Serial blood collections will be necessary for most clinical trials. For drug or interventional studies, results will be most meaningful when baseline concentrations of cardiac markers are determined and compared to results of samples collected after initiation of therapy. In this situation, the investigator may choose to discard pre-established reference ranges or cutoff concentrations, and instead make use of some difference value that is deemed to be statistically significant. For detection of cardiac events or side effects, measurement of the change in concentrations is likely to be more sensitive than absolute concentrations at a single time point. For infarct sizing, myosin light or heavy chains may be the most useful, and will require testing at very regular intervals, e.g., every 4 hours, for several days.

Table 3. Antiquated Cardiac Markers and Optimum Timing Windows for Current Markers used in Clinical Trials

Marker	Timing
Antiquated markers	
aspartate aminotransferase	
lactate dehydrogenase isoenzymes	
creatine kinase-MB activity assays	
Markers of ischaemia and inflammation	2-6 h
C-reactive protein	
soluble fibrin monomers	
thrombus precursor protein	
P-selectin	
glycogen phosphorylase BB isoenzyme	
Early markers of necrosis	3-12 h
myoglobin	
carbonic anhydrase III	
CK-MB isoforms	
fatty acid binding protein	
Definitive AMI markers	
CK-MB mass assays	9-72 h
cardiac troponin T	9-200 h
cardiac troponin I	9-150 h
Infarct sizing	
Myosin light or heavy chains	serially for several days

With regards to actual analysis, it is common and advantageous to have one "core" laboratory site that is assigned to conduct all testing. Fortunately, most cardiac markers are stable in serum or plasma when stored and shipped frozen. Use of a single laboratory eliminates the biases that are likely to be introduced by use of different instrumentation, cutoff concentrations, and laboratory practices. Laboratories conducting large multicentre trials should request enough blood so that they can be separated into one milliliter aliquots and separately frozen. This will enable each aliquot to be tested in duplicate, or triplicate if necessary. When new markers or indications for existing markers become available, pilot phase I or IIa studies can be retrospectively conducted on these stored samples.

Assay performance must also be considered. The precision of the assays may be the limiting factor. Commercial assays on automated immunoassay analyzers will be superior to prototype assays on manual platforms (e.g., ELISA). Older analysis procedures such as electrophoresis, column chromatography, and immunoinhibition are not as precise or sensitive as mass assays, and for cardiac markers, they should be discarded. The specificity of cardiac markers will be critical for detecting endpoints of clinical trials. Non-specific assays such as aspartate aminotransferase, lactate dehydrogenase with isoenzymes, and even total CK and CK-MB may no longer have a role in prospective clinical trials.

Evaluation of cardiac ischaemia

In the past, the existing cardiac markers have not been sufficiently sensitive to be of use in detection of ischaemic changes. Although the question has not been fully resolved, many investigators feel that release of cytosolic protein markers such as creatine kinase reflect irreversible injury thereby making them unusable for detecting reversible ischaemic changes [14,15]. In a recent study, however, cardiac troponin I was shown to be released in ischaemic tissue following stenosis of coronary arteries in a pig model [16]. The mechanism of troponin I release in ischaemic tissue is unclear. It is possible the troponin originates from the free cytoplasmic pool and not the structural component. Glycogen phosphorylase BB isoenzyme is a marker that may be more effective than troponin for ischaemia because this enzyme directly participates in the glycogenolysis that occurs with oxygen deficits [17]. Therefore, release of this isoenzyme might occur without irreversible damage to the myocytes.

An area whereby ischaemic markers may be of tremendous value is in new treatment strategies for patients with unstable angina pectoris. Unstable angina encompasses a wide rage of clinical diseases ranging from ischaemia to non-Q wave acute myocardial infarction [18]. Aspirin, heparin, nitrates, and beta-blockers have been proposed as standard medical therapy for patients with unstable angina [19]. For patients with non-Q wave AMI, traditional and new biochemical markers such as CK-MB and cardiac troponin have and will continue to be used for diagnosis and risk assessment. However, new therapies are being directed to minimize the extent of injuries caused by plaque disruption. They have largely focused on inhibiting both thrombosis and activation of platelets, with drugs such as hirudin, hirulog, and argatroban [20]. These agents may be more effective than heparin because they inhibit thrombin that is both clot bound, as well as that present free in the circulation. In clinical studies, hirudin, argatroban, and hirulog have been examined in patients with unstable angina [21-23]. The primary endpoints included the demonstration that these agents inhibited clot-bound thrombin, and to document the incidence of untoward events such as death, reinfarction, recurrent ischaemia, and bleeding.

Antagonists to the platelet receptor glycoprotein (GP) IIb/IIIa is another area of active clinical research for patients with acute coronary syndromes. Abciximab (ReoPro) is a Fab monoclonal antibody fragment that binds to the GP IIb/IIIa receptor thereby inhibiting platelet adhesion to endothelial surfaces. Clinical trials have initially focused on the protective effect of ReoPro and other antagonists such as Integrelin on patients undergoing PTCA. The EPIC and IMPACT trials demonstrated a reduction in the incidence of death, AMI, and restenosis [12,24]. In a preliminary study of refractory angina, ReoPro was shown to reduce the incidence of recurring ischaemic events as measured by electrocardiography, and of TIMI coronary artery blood flow, as measured by angiography [25].

Biochemical markers of ischaemia are not currently part of any of these clinical trials. However, if reduction of ischaemic episodes is a desired objective of these new drugs, then analysis for glycogen phosphorylase BB, thrombin precursor protein, and p-selectin should be seriously considered, once the pathophysiologic and clinical utility of these markers have been separately validated, and commercial assays made available.

Detection of minor myocardial injury following use of cardiotoxic drugs

Sensitive cardiac markers can be used to determine the presence of cardiotoxicities in clinical drug trials. For example, anthracycline drugs such as doxorubicin and daunorubicin are effective antineoplastic drugs in pediatric patients with tumors, such as malignant lymphoma [26]. Many of these drugs can have serious cardiotoxicities. The mechanism is thought to be induced by the formation of free radicals [27]. In clinical studies, serum cardiac troponin T has been used to determine the extent of cardiac injuries when doxorubicin was used to treat various malignancies. In some studies, no increases were observed in studies of pediatric [28] and adult [29] malignancies, indicating the absence of significant cardiotoxicities. In others, however, mild increases in cTnT were observed in a subset of patients [30].

Other chemotherapeutic drugs have been reported to have toxic effects that include the myocardium. 5-fluorouracil (5-FU) inhibits DNA synthesis and has been used for many years for solid tumors such as breast and colorectal cancers. In one study of 231 patients, six (2.6%) developed significant cardiotoxicities [31]. Symptoms, including angina, hypotension, dyspnea, tachycardia, and arrhythmias are thought to be caused by coronary artery spasm and do not contribute to irreversible injury [32]. However, there has been one case report in which use of 5-FU led to an AMI, suggesting that irreversible injury can also occur [33]. Paclitaxel is a antimitotic cytotoxic agent also used in patients with metastatic breast cancer, and can produce cardiotoxicities. In one study of 27 patients with ovarian cancer, cardiotoxicity was reported in 14.8% of cases with the combined use of paclitaxel and carboplatin [34]. In a case report of a paclitaxel-induced fatality, ultrastructural analysis suggested that the drug causes loss of myofibrils, and is similar to anthracycline-induced myocardial injury [35]. Cardiac markers are not routinely used to evaluate drug toxicities affecting the heart. With the development of sensitive and specific markers for ischaemia and necrosis, such as cardiac troponins, future drug trials for antineoplastic agents should include a biochemical assessment of myocardial damage.

Cardiac markers in conjunction with revascularization

Arguably the most important advance in the treatment of acute myocardial infarction has been the introduction of the medical and surgical revascularization techniques of intravenous thrombolytic therapy and percutaneous transluminal angioplasty (PTCA). In the initial trials, the outcomes for clinical trials for IV therapy have focused on mortality, and rate of coronary artery patency vs. placebo (Table 2). After thrombolytic therapy became accepted as standard practice, subsequent clinical trials focused on comparing success rates against different IV agents, and the success of the combination of IV therapy with PTCA. The accepted standard for determining coronary artery patency is angiography. New thrombolytic agents have been developed, such as reteplase (a deletion mutant of wild-type t-PA), and will be the focus of future clinical trials [36].

Coronary blood flow is graded according to a scale established by the TIMI investigators [4]: grades 0 and 1 are no flow to the coronary beds distal to the obstruction, while grades 2 and 3 indicate partial and complete perfusion, respectively, of previously affected areas. Serial measurement of cardiac markers have been proposed as a mechanism for the non-invasive determination of coronary blood flow (see Chapter 11). Many studies have suggested that blood is collected just prior to initiation of therapy and either 90 or 120 minutes thereafter [37,38]. Detection of the "washout phenomenon", i.e., a bolus-like release of enzymes and proteins, indicates the presence of successful reperfusion. Although serial measurements of markers may become part of routine clinical practice, it seems unlikely that serologic testing will be definitive enough to replace angiography as the "gold standard" for determination of success of thrombolytic therapy in clinical trials.

Although the very objective of revascularization therapy is to restore coronary artery blood flow, research in experimental animal models have shown that when such therapy is successful, irreversible damage to these arteries can, and does, occur. This phenomenon is termed "reperfusion injury." The mechanism of reperfusion injury is thought to be a combination of free radical formation, activation of complement, and a disturbance in calcium homeostasis [39]. Therapeutic drug trials are underway to reduce the magnitude of reperfusion injury. Clinical and experimental trials have focused on the use of magnesium [39], liposomal prostaglandin E1 [40], N-acetylcysteine [41], and poloxamer 188 [42] to reduce infarct sizing, improve left ventricular function and regional wall motion.

Reperfusion injury also occurs following coronary artery bypass graft (CABG) surgery. Traditional cardioplegic protective agents have included tepid hyperkalemic crystalloids. Clinical trials are being conducted on new protective agents such as inotropic agents (e.g., dopamine, dobutamine, etc.), adenosine, a potassium channel opener [43], L-arginine, which augments nitric oxide release to inhibit neutrophilic injury [44], and allopurinol which can attenuate free radical generation [45]. There has been a lot of recent interest in acadesine, as it can increase the availability of adeonsine in ischaemic tissues. In the Muticentre Study of Perioperative Ischaemia Research Group, acadesine was shown to reduce early cardiac death, MI, and combined adverse cardiovascular outcomes when administered before and during CABG surgery [46].

Biochemical markers may be useful in estimating the effect of therapy designed to limit ischaemic reperfusion injury in controlled animal studies and human clinical trials. In the past, infarct sizing studies have been based on serial measurements of creatine kinase and CK-MB. These markers do not produce accurate measurements of infarct size following reperfusion because, per gram of infarcted tissue, there is a higher recovery of enzymes in the blood of patients with successful reperfusion, compared with those with unsuccessful reperfusion [47]. It is not known how well CK-MB will perform when comparing subsets of patients on whom successful reperfusion has been achieve Nevertheless, markers that are less subject to influence by reperfusion status may be more appropriate such as myosin light and heavy chains [48]. Validation studies for experimental markers will need to be conducted in controlled animal models prior to infarct size estimates made on humans.

Therapeutic trials of congestive heart failure (CHF)

Among the long term consequences of repeated episodes of ischaemic injury is development of congestive heart failure. Clinical practice guidelines have been developed for management of patients with CHF, which has an estimated annual mortality rate of 10% [49]. CHF is classified by severity using guidelines established by Killip *et al.*, and the New York Heart Association [50,51]. Practice guidelines suggest the use of angiotensin converting enzyme (ACE) inhibitors and diuretics if there are signs of significant volume overload. Digoxin can be used to improve contractile force if the case is severe. Patients should be evaluated for left ventricular function with echocardiography and radionuclide ventriculography to measure ejection fraction. There has been interest recently in the use of beta-blockers as adjunctive therapy for CHF patients, although the improvements in mortality have not yet been documented [52].

Clinical trials for digoxin and ACE inhibitors in patients with heart failure have focused on the rate of clinical deterioration, exercise tolerance, ejection fraction, CHF classifications, and quality of life [53,54]. Biochemical markers are not currently used in these assessments. Recently, however, brain natriuretic peptide (BNP) has been suggested as a marker of CHF. BNP is a 32-amino acid peptide that originates from the brain and ventricles of the heart and is an important hormone for regulation of fluid volume, sodium balance and blood pressure [55]. Blood concentrations of BNP in patients with CHF are significantly increased over normal values. Although there is overlap between groups, there is a correlation between the extent of abnormal concentrations and the severity of disease [56]. When new clinical trials are being planned, it may be possible to use blood concentrations of BNP before and after initiation of new therapy, to determine if there is a decrease in the concentration of BNP, suggestive of an amelioration of the disease.

Research in stunning and hibernating myocardium

Brain natriuretic peptide, and possibly other cardiac markers, might also be useful in future research and clinical studies involving myocardial stunning and hibernation. Myocardial tissue is said to be stunned when there is left ventricular contractile dysfunction, despite the presence of viable myocardial cells [57]. This phenomenon occurs immediately after revascularization of ischaemic areas (e.g., thrombolytic therapy, angioplasty, bypass surgery) and can last for 24-48 hours. In the hibernating myocardium, there is a persistent dysfunction of viable tissue caused by a reduction in coronary blood flow, such as that seen in patients with acute coronary syndromes. The hibernating myocardium may be a compensatory mechanism to reduce the oxygen needs of ischaemic tissue. Restoration of blood flow leads to a return of contractile function. The echocardiogram is used to detect regional wall motion abnormalities that are caused by stunned and hibernating myocardium. However, irreversible myocardial necrosis resulting from acute myocardial infarction will also produce these abnormalities. These cells, however, cannot be recovered. Positron emission tomography can be used to distinguish between dysfunctional but viable tissue from that which is permanently damaged [58]. The uptake of substances such as fluorodeoxyglucose indicates active metabolites by myocytes that have latent contractile capabilities.

Cardioprotective agents used before and after bypass surgery are logical candidate pharmacologic therapy for stunned myocardium. In addition, other drug classes have been used to treat or demonstrate stunning and hibernation including nitrates, calcium antagonists, ACE inhibitors, and beta-blockers [58]. Because neither stunning nor hibernating myocardium are associated with irreversible injury, conventional serum markers such as CK-MB and troponin are within the normal range. To evaluate the success of protective agents, early ischaemic markers, such as glycogen phosphorylase BB, or tests of left ventricular function such as BNP may have the most to offer in clinical trials.

References

1. Spilker B. Guide to Clinical Trials. New York: Raven Press, 1991.
2. Gruppo Italiano per lo Studio della Streptochinasi nell'Infarcto Miocardico (GISSI): Effectiveness of intravenous thrombolytic treatment in acute myocardial infarction. Lancet 1986:1:397-402.
3. Chesebro JH, Knatterud G, Roberts R, *et al.* Thrombolysis in Myocardial Infarction (TIMI) Trial, Phase I: a comparison between intravenous tissue plasminogen activator and intravenous streptokinase. Circulation 1987;76:142-54.
4. TIMI Study Group. Comparison of invasive and conservative strategies after treatment with intravenous tissue plasminogen activator in acute myocardial infarction. N Engl J Med 1989;320:618-27.
5. Topol EJ, Califf RM, Kereiakes DJ, George BS. Thrombolysis and angioplasty in myocardial infarction (TAMI) trial. J Am Coll Cardiol 1987:10(5 Suppl B):65-74B.
6. The GUSTO Investigators. An international randomized trial comparing four thrombolytic strategies for acute myocardial infarction. N Engl J Med 1993;329:673-82.
7. Second International Study of Infarct Survival Collaborative Group. Randomized trial of intravenous streptokinase, oral aspirin, both, or neither among 17,187 cases of suspected acute myocardial infarction: ISIS-2. Lancet 1988;2(8607):349-60.
8. TIMI IIIB Investigators. Effects of tissue plasminogen activator and a comparison of early invasive and conservative strategies in unstable angina and non-Q-wave infarction: results of the TIMI IIIB trial. Circulation 1994;89:1545-56.
9. Bell RL, Furb JD, Friedman LM, McIntyre KM, Payton-Ross C. Enhancement of visit adherence in the national beta-blocker heart attack trial. Cont Clin Trials 1985;6:89-101.
10. The Captopril-Digoxin Multicenter Research Group. Comparative effects of therapy with captopril and digoxin in patients with mild to moderate heart failure. JAMA 1988;259:539-44.
11. Cohort of Rescue Angioplasty in Myocardial Infarction (CORAMI) Study group. Outcome of attempted rescue coronary angiography after failed thrombolysis for acute myocardial infarction. Am J Cardiol 1994;74:172-4.
12. The EPIC Investigators. Use of a monoclonal antibody directed against the platelet glycoprotein IIb/IIIa receptor in high-risk coronary angioplasty. N Engl J Med 1994;330:956-61.
13. Wu AHB. Use of cardiac markers as assessed by outcomes analysis. Clin Biochem 1997,30:339-50.
14. Sakai K, Gebhard MM, Spieckermann PG, Bretschneider HJ. Enzyme release resulting from total ischaemia and reperfusion in the isolated, perfused guinea pig heart. J Mol Cell Cardiol 1973;5:359-407.
15. Ishikawa Y, Saffitz JE, Mealman TL, Grace AM, Roberts R. Reversible myocardial ischaemic injury is not associated with increased creatine kinase activity in plasma. Clin Chem 1997;43:467-75.
16. Feng YJ, Chen C, Fallon JT, Ma L, Waters DD, Wu AHB. Comparison of cardiac troponin I, creatine kinase-MB, and myoglobin for detection of acute myocardial necrosis in a swine myocardial ischaemic model. Am J Clin Pathol 1997; in press.
17. Rabitzsch G, Mair J, Lechleitner P, *et al.* Immunoenzymometric assay of human glycogen phosphorylase isoenzyme BB in diagnosis of ischaemic myocardial injury. Clin Chem 1995;41:966-78.
18. Braunwald E, Mark DB, Jones RH, *et al.* Unstable angina: diagnosing and management, Clinical Practice Guideline, No. 10 (amended). U.S. Department of Health and Human Services Public Health Service, May, 1994.
19. Waters DD, Lam J, Theroux P. Newer concepts in the treatment of unstable angina pectoris. Am J Cardiol

1991;68:34-41C.
20. Ambrose JA, Weinrauch M. Thrombosis in ischaemic heart disease. Arch Intern Med 1996;145:1382-94.
21. Rao AK, Sun L, Chesebro JH, *et al.* Distinct effects of recombinant desulfatohirudin (Revasc) and heparin on plasma levels of fibrinopeptide A and prothrombin fragment F1.2 in unstable angina. A multicenter trial. Circulation 1996;94:2389-95.
22. Gold HK, Torres FW, Garaedian HD, *et al.* Evidence for a rebound coagulation phenomenon after cessation of a 4-hour infusion of a specific thrombin inhibitor in patients with unstable anbina pectoris. J Am Coll Cardiol 1993;21:1039-47.
23. Fuchs J, Cannon CP. Hirulog in the treatment of unstable angina. Results of the thrombin inhibition in myocardial ischaemia (TIMI) 7 trial. Circulation 1995;92:727-33.
24. Tchen JE, Harrington RA, Kottke-Marchant K, *et al.* Multicenter randomized, double-blind, pacebo-controlled trial of the platelet integrin glycoprotein IIb/IIIa blocker Integrelin in elective coronary intervention. IMPAC Investigators. Circulation 1995;91:2151-7.
25. Simoons ML, Jande Boer MJ, van den Brand MJBM *et al.* Randomized trial of a GPIIb/IIIa platelet receptor blocker in refractory unstable angina. Circulation 1994;89:596-603.
26. Muggia FM, Green MD. New antracycline anti tumor antibiotics. Crit Rev Oncol Hematol 1991;11:43-64.
27. Von Hoff DD, Layard MW, Basa P, *et al.* Risk factors for doxorubicin-induced congestive heart failure. Ann Intern Med 1979;91:710-7.
28. Fink FM, Genser N, Fink C, *et al.* Cardiac troponin T and creatine kinase MB mass concentrations in children receiving anthracycline chemotherapy. Med Ped Oncol 1995;25:185-9.
29. Raderer M, Kornek G, Weinlander G, Kastner J. Serum troponin T levels in adults undergoing anthracycline therapy. J Nat Cancer Instit 1997;89:171.
30. Ottlinger ME, Sallan SE, Rifai N, Sacks DB, Lipshultz SE. Myocardial damage in doxorubicin-treated children: a study of serum cardiac troponin-T measurements (abstract). ASCO Annual Meeting, Los Angeles, CA, 1995.
31. Weidmann B, Jansen W, Heider A, Niederle N. 5-fluorouracil cardiotoxicity with left ventricular dysfunction under different dosing regimens. Am J Cardiol 1995;75:194-5.
32. Kleiman NS, Lehane DE, Geyer CE Jr, Paratt CM, Young JB. Prinzmetal's angina during 5-fluorouracil chemotherapy. Am J Med 1987;82:566-8.
33. Mizuno Y, Hokamura Y, Kimura T, Kimura Y, Kaikita K, Yasue H. A case of 5-fluorouracil cardiotoxicity simulating acute myocardial infarction. Jap Circulation J 1995;59:303-7.
34. Bolis G, Scarfone G, Villa A. Acerboni S, Siliprandi V, Guarnerio P. A phase I trial with fixed-dose carboplatin and escalating doses of paclitaxel in advanced ovarian cancer. Sem Oncol 1997;24(1 Suppl 2): S2-23-5.
35. Shek TW, Luk IS, Cheung KL. Paclitaxel-induced cardiotoxicity. An ultrastructural study. Arch Path Lab Med 1996;120:89-91.
36. International Joint Efficacy Comparison of Thrombolytics. Randomized, Double-blind Comparison of Reteplase Double-bolus Administration with Streptokinase in Acute Myocardial Infarction (INJECT): Trial to Investigate Equivalence. Lancet 1995;346:329-336.
37. Zabel M, Hohnloser SH, Koster W, Prinz M, Kasper W, Just H. Analysis of creatine kinase, CK-MB, myoglobin, and troponin T time-activity curves for early assessment of coronary artery reperfusion after intravenous thrombolysis. Circulation 1993;87:1542-50.
38. Apple FS, Henry TD, Berger CR, Landt YA. Early monitoring of serum cardiac troponin I for assessment of coronary reperfusion following thrombolytic therapy. Am J Clin Pathol 1996;105:6-10.
39. Steurer G, Yang P, Rao V, *et al.* Acute myocardial infarction, reperfusion injury, and intravenous magnesium therapy: basic concepts and clinical implications. Am Heart J 1996;132:478-82.
40. Feld S, Li G, Wu A, *et al.* Reduction of canine infarct size by bolus intravenous administration of liposomal prostaglandin E1: comparison with control, pacebo liposomes, and continuous intravenous infusion of prostaglandin E. Am Heart J 1996;132:747-57.
41. Sochman J, Vrbska J, Musilova B, Rocek M. Infarct size limitation: acute N-acetylcysteine defense (ISLAND trial: preliminary analysis and report after the first 30 patients. Clin Cardiol 1996;19:94-100.
42. Schaer GL, Spaccavento LJ, Browne KF, *et al.* Beneficial effects of RheothRx injection inpatients receiving thrombolytic therapy for acute myocardial infarction. Results of a randomized, double-blind, placebo-controlled trial. Circulation 1996;94:298-307.
43. Fremes SE, Levy SI, Christakis GT, *et al.* Phase 1 human trial of adeonsine-potassium cardioplegia. Circulation 1996;94(9Suppl 1):II370-5.
44. Vinten-Johansen J, Sato H, Zhao ZQ. The role of nitric oxide and NO-=donor agents in myocardial

protection from surgical ischaemic-reperfusion injyry. Int J Cardiol 1995;50:273-81.

45. Movahed A, Nair KG, Ashavaid TF, Kumar P. Free radical generation and the role of allopurinol as a cardioportective agent during coronary artery bypass grafting surgery. Can J Cardiol 1996;12:138-44.
46. Mangano DT. Effects of acadesine on myocardial infarction, stroke, and death following surgery. A meta-analysis of the 5 international randomized trials. The Multicenter Study of Perioperative Ischaemia (McSPI) Research Group. JAMA 1997;227:325-32.
47. Vatner SF, Baig H, Manders WT, Maroko PR. Effects of coronary artery reperfusion on myocardial infarct size calculated from creatine kinase. J Clin Invest 1978;61:1048-56.
48. Katus HA, Diederich KW, Uellner M, Remppis A, Schuler G, Kubler W. Myosin light chains release in acute myocardial infarction: non-invasive estimation of infarct size. Cardiovasc Res 1988;22:456-63.
49. Konstam MA, Dracup K, Baker DW, *et al.* Heart failure: evaluationa dn care of patients with left-ventricular systolic dysfunction. Clinical Practice Guideline, No. 11 (amended). U.S. Department of Health and Human Services Public Health Service, June, 1994.
50. Functional capacity and objective assessment. In: Nomenclature and criteria for diagnosis of diseases of the heart and great vessels, 9th ed. Dolgin M, ed., Little, Brown and Co., New York, 1994:253-55.
51. Killip T, Kimball J. Treatment of myocardial infarction in a coronary care unit. A two year expeience with 250 patients. Am J Cardiol 1967;20:457-64.
52. Sharpe N. Beta-blockers in heart failure. Future directions. Europ Heart J 1996;17 Suppl B:39-42.
53. DiBianco R, Shabetai R, Kostuk W, *et al.* A comparison of oral milrinone, digoxin, and their combination in the treatment of patients with chronic heart failure. N Engl J Med 1989;320:677-83.
54. Packer M, Gheorghiade M, Young D, *et al.* Withdrawal of digoxin from patients with chronic heart failure treated with angiotensin-cinverting-enzyme inhibitors. N Engl J Med 1993;329:1-7.
55. Wei CM, Heublein DM, Perrella MA, *et al.* Natriuretic peptide system in human heart failure. Circulation 1993;88:1004-9.
56. Mukoyama M, Nakao K, Hosoda K, *et al.* Brain natriuretic peptide as a novel cardiac hormone in humans. Evidence for an exquisite dual natriuretic peptide system, atrial natriuretic peptide and brain natriuretic peptide. J Clin Invest 1991;87:1402-12.
57. Klooner RA, Przykleng K. Stunned an hibernating myocardium. Ann Rev Med 1991;42:1-8.
58. Kloner RA, Przyklenk K, Rahimtoola SH, Braunwald E. Myocardial stunning and hibernation: mechanisms and clinical implcationse. In: Heart Disease Update, E. Braunwald, ed., 241-56.

Chapter 17

PRE-CLINICAL APPLICATION OF MARKERS OF MYOCARDIAL DAMAGE

David W. Holt

There is increasing interest in the specific detection of myocardial damage in pre-clinical studies. In addition to basic research studies involving animals this interest stems from the need to rule-out the potential for myocardial damage of new drug molecules at an early stage in their development. Specific detection of myocardial damage in animals may also be of use in the diagnosis and treatment of pathological conditions affecting the hearts of agricultural or domestic animals.

Until recently, the analytes available in this setting lacked both specificity and sensitivity for myocardial damage, and were poorly validated in animal species. With the introduction of assays for the cardiac troponins, and for the separation of creatine kinase MB isoforms, the potential for studies in this field has been substantially enhanced.

This chapter will review some of the data which suggest that assays optimised for the measurement of cardiac marker proteins in samples from humans can be used in samples from some animal species with some confidence.

Analytes available

For most experimental studies the aim is, as in the human setting, to detect whether there are biochemical signs of myocardial damage by performing single or serial measurements of marker analytes in serum or plasma. For such measurements to be applicable in regulatory studies associated with drug development there needs to be evidence that the significance of the marker analytes detected in animal species is the same or similar to that when the marker appears in the human circulation.

Some conventional analytes used in clinical medicine to detect myocardial damage are applicable for use with samples of non-human origin. Thus, measurement of the enzymatic activity of creatine kinase (CK), the isoenzymes of CK, the isoenzymes of lactate dehydrogenase (LDH), or hydoxybutyrate dehydrogenase (HBDH) are all feasible, but they lack both specificity and sensitivity for myocardial injury.

Of the newer analytes with a higher specificity or sensitivity for cardiac muscle damage, measurements of CK MB isoenzyme mass (CK-MB) are not practicable because there is poor species conservation of the molecule. Thus, the immunoassays developed for human application are too selective for epitopes on the human form of CK-MB and will not even cross-react to an appreciable extent with CK-MB from higher primates. Currently, there are no commercial assays for the BB isoenzyme of glycogen phosphorylase or for cardiac fatty acid binding protein, and there are not data to indicate that they would be of value in pre-clinical studies.

There is good reason to suppose that clinical assays for the cardiac forms of troponin T (cTnT) and troponin I (cTnI) can be used for samples from pre-clinical studies. There is species conservation in the structure of these proteins, certainly in the higher vertebrates, and there is growing evidence that the measurement of these analytes can distinguish between troponins of cardiac and skeletal origin, in species other than man.[1-5] In addition, there is evidence from this laboratory that separation of the isoforms of CK-MB provides a sensitive and selective indication of myocardial damage which can be used in association with measurements of the troponins in a variety of animal species.[6]

The remainder of this chapter will summarise some of the potential problems in applying measurements of marker proteins to pre-clinical studies and will also document some practical applications of the assays.

Potential problems

Assay related

Antibody cross-reactivity. Whilst immunoassays which have been optimised for the measurement of cTnT and cTnI in clinical samples may give positive results in samples from experimental animals, they may not be cross-reacting with these analytes on a 1:1 basis. Nor does detection automatically imply cardiac, rather than skeletal, muscle damage. The need is for studies which demonstrate that antibodies to cardiac TnT and TnI cross-react only with troponin molecules of cardiac origin in the target species, and which establish the significance of the numerical values obtained.

Whilst only one commercial assay is currently available for cTnT,[2,7] a number of cTnI assays are available.[8-14] There are differences between the assays in the epitopes on the cTnI molecule recognised by the anti-cTnI antibody, and there appear to be some differences related to the calibration of the assay. For samples from human subjects the diagnostic cut-off values are not the same for each assay. It should be remembered that validation of one of these assays in an animal species does not imply that all of these assays will give comparable results.

Analyte stability. It is common practice to store serum or plasma samples deep-frozen, in the range -20° to -70°C, whilst tissue samples are often snap-frozen in liquid nitrogen. It is advisable to demonstrate the stability of the marker proteins chosen for experimental use during the period of intended storage, as well as their stability through at least two freeze/thaw cycles and at ambient temperature during a working day. Again, stability in one assay format does not imply stability of an analyte in a measurement system devised by another manufacturer.

Reference ranges. Individual, species-related, reference ranges should be determined for the marker proteins. When control data are not available for a particular species, at the very least, data should be generated from a control group studied in parallel with the test group of animals. If this is not possible, because animals are not available to form a control group, frequent samples should be collected before the start of the study, to establish base-line measurements.

The importance of determining reference ranges for a particular species is underlined by recent data showing that, even in apparently healthy individuals, there is substantial within-subject variability in the measurement of such parameters as CK-MB mass.[15] Thus, for animal studies, within-species variability for each analyte should be determined; the use of mean data derived in humans is entirely inappropriate without adequate control studies. There are also data to suggest that poor laboratory performance of assays for, even, the conventional analytes could introduce marked variability in the calculation of reference ranges.[16]

In determining reference ranges the dynamic range of the assay should also be borne in mind. Measured concentrations should be within the range of the calibrators, or concentrations outside this range should be validated with appropriate control samples. When large sample dilutions are needed, for instance if analytes are being measured in tissue extracts, the assayed sample should correspond as closely as possible with the matrix recommended by the manufacturer of the assay.[17] Failure to optimise the matrix can lead to spurious results. For instance, tap water can give a positive result if measured for cTnT (Collinson P., personal communication).

Tissue distribution of the analytes

The marker proteins under discussion differ in their distribution from one species to another. The potential differences are well illustrated in Figure 1, which shows the electrophoretic separation of CK isoenzymes in the cytosolic fraction of human and canine ventricular myocardium. There is a markedly higher proportion of the MB isoenzyme in the sample of human origin, compared with the canine sample. In some species, such as swine, rabbits and horses, the activity of CK-MB in plasma is very low or undetectable.[5]

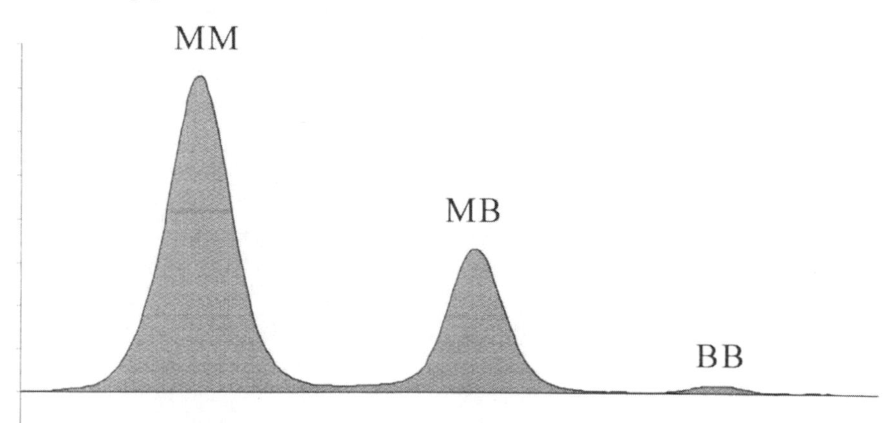

Figure 1a. Electrophoretic separation of CK isoenzymes in a cytosolic fraction of human myocardial muscle. The proportions for each isoenzyme were 69.8% (MM), 28.6% (MB) and 1.6% (BB). Data: Analytical Unit, unpublished observation.

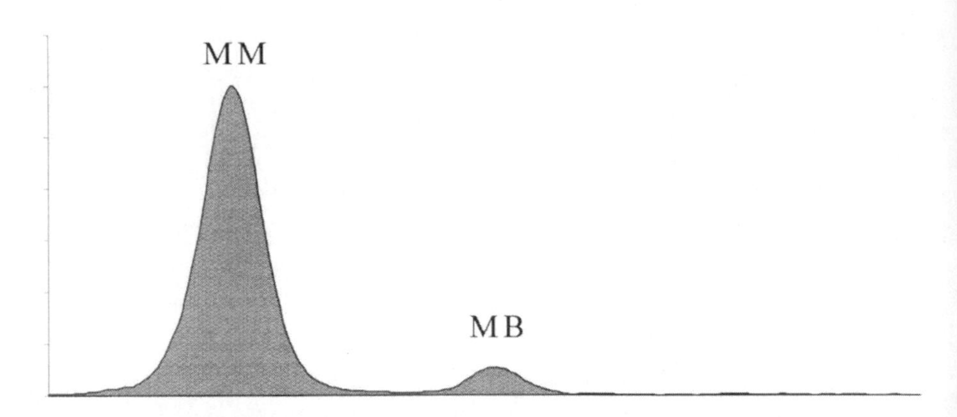

Figure 1b. Electrophoretic separation of CK isoenzymes in a cytosolic fraction of beagle dog myocardial muscle. The proportions for each isoenzyme were 91.1% (MM) and 8.9% (MB). Data: Analytical Unit, unpublished observation.

Similarly, there are between-species differences in the cytosolic fraction of cTnT. In human myocardial samples it is approximately 6-8% of the total myocardial content.[18,19] In Wistar rats the cytosolic fraction was found to be identical with that of humans, but in dogs the cytosolic fraction was only 2%.[19,20] These findings have implications for the kinetics of release of cTnT following myocardial damage or revascularisation.[20]

Underlying pathology

The distribution of marker proteins may vary between-species and may be influenced by disease state. Voss *et al* [19] noted that the distribution of cTnT and CK-MB differed between the right and left ventricles in the dog; cTnT was higher and CK-MB was lower in the right ventricle. These differences impacted on the calculation of infarct size from measurements of these analytes, depending on the location of the infarct.
Ricchiuti *et al* [21] found differences in the tissue content of cTnT and cTnI in porcine left ventricle from control animals and those which had suffered myocardial infarction following ligation. Tissue remote from the site of infarction contained significantly less cTnT and cTnI than tissue from the control animals, suggesting that measurement of these proteins in the circulation could correlate with their chronic loss from injured myocardium.

O'Brien [22] noted that myofibrillar cTnT was 25% lower in Doberman Pinschers with advanced congestive heart failure secondary to idiopathic dilated cardiomyopathy, compared with mixed-breed control dogs; cytosolic CK-MB was 50% lower. Thus, tissue measurements of these analytes in diseased animals must be related to healthy controls. It is interesting to note that, in humans, changes in the CK-MB isoform ratio have been related to signs of deterioration in patients with idiopathic dilated cardiomyopathy.[23]

There are two current, unresolved, controversies concerning the measurement of the cardiac troponins in samples from humans, both of which should be borne in mind when applying the measurement to pre-clinical studies.

Firstly, there is the possibility that cTnT may be expressed in regenerating human skeletal muscle, such as from patients with polymyositis or Duchenne muscular dystrophy.[24] There are similar data in regenerating rat muscle fibres following cold injury.[25] Whilst it may be possible to detect cTnT in these clinical settings, using immunohistochemistry and Western blot techniques, the significance of these finding in relation to serum or plasma measurements of cTnT is not clear. However, spurious measurements of cTnT in serum or plasma should be excluded in animals bred specifically for research on these diseases.

Secondly, the influence of renal failure on the measurement of both cTnT and cTnI has been reported. Several groups have noted that both analytes can be present at concentrations above the cut-off values in the serum or plasma of patients with impaired renal function, but without overt signs of cardiac muscle damage.[26-28] The implication is that the detection of these analytes is due to their extracardiac expression in patients with renal disease, and that this phenomenon diminishes the cardiac specificity of the measurements. Whilst improvements to the cardiac specificity of the cTnT assay have reduced the incidence of apparent false positive results, there are a significant number of patients with cTnT values above the normal cut-off but without signs of cardiac involvement.[7] More recent findings, in which patients with end-stage renal disease were subjected to intensive cardiovascular diagnostic tests, suggests that those with a raised cTnT may, indeed, have increased cardiovascular risk factors.[29] Thus, the measurement may be of diagnostic value. No matter how this matter is resolved in clinical practice, care should be taken to exclude the influence of renal function in the interpretation of data obtained from animals, especially if the object of the study is to detect or rule-out minor myocardial damage.

In animal studies it is very important that markers of muscle damage can differentiate between skeletal and cardiac muscle damage. Skeletal muscle damage is relatively common in laboratory animals, due to such factors as struggling during handling, and injuries caused by cages or restraints. The conventional analytes, such as CK may be grossly elevated at a time when more selective analytes show no evidence of cardiac muscle damage. Figure 2 illustrates two points in this respect.

A cynomolgus monkey was exposed to a test compound that had been implicated in the development of histological signs of myocardial damage. Measurement of total CK in this animal was in excess of 4000U/L, but there were no signs of cardiac muscle damage at autopsy. Two specific markers of myocardial damage - the CK-MB2/MB1 isoform ratio remained within normal limits noted in control animals and cTnI (Sanofi-Pasteur) concentration was below the cut-off value for the assay. However, measurements of cTnT became very elevated during the study period. The cTnT assay used at the time was the first generation assay, which was known to give false positive results in the presence of high concentrations of skeletal TnT. Taking into account the other two marker protein results it was concluded that the apparently high cTnT concentrations were not due to cardiac muscle damage. Samples were not available for testing with the second generation assay, which is not affected by high concentrations of skeletal muscle TnT.[30]

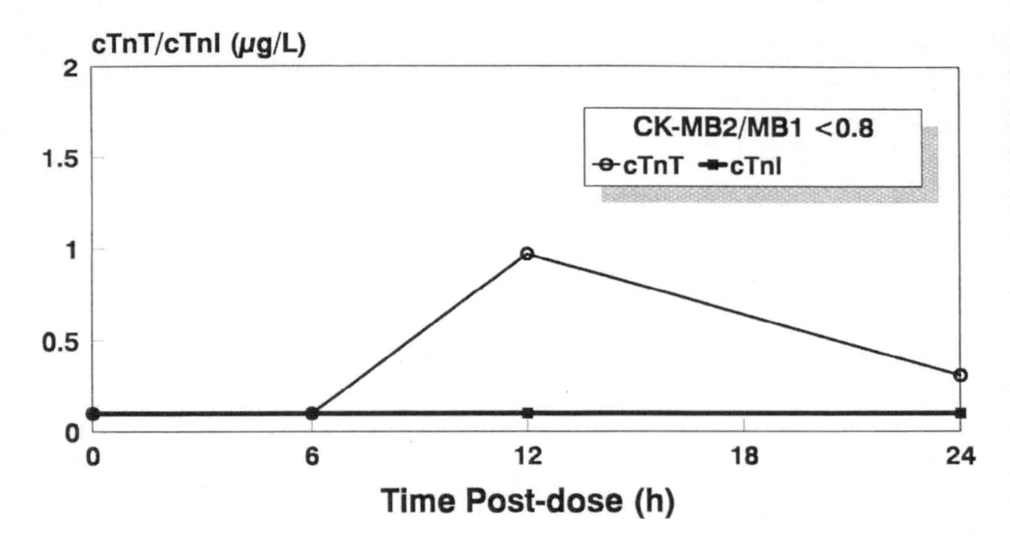

Figure 2. Measurements of cTnT and cTnI in serum samples from a cynomolgus monkey following administration of a test compound which was not associated with histological signs of myocardial injury. During the course of these observations the CK-MB2/MB1 isoform ratio was below 0.8 at all time points. Data: Analytical Unit, unpublished observation.

Similarly, it is interesting to note a series of studies in which serum cTnT was measured in broiler chickens as a marker of ascites, secondary to right ventricular hypertrophy.[31] The authors reported significantly higher cTnT concentrations in the affected birds, compared with controls, but the range of values, even in the control animals, was wide and above the normal cut-off value. Subsequently, measurements made in healthy birds up to 28 days following hatching were as high as 7.9μg/L at 28 days.[32] These measurements, too, were made with the first generation cTnT assay and it seems likely that the findings were influenced by skeletal muscle damage early in life, highlighting the need to exclude confounding variables when interpreting these measurements.

Finally, Bernardi *et al* [33] noted that measurements of the conventional markers CK, LDH and HBDH were elevated in samples collected from orbital sites in rats, compared with samples collected from the abdominal aorta. The data were consistent with soft tissue injury, underlining the need for both assessing the influence of sampling techniques and identifying markers with a high specificity for cardiac muscle damage if a differential diagnosis between skeletal and cardiac muscle damage is to be achieved.

Applications

There have been three important recent reports in the literature which confirm the validity of using measurements of the cardiac troponins as markers of myocardial damage in animals.

Firstly, Chocron *et al* [34] measured the release of cTnI from Langendorff perfused Wistar rat hearts. Three experimental groups were studied: 1. control hearts perfused after excision, 2. hearts perfused after immersion in St Thomas's Hospital solution for 3 hours, 3. hearts perfused after immersion in St Thomas's Hospital solution for 6 hours. Hearts were allowed to stabilise for 30 minutes, after which serial measurements of cTnI (Sanofi Pasteur), CK-MB and LDH were made for 2 hours.

Concentrations of cTnI increased in proportion to the period of ischaemia, and the period of ischaemia was related to histological signs of myocardial cell damage. The activities of the two conventional markers were not related to the period of ischaemia. Thus, cTnI was shown to be an early and sensitive marker of ischaemic myocardial injury, with distinct advantages over the conventional markers.

Secondly, Vorderwinkler *et al* [35] also established Langendorff perfusion of Wistar rat hearts which, after 30 minutes of equilibration, were subjected to 60 minutes of hypoxic perfusion, then 75 minutes of re-oxygenation. Measurements of both cTnT and cTnI (Sanofi Pasteur) showed a significant and parallel rise within 5 minutes of re-oxygenation, acting as sensitive markers under conditions simulating those of myocardial infarction followed by revascularisation. In this instance the release of the troponins was similar to the release of the conventional markers CK-MB and LDH.

In contrast to the latter findings, Hjelms *et al* [36] demonstrated the poor diagnostic sensitivity for the non-specific marker CK. Three groups of dogs were studied - control, those subjected to irreversible myocardial ischaemic damage leading to transmural myocardial infarction, and those subjected to short periods of ischaemia producing no evidence of infarction. Whilst CK activity was elevated in all three groups there was no significant difference in the cumulative release over 8 hours for the two groups exposed to ischemic stress.

Thirdly, O'Brien *et al* [5] have made measurements of cTnT as a marker of myocardial damage in a variety of laboratory animals. In mixed-breed dogs and Sprague-Dawley rats there was a significant trend for cTnT to increase with the time of reperfusion, following coronary occlusion. Concentrations of cTnT also correlated with infarct size within 3 hours of occlusion. These authors also demonstrated that the assay could be applied to drug-induced myocardial damage in animals by measuring cTnT in mice treated with doxorubicin for 5 days. There was a more than 10 fold increase in cTnT in the treated mice compared with the controls, although it should be noted that concentrations of cTnT in the control mice were very high. For this experiment the first generation cTnT assay was used.

Isoprenaline-induced, and histologically confirmed, myocardial damage has also been detected in Sprague-Dawley rats by means of cTnT.[37] There were marked elevations of cTnT within 6 hours of treatment and concentrations were still significantly elevated 48 hours after treatment. The conventional markers CK and LDH were of little diagnostic value.

The practical value of using animals in preliminary studies to determine the clinical value of cTnI measurements has been shown by Smith *et al.* [38] This group demonstrated that cTnI was a useful non-invasive marker for the detection of myocarditis by first making measurements in mice with autoimmune myocarditis. The marker was elevated in 24 of 26 test mice but in none of the control mice.

In this laboratory, separation of CK-MB isoforms has been shown to be a sensitive technique for the detection of drug-induced myocardial damage in beagle dogs and has been used in association with measurements of cTnT.

In control dogs the mean ratio CK-MB2/MB1 was 0.99, very similar to the ratio found in humans without signs of myocardial injury. Figure 3a shows the rapid rise in this ratio shortly after the administration of a drug which produced histological signs of myocardial injury; there was no significant change in the ratio in samples from a control animal. In Figure 3b the changes in serum cTnT are shown in the same test animal. Both markers are rapid and sensitive indicators of myocardial injury but there are clear differences in the kinetics of their release and clearance. The CK-MB2-MB1 ratio returns within normal limits relatively quickly, making this a suitable marker for any subsequent episode of myocardial injury.[39]

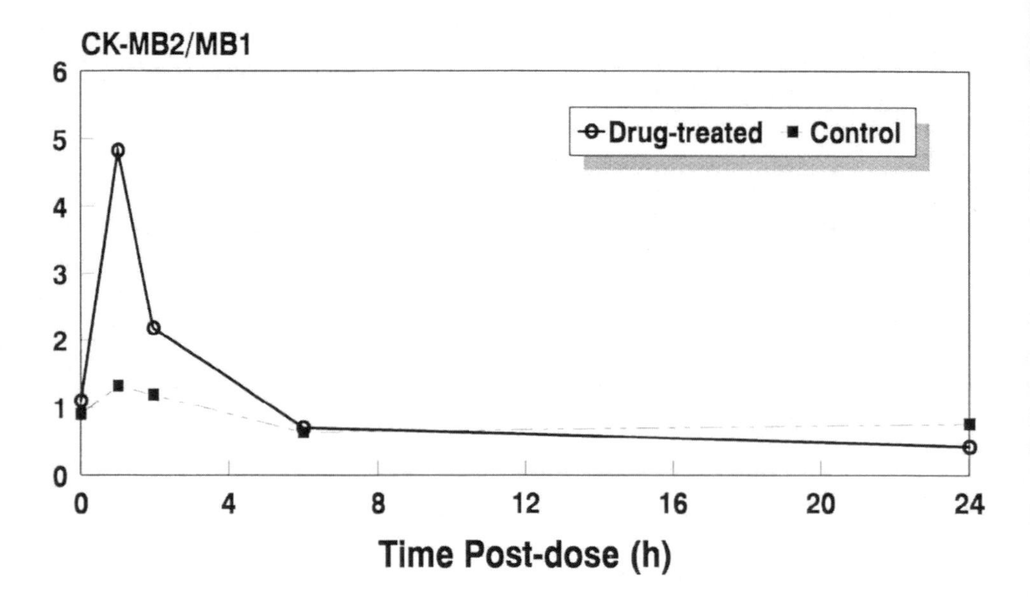

Figure 3a. Measurement of the CK-MB2/MB1 isoform ratio in a control beagle dog and a beagle dog exposed to a test compound associated with histological signs of myocardial injury. Data: Analytical Unit, unpublished observation.

Further work in this laboratory suggests that separation of CK-MB isoforms is also applicable as a sensitive marker of myocardial damage in monkey, rat, and mini pig. Work is continuing in this field to establish the skeletal muscle and cardiac muscle distribution of the CK isoforms in these species.

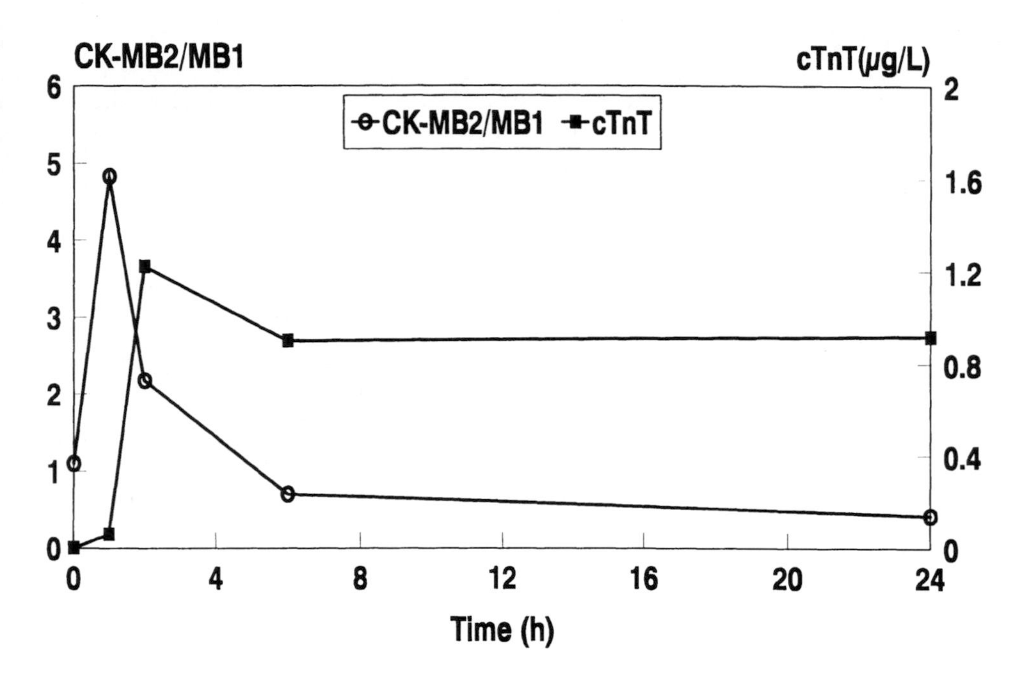

Figure 3b. Measurement of the CK-MB2/MB1 isoform ratio and cTnT in a beagle dog exposed to a test compound associated with histological signs of myocardial injury. Data: Analytical Unit, unpublished observation.

Conclusions

This chapter has set-out to address some of the problems associated with the specific detection of myocardial damage in animals. Whilst there are caveats which must be placed on the use of the tests described it is clear that the existing assays for cTnT and cTnI, together with the separation of CK-MB isoforms, can be applied in a broad range of species.

The data so far suggest that cTnT and cTnI can be used in several common laboratory animal species to detect ischaemic or drug-induced myocardial damage and it is likely that they could be used to monitor myocardial damage associated with invasive procedures.[40] They represent a significant advance over the conventional markers in terms of specificity and sensitivity.

The application of these markers in the diagnosis or monitoring of diseases in animals is limited but there are some encouraging initial results.

Rigorous validation of the assays is needed for regulatory studies, paying attention to within-species and between-species differences in the distribution of the analytes, and matrix effects when they are measured in assay systems designed for use with human serum or plasma samples.

Acknowledgements

I am grateful to Salim Fredericks, Gurcharin Merton and Mojgan Hossein-Nia for skilled technical assistance and thoughtful input on the experimental design of studies to assess the value of myocardial marker proteins in animal studies.

References

1. Hastings KEM. Strong evolutionary conservation of broadly expressed protein isoforms in the troponin I gene family and other vertebrate gene families. J Mol Evol 1996; 42:631-640.
2. Katus HA, Looser S, Hallermayer K, Remppis A, Scheffold T, Borgya A, *et al.* Development and in vitro characterization of a new immunoassay of cardiac troponin T. Clin Chem 1992; 38:386-393.
3. Leszyk J, Dumaswala R, Potter JD, Collins JH. Amino acid sequence of bovine cardiac troponin I. Biochem 1988; 27:2821-2827.
4. Leszyk J, Dumaswala R, Potter JD, Gusev NB, Verin AD, Tobacman LS, *et al.* Bovine cardiac troponin T: amino acid sequences of the two isoforms. Biochem 1987; 26:7035-7042.
5. O'Brien PJ, Dameron GW, Beck ML, *et al.* Cardiac troponin T is a sensitive, specific biomarker of cardiac injury in laboratory animals. Lab Animal Sci 1997; 47:486-495.
6. Hossein-Nia M, Suter KE, Heining P, Zwanenburg SB, Holt DW. Creatine konase MB isoforms and troponin T and I: sensitive markers of myocardial damage in pre-clinical studies. Clin Chem 1996; 42:S241 Abstract.
7. Muller-Bardorff M, Hallermayer K, Schroder A, Ebert C, Borgya A, Gerhardt W, *et al.* Improved troponin T ELISA specific for cardiac troponin T isoform: assay development and analytical and clinical validation [see comments]. Clin Chem 1997; 43:458-466.
8. Larue C, Defacque-Lacquement H, Calzolari C, Le Nguyen D, Pau B. New monoclonal antibodies as probes for human cardiac troponin I: epitopic analysis with synthetic peptides. Mol Immunol 1992; 29:271-278.
9. Bodor GS, Porter S, Landt Y, Ladenson JH. Development of monoclonal antibodies for an assay of cardiac troponin-I and preliminary results in suspected cases of myocardial infarction. Clin Chem 1992; 38:2203-2214.
10. McLellan B. Stratus cardiac specific troponin-I. Immuno-Analyse et Biologie Specialisee 1995; 10: 175-177.
11. Adams III, JE, Sicard GA, Allen BT, Bridwell KH, Lenke LG, Davila- Roma, *et al.* Diagnosis of perioperative myocardial infarction with measurement of cardiac troponin I. N Engl J Med 1994; 330: 670-674.
12. Bhayana V, Henderson AR. Biochemical markers of myocardial damage. [Review] [208 refs]. Clin Biochem 1995; 28:1-29.
13. Davies E, Gawad Y, Takahashi M, *et al.* Analytical performance and clinical utility of a sensitive immunoassay for determination of human cardiac troponin I. Clin Biochem 1997; 30:479-490.
14. Christenson RH, Apple FS, Morgan DL, *et al.* Cardiac troponin I measurement with the ACCESS immunoassay system: analytical and clinical performance characteristics. Clin Chem 1998; 44:52-60.
15. Ross SM, Fraser CG. Biological variation of cardiac markers: analytical and clinical considerations. Ann Clin Biochem 1998; 35:80-84.
16. Henderson AR, Krishnan S, Webb S, Cheung CM, Nazir DJ, Richardson H. Proficiency testing of creatine kinase and creatine kinase-2: the experience of the Ontario Laboratory Proficiency Testing Program. Clin Chem 1998; 44:124-133.
17. Fredericks S, Merton GK, Lerena M, Holt DW. Markers of myocardial damage. Clin Chem 1998; 44: (In Press)
18. Katus HA, Remppis A, Scheffold T, Diederich KW, Kuebler W. Intracellular compartmentation of cardiac troponin T and its release kinetics in patients with reperfused and nonreperfused myocardial infarction. Am J Cardiol 1991; 67:1360-1367.
19. Voss EM, Sharkey SW, Gernert AE, Murakami MM, Johnston RB, Hsieh CC, *et al.* Human and canine cardiac troponin T and creatine kinase-MB distribution in normal and diseased myocardium. Infarct sizing using serum profiles. Arch Path Lab Med 1995; 119:799-806.

20. Remppis A, Scheffold T, Greten J, Haass M, Greten T, Kubler W, *et al.* Intracellular compartmentation of troponin T: release kinetics after global ischemia and calcium paradox in the isolated perfused rat heart. J Mol Cell Cardiol 1995; 27:793-803.
21. Ricchiuti V, Zhang J, Apple FS. Cardiac troponin I and T alterations in hearts with severe left ventricular remodeling. Clin Chem 1997; 43:990-995.
22. O'Brien PJ. Deficiencies of myocardial troponin-T and creatine kinase MB isoenzyme in dogs with idiopathic dilated cardiomyopathy. Am J Vet Res 1997; 58:11-16.
23. Hossein-Nia M, Baig K, Goldman JH, Keeling PJ, Caforio ALP, Holt DW, *et al.* Creatine kinase isoforms as circulating markers of deterioration in idiopathic dilated cardiomyopathy. Clin Cardiol 1997; 20:55-60.
24. Bodor GS, Survant L,' Voss EM, Smith S, Porterfield D, Apple FS. Cardiac troponin T composition in normal and regenerating human skeletal muscle [see comments]. Clin Chem 1997; 43:476-484.
25. Saggin L, Gorza L, Ausoni S, Schiaffino S. Cardiac troponin T in developing, regenerating and denervated rat skeletal muscle. Development 1990; 110:547-554.
26. Bhayana V, Gougoulias T, Cohoe S, Henderson AR. Discordance between results for serum troponin T and troponin I in renal disease. Clin Chem 1995; 41:312-317.
27. Li D, Jialal I, Keffer J. Greater frequency of increased cardiac troponin T than increased cardiac troponin I in patients with chronic renal failure. Clin Chem 1996; 42:114-115.
28. Hafner G, Thome-Kromer B, Schaube J, Kupferwasser I, Ehrenthal W, Cummins P, *et al.* Cardiac troponins in serum in chronic renal failure. Clin Chem 1994; 40:1790-1791.
29. Haller C, Zehelein J, Remppis A, Muller-Bardorff M, Katus HA. Cardiac troponin T in patients with end stage renal disease: absence of expression in skeletal muscle. Clin Chem 1998; 44: (In Press)
30. Hossein-Nia M, Nisbet J, Merton GK, Holt DW. Spurious rises of cardiac troponin T. Lancet 1995; 346:1558
31. Maxwell MH, Robertson GW, Moseley D. Potential role of serum troponin T in cardiomyocyte injury in the broiler ascites syndrome. Br Poultry Sci 1994; 35:663-667.
32. Maxwell MH, Robertson GW, Moseley D. Serum troponin T concentrations in two strains of commercial broiler chickens aged one to 56 days. Res Vet Sci 1995; 58:244-247.
33. Bernardi C, Moneta D, Brughera M, Di Salvo M, Lamparelli D, Mazue G, *et al.* Haematology and clinical chemistry in rats: Comparison of different blood collection sites.. Comp Haematology Int 1996; 6:160-166.
34. Chocron S, Alwan K, Toubin G, Kantelip B, Clement F, Kantelip JP, *et al.* Effects of myocardial ischemia on the release of cardiac troponin I in isolated rat hearts. J Thoracic Cardiovasc Surg 1996; 112:508-513.
35. Vorderwinkler KP, Mair J, Puschendorf B, Hempel A, Schluter KD, Piper, *et al.* Cardiac troponin I increases in parallel to cardiac troponin T, creatine kinase and lactate dehydrogenase in effluents from isolated perfused rat hearts after hypoxia-reoxygenation-induced myocardial injury. Clinica Chimica Acta 1996; 251:113-117.
36. Hjelms E, Hansen BF, Waldorff S, Steiness E. Evaluation of increased serum creatine kinase as an indicator of irreversible myocardial damage in dogs. Scandinavian J Thoracic Cardiovasc Surg 1987; 21:165-168.
37. Bleuel H, Deschl U, Bertsch T, Bolz G, Rebel W. Diagnostic efficiency of troponin T measurements in rats with experimental myocardial cell damage. Exp Toxicol Path 1995; 47:121-127.
38. Smith SC, Ladenson JH, Mason JW, Jaffe AS. Elevations of cardiac troponin I associated with myocarditis. Experimental clinical correlates. Circulation 1997; 95:163-168.
39. Anderson JR, Hossein-Nia M, Kallis P, Pye M, Holt DW, Murday AJ, *et al.* Comparison of two strategies for myocardial management during coronary artery operations. Ann Thoracic Surg 1994; 58:768-72; discussion 772-3.
40. Katritsis D, Hossein-Nia M, Anastasakis A, Poloniecki I, Holt DW, Camm AJ, *et al.* Use of troponin-T concentration and creatinine kinase isoforms for quantitation of myocardial injury induced by radiofrequency catheter ablation. Eur Heart J 1997; 18:1007-1013.

INDEX

Developments in Cardiovascular Medicine

109. J.P.M. Hamer: *Practical Echocardiography in the Adult.* With Doppler and Color-Doppler Flow Imaging. 1990 ISBN 0-7923-0670-8
110. A. Bayés de Luna, P. Brugada, J. Cosin Aguilar and F. Navarro Lopez (eds.): *Sudden Cardiac Death.* 1991 ISBN 0-7923-0716-X
111. E. Andries and R. Stroobandt (eds.): *Hemodynamics in Daily Practice.* 1991 ISBN 0-7923-0725-9
112. J. Morganroth and E.N. Moore (eds.): *Use and Approval of Antihypertensive Agents and Surrogate Endpoints for the Approval of Drugs affecting Antiarrhythmic Heart Failure and Hypolipidemia.* Proceedings of the 10th Annual Symposium on New Drugs and Devices (1989). 1990 ISBN 0-7923-0756-9
113. S. Iliceto, P. Rizzon and J.R.T.C. Roelandt (eds.): *Ultrasound in Coronary Artery Disease.* Present Role and Future Perspectives. 1990 ISBN 0-7923-0784-4
114. J.V. Chapman and G.R. Sutherland (eds.): *The Noninvasive Evaluation of Hemodynamics in Congenital Heart Disease.* Doppler Ultrasound Applications in the Adult and Pediatric Patient with Congenital Heart Disease. 1990 ISBN 0-7923-0836-0
115. G.T. Meester and F. Pinciroli (eds.): *Databases for Cardiology.* 1991 ISBN 0-7923-0886-7
116. B. Korecky and N.S. Dhalla (eds.): *Subcellular Basis of Contractile Failure.* 1990 ISBN 0-7923-0890-5
117. J.H.C. Reiber and P.W. Serruys (eds.): *Quantitative Coronary Arteriography.* 1991 ISBN 0-7923-0913-8
118. E. van der Wall and A. de Roos (eds.): *Magnetic Resonance Imaging in Coronary Artery Disease.* 1991 ISBN 0-7923-0940-5
119. V. Hombach, M. Kochs and A.J. Camm (eds.): *Interventional Techniques in Cardiovascular Medicine.* 1991 ISBN 0-7923-0956-1
120. R. Vos: *Drugs Looking for Diseases.* Innovative Drug Research and the Development of the Beta Blockers and the Calcium Antagonists. 1991 ISBN 0-7923-0968-5
121. S. Sideman, R. Beyar and A.G. Kleber (eds.): *Cardiac Electrophysiology, Circulation, and Transport.* Proceedings of the 7th Henry Goldberg Workshop (Berne, Switzerland, 1990). 1991 ISBN 0-7923-1145-0
122. D.M. Bers: *Excitation-Contraction Coupling and Cardiac Contractile Force.* 1991 ISBN 0-7923-1186-8
123. A.-M. Salmasi and A.N. Nicolaides (eds.): *Occult Atherosclerotic Disease.* Diagnosis, Assessment and Management. 1991 ISBN 0-7923-1188-4
124. J.A.E. Spaan: *Coronary Blood Flow.* Mechanics, Distribution, and Control. 1991 ISBN 0-7923-1210-4
125. R.W. Stout (ed.): *Diabetes and Atherosclerosis.* 1991 ISBN 0-7923-1310-0
126. A.G. Herman (ed.): *Antithrombotics.* Pathophysiological Rationale for Pharmacological Interventions. 1991 ISBN 0-7923-1413-1
127. N.H.J. Pijls: *Maximal Myocardial Perfusion as a Measure of the Functional Significance of Coronary Arteriogram.* From a Pathoanatomic to a Pathophysiologic Interpretation of the Coronary Arteriogram. 1991 ISBN 0-7923-1430-1
128. J.H.C. Reiber and E.E. v.d. Wall (eds.): *Cardiovascular Nuclear Medicine and MRI.* Quantitation and Clinical Applications. 1992 ISBN 0-7923-1467-0
129. E. Andries, P. Brugada and R. Stroobrandt (eds.): *How to Face "the Faces' of Cardiac Pacing.* 1992 ISBN 0-7923-1528-6
130. M. Nagano, S. Mochizuki and N.S. Dhalla (eds.): *Cardiovascular Disease in Diabetes.* 1992 ISBN 0-7923-1554-5
131. P.W. Serruys, B.H. Strauss and S.B. King III (eds.): *Restenosis after Intervention with New Mechanical Devices.* 1992 ISBN 0-7923-1555-3
132. P.J. Walter (ed.): *Quality of Life after Open Heart Surgery.* 1992 ISBN 0-7923-1580-4
133. E.E. van der Wall, H. Sochor, A. Righetti and M.G. Niemeyer (eds.): *What's new in Cardiac Imaging?* SPECT, PET and MRI. 1992 ISBN 0-7923-1615-0
134. P. Hanrath, R. Uebis and W. Krebs (eds.): *Cardiovascular Imaging by Ultrasound.* 1992 ISBN 0-7923-1755-6
135. F.H. Messerli (ed.): *Cardiovascular Disease in the Elderly.* 3rd ed. 1992 ISBN 0-7923-1859-5
136. J. Hess and G.R. Sutherland (eds.): *Congenital Heart Disease in Adolescents and Adults.* 1992 ISBN 0-7923-1862-5
137. J.H.C. Reiber and P.W. Serruys (eds.): *Advances in Quantitative Coronary Arteriography.* 1993 ISBN 0-7923-1863-3

Developments in Cardiovascular Medicine

138. A.-M. Salmasi and A.S. Iskandrian (eds.): *Cardiac Output and Regional Flow in Health and Disease.* 1993 ISBN 0-7923-1911-7
139. J.H. Kingma, N.M. van Hemel and K.I. Lie (eds.): *Atrial Fibrillation, a Treatable Disease?* 1992 ISBN 0-7923-2008-5
140. B. Ostadel and N.S. Dhalla (eds.): *Heart Function in Health and Disease.* Proceedings of the Cardiovascular Program (Prague, Czechoslovakia, 1991). 1992 ISBN 0-7923-2052-2
141. D. Noble and Y.E. Earm (eds.): *Ionic Channels and Effect of Taurine on the Heart.* Proceedings of an International Symposium (Seoul, Korea, 1992). 1993 ISBN 0-7923-2199-5
142. H.M. Piper and C.J. Preusse (eds.): *Ischemia-reperfusion in Cardiac Surgery.* 1993 ISBN 0-7923-2241-X
143. J. Roelandt, E.J. Gussenhoven and N. Bom (eds.): *Intravascular Ultrasound.* 1993 ISBN 0-7923-2301-7
144. M.E. Safar and M.F. O'Rourke (eds.): *The Arterial System in Hypertension.* 1993 ISBN 0-7923-2343-2
145. P.W. Serruys, D.P. Foley and P.J. de Feyter (eds.): *Quantitative Coronary Angio- graphy in Clinical Practice.* With a Foreword by Spencer B. King III. 1994 ISBN 0-7923-2368-8
146. J. Candell-Riera and D. Ortega-Alcalde (eds.): *Nuclear Cardiology in Everyday Practice.* 1994 ISBN 0-7923-2374-2
147. P. Cummins (ed.): *Growth Factors and the Cardiovascular System.* 1993 ISBN 0-7923-2401-3
148. K. Przyklenk, R.A. Kloner and D.M. Yellon (eds.): *Ischemic Preconditioning: The Concept of Endogenous Cardioprotection.* 1993 ISBN 0-7923-2410-2
149. T.H. Marwick: *Stress Echocardiography.* Its Role in the Diagnosis and Evaluation of Coronary Artery Disease. 1994 ISBN 0-7923-2579-6
150. W.H. van Gilst and K.I. Lie (eds.): *Neurohumoral Regulation of Coronary Flow.* Role of the Endothelium. 1993 ISBN 0-7923-2588-5
151. N. Sperelakis (ed.): *Physiology and Pathophysiology of the Heart.* 3rd rev. ed. 1994 ISBN 0-7923-2612-1
152. J.C. Kaski (ed.): *Angina Pectoris with Normal Coronary Arteries: Syndrome X.* 1994 ISBN 0-7923-2651-2
153. D.R. Gross: *Animal Models in Cardiovascular Research.* 2nd rev. ed. 1994 ISBN 0-7923-2712-8
154. A.S. Iskandrian and E.E. van der Wall (eds.): *Myocardial Viability.* Detection and Clinical Relevance. 1994 ISBN 0-7923-2813-2
155. J.H.C. Reiber and P.W. Serruys (eds.): *Progress in Quantitative Coronary Arteriography.* 1994 ISBN 0-7923-2814-0
156. U. Goldbourt, U. de Faire and K. Berg (eds.): *Genetic Factors in Coronary Heart Disease.* 1994 ISBN 0-7923-2752-7
157. G. Leonetti and C. Cuspidi (eds.): *Hypertension in the Elderly.* 1994 ISBN 0-7923-2852-3
158. D. Ardissino, S. Savonitto and L.H. Opie (eds.): *Drug Evaluation in Angina Pectoris.* 1994 ISBN 0-7923-2897-3
159. G. Bkaily (ed.): *Membrane Physiopathology.* 1994 ISBN 0-7923-3062-5
160. R.C. Becker (ed.): *The Modern Era of Coronary Thrombolysis.* 1994 ISBN 0-7923-3063-3
161. P.J. Walter (ed.): *Coronary Bypass Surgery in the Elderly.* Ethical, Economical and Quality of Life Aspects. With a foreword by N.K. Wenger. 1995 ISBN 0-7923-3188-5
162. J.W. de Jong and R. Ferrari (eds.), *The Carnitine System.* A New Therapeutical Approach to Cardiovascular Diseases. 1995 ISBN 0-7923-3318-7
163. C.A. Neill and E.B. Clark: *The Developing Heart: A "History' of Pediatric Cardiology.* 1995 ISBN 0-7923-3375-6
164. N. Sperelakis: *Electrogenesis of Biopotentials in the Cardiovascular System.* 1995 ISBN 0-7923-3398-5
165. M. Schwaiger (ed.): *Cardiac Positron Emission Tomography.* 1995 ISBN 0-7923-3417-5
166. E.E. van der Wall, P.K. Blanksma, M.G. Niemeyer and A.M.J. Paans (eds.): *Cardiac Positron Emission Tomography.* Viability, Perfusion, Receptors and Cardiomyopathy. 1995 ISBN 0-7923-3472-8
167. P.K. Singal, I.M.C. Dixon, R.E. Beamish and N.S. Dhalla (eds.): *Mechanism of Heart Failure.* 1995 ISBN 0-7923-3490-6
168. N.S. Dhalla, P.K. Singal, N. Takeda and R.E. Beamish (eds.): *Pathophysiology of Heart Failure.* 1995 ISBN 0-7923-3571-6
169. N.S. Dhalla, G.N. Pierce, V. Panagia and R.E. Beamish (eds.): *Heart Hypertrophy and Failure.* 1995 ISBN 0-7923-3572-4

Developments in Cardiovascular Medicine

170. S.N. Willich and J.E. Muller (eds.): *Triggering of Acute Coronary Syndromes*. Implications for Prevention. 1995 ISBN 0-7923-3605-4
171. E.E. van der Wall, T.H. Marwick and J.H.C. Reiber (eds.): *Advances in Imaging Techniques in Ischemic Heart Disease.* 1995 ISBN 0-7923-3620-8
172. B. Swynghedauw: *Molecular Cardiology for the Cardiologist.* 1995 ISBN 0-7923-3622-4
173. C.A. Nienaber and U. Sechtem (eds.): *Imaging and Intervention in Cardiology.* 1996 ISBN 0-7923-3649-6
174. G. Assmann (ed.): *HDL Deficiency and Atherosclerosis.* 1995 ISBN 0-7923-8888-7
175. N.M. van Hemel, F.H.M. Wittkampf and H. Ector (eds.): *The Pacemaker Clinic of the 90's.* Essentials in Brady-Pacing. 1995 ISBN 0-7923-3688-7
176. N. Wilke (ed.): *Advanced Cardiovascular MRI of the Heart and Great Vessels.* Forthcoming. ISBN 0-7923-3720-4
177. M. LeWinter, H. Suga and M.W. Watkins (eds.): *Cardiac Energetics: From Emax to Pressure-volume Area.* 1995 ISBN 0-7923-3721-2
178. R.J. Siegel (ed.): *Ultrasound Angioplasty.* 1995 ISBN 0-7923-3722-0
179. D.M. Yellon and G.J. Gross (eds.): *Myocardial Protection and the* K_{ATP} *Channel.* 1995 ISBN 0-7923-3791-3
180. A.V.G. Bruschke, J.H.C. Reiber, K.I. Lie and H.J.J. Wellens (eds.): *Lipid Lowering Therapy and Progression of Coronary Atherosclerosis.* 1996 ISBN 0-7923-3807-3
181. A.-S.A. Abd-Eyattah and A.S. Wechsler (eds.): *Purines and Myocardial Protection.* 1995 ISBN 0-7923-3831-6
182. M. Morad, S. Ebashi, W. Trautwein and Y. Kurachi (eds.): *Molecular Physiology and Pharmacology of Cardiac Ion Channels and Transporters.* 1996 ISBN 0-7923-3913-4
183. M.A. Oto (ed.): *Practice and Progress in Cardiac Pacing and Electrophysiology.* 1996 ISBN 0-7923-3950-9
184. W.H. Birkenhäger (ed.): *Practical Management of Hypertension.* Second Edition. 1996 ISBN 0-7923-3952-5
185. J.C. Chatham, J.R. Forder and J.H. McNeill (eds.): *The Heart in Diabetes.* 1996 ISBN 0-7923-4052-3
186. J.H.C. Reiber and E.E. van der Wall (eds.): *Cardiovascular Imaging.* 1996 ISBN 0-7923-4109-0
187. A-M. Salmasi and A. Strano (eds.): *Angiology in Practice.* 1996 ISBN 0-7923-4143-0
188. M.W. Kroll and M.H. Lehmann (eds.): *Implantable Cardioverter Defibrillator Therapy: The Engineering – Clinical Interface.* 1996 ISBN 0-7923-4300-X
189. K.L. March (ed.): *Gene Transfer in the Cardiovascular System.* Experimental Approaches and Therapeutic Implications. 1996 ISBN 0-7923-9859-9
190. L. Klein (ed.): *Coronary Stenosis Morphology: Analysis and Implication.* 1997 ISBN 0-7923-9867-X
191. J.E. Pérez and R.M. Lang (eds.): *Echocardiography and Cardiovascular Function: Tools for the Next Decade.* 1997 ISBN 0-7923-9884-X
192. A.A. Knowlton (ed.): *Heat Shock Proteins and the Cardiovascular System.* 1997 ISBN 0-7923-9910-2
193. R.C. Becker (ed.): *The Textbook of Coronary Thrombosis and Thrombolysis.* 1997 ISBN 0-7923-9923-4
194. R.M. Mentzer, Jr., M. Kitakaze, J.M. Downey and M. Hori (eds.): *Adenosine, Cardioprotection and Its Clinical Application.* 1997 ISBN 0-7923-9954-4
195. N.H.J. Pijls and B. De Bruyne: *Coronary Pressure.* 1997 ISBN 0-7923-4672-6
196. I. Graham, H. Refsum, I.H. Rosenberg and P.M. Ueland (eds.): *Homocysteine Metabolism: from Basic Science to Clinical Medicine.* 1997 ISBN 0-7923-9983-8
197. E.E. van der Wall, V. Manger Cats and J. Baan (eds.): *Vascular Medicine – From Endothelium to Myocardium.* 1997 ISBN 0-7923-4740-4
198. A. Lafont and E. Topol (eds.): *Arterial Remodeling.* A Critical Factor in Restenosis. 1997 ISBN 0-7923-8008-8

200. W.C. De Mello and M.J. Janse (eds.): *Heart Cell Communication in Health and Disease.* 1997 ISBN 0-7923-8052-5
201. P.E. Vardas (ed.): *Cardiac Arrhythmias Pacing and Electrophysiology.* The Expert View. 1998 ISBN 0-7923-4908-3

Developments in Cardiovascular Medicine

201. P.E. Vardas (ed.): *Cardiac Arrhythmias Pacing and Electrophysiology.* The Expert View. 1998 ISBN 0-7923-4908-3
202. E.E. van der Wall, P.K. Blanksma, M.G. Niemeyer, W. Vaalburg and H.J.G.M. Crijns (eds.): *Advanced Imaging in Coronary Artery Disease.* PET, SPECT, MRI, IVUS, EBCT. 1998 ISBN 0-7923-5083-9
203. R.L. Wilenski (ed.): *Unstable Coronary Artery Syndromes, Pathophysiology, Diagnosis and Treatment.* 1998 ISBN 0-7923-8201-3
204. J.H.C. Reiber and E.E. van der Wall (eds.): *What's New in Cardiovascular Imaging?* 1998 ISBN 0-7923-5121-5
205. J.C. Kaski and D.W. Holt (eds.): *Myocardial Damage.* Early Detection by Novel Biochemical Markers. 1998 ISBN 0-7923-5140-1

Previous volumes are still available

KLUWER ACADEMIC PUBLISHERS – DORDRECHT / BOSTON / LONDON